ADVANCES IN ELECTRONIC CIRCUIT PACKAGING

Volume 2

ADVANCES IN
Volume 2
ELECTRONIC CIRCUIT PACKAGING

Proceedings of the Second International Electronic Circuit
Packaging Symposium,
sponsored by the University of Colorado and EDN (Electrical Design News),
held at Boulder, Colorado

Edited by Gerald A. Walker, Assistant Editor, Electrical Design News

SPRINGER SCIENCE+BUSINESS MEDIA, LLC

ISBN 978-1-4899-7297-2   ISBN 978-1-4899-7311-5 (eBook)
DOI 10.1007/978-1-4899-7311-5

FOREWORD

The Proceedings of The Second Electronic Circuit Packaging Symposium provides diverse examinations of an ever-growing problem—the assembly of electrical circuitry so that the final product fulfills its intended mission. Although approaches and techniques differ, the goal is the same: combine components into a package that will pass size, environmental, appearance, and/or economic "tests."

Subjects treated in this "Proceedings" range from large, rack-type units packaged for use in Polaris-firing submarines to microminiaturized, solid-circuit semiconductor networks. They vary from the shock and acceleration problems of landing operating instrumentation on the moon to the temperature and pressure problems of oil-well logging instruments lowered several miles below the earth. And yet, universal approaches to improved materials, interconnections, and component placement are significant extensions of the state-of-the art of all electronic packaging.

This volume, therefore, is a collection of varying means to a common goal. It is written by mechanical, electrical, and electronic engineers along with physicists, chemists, and mathematicians, but it is directed to the newest of all designers, the electronic circuit packaging engineer.

It should be noted that in some cases, the discussion period is even more fruitful and revealing than the paper itself. This is due in part to the advance distribution of each paper. During the meeting, the authors highlighted their presentations with slides and points of special interest. This procedure eliminated the need for verbatim reading of the papers and resulted in the lively discussions included with each paper.

Of course, the success of the second Symposium must be attributed to the speakers and to those who attended. Their comments and evaluations have clearly indicated a need for such conferences. Plans are now being made for the third Symposium and your interest in packaging can be manifested in your participation—either as a speaker or as an attendee — at future Symposia.

Gerald A. Walker
Assistant Editor
Electrical Design News
November 1, 1961

CONTENTS

Opening Address

WHERE WE STAND

Colonel Paul E. Worthman

I want to thank each of you, each of you participants and guests in the Electronic Circuit Packaging Symposium, for inviting the Air Force, and particularly this representative, to be here with you this morning. We like to tell the story of space in the Air Force. We like an opportunity like this. My Commander, as many of you know, is Major General O. J. Ritland. He sends his greetings to you as well as the greetings of all of us in the Space Systems Division.

I am meeting with you this morning essentially to tell you about three things. And first, I am going to point out some of the highlights of the work of the Ballistic Systems Division. You know, we have two divisions in Los Angeles now, working for the Air Force. One of them is the Ballistic Systems Division and the other one is the Space Systems Division. We have "spun off," a familiar expression, the Space Systems Division from the old AFBMD, which is a sign in itself of the importance of space work. The second thing I want to tell you is something about that division, the Space Systems Division. And finally, I want to make a few comments on the directions that our work has taken—how we see the future. I consider everybody in this room to be involved directly—or indirectly—in the space business and, as you have heard, I consider this to be quite literally a report to the stockholders.

We have been working very diligently, since 1954, on the development of ballistic weapons systems. New weapons are now moving into the operational inventory. They are new weapons for a new age. These weapons are complex, they are costly, and often they are decidedly contrary. But if they serve out their lifetimes without ever stirring in anger, like the B-36, they will have more than paid for every bit of the energy and every bit of the national resource that went into them.

The first large missile that was delivered to the nation's operational inventory was Thor. Now this is an IRBM (intermediate range ballistic missile). It has been operationally deployed, and when I say it's part of this nation's inventory, I'm speaking very broadly, freely speaking of NATO and of the British because, as you know, the operational Thors are deployed in the United Kingdom in the hands of the RAF. I think the Thor's dependability and reliability, two words that mean a great deal to you folks, have been quite amply demonstrated. The last Discoverer flight was boosted by Thor No. 121. Of these 121 Thor flights, 90 have been completely satisfactory. So I think the Thor has honestly earned the reputation of being the "Workhorse of the Space Age."

The first Atlas Intercontinental Ballistic Missile was turned over to General Powers' Strategic Air Command a little over two years ago. And today, in SAC, Atlas installations are complete at Vandenberg Air Force Base, just north of L.A.; near Cheyenne, Wyoming, at our Warren Air Force Base; and at Offut Air Force Base itself, near Omaha, Nebraska. The Atlas, in 91 flights, both before and after it became operational, has proved itself both reliable and dependable. Many of you will recall the presidential statement, made last winter, to the effect that the Atlas, fired for mark, is "striking within two miles of the target." Since delivering the first Atlas to the Strategic Air Command, the Ballistic Systems Division has continued to improve the capabilities of range, payload, and accuracy, and our present ICBM's are designed to behave in a much more sophisticated manner than the early ones. If you recall last month's 9050-mile flight of an Atlas from Cape Canaveral into the Indian Ocean around the bend to South Africa, I think you get sort of an index as to the nature, the scope, and the magnitude of these improvements.

The Titan is a sister ICBM to the Atlas, and it is progressing quite nicely. There have been 28 completely successful launchings of Titan out of 38 attempts. All the flights now are for the full 5500-nautical-mile range. From now on, the flight test program will point toward sharpening the accuracy, confirming the reliability, and establishing the Titan as a deployable missile.

The third ICBM is Minuteman, and Minuteman belongs to an entirely new generation of ballistic missiles. It has a number of unique advantages over both the Atlas and the Titan. The first of these, of great importance to us, is that it generates its thrust from solid propellants and these, of course, are much simpler to store, much easier to handle than the liquid oxygen and the kerosene which we use in Atlas and in Titan. Second, since Minuteman is so much easier to launch, it has a much quicker reaction time, again, a very important operational consideration. And the most dramatic difference of all, to me, is the fact that it is cheaper in all respects than its predecessors. I know of no other weapon system which replaces something prior to it and costs less.

The development on Minuteman has progressed to the point now where all of the major unknowns have been resolved and on the first of February of this year the first flight was made out of Cape Canaveral. It was a double first because this was the first time an ICBM had ever been fired on round one with all stages lighting; that is, it intentionally triggered off all stages. You know the earlier Atlases and Titans really just flew booster stages with dummy second stages for quite awhile in their flight history. This particular bird lighted all stages and flew a very accurate trajectory 4800 nautical miles down range. It's beginning to look now (these statements are always a little dangerous), but it's beginning to look now, as though the operational date for 1963 is going to move back into 1962.

The scope of the nation's ballistic missile program is something that comes into perspective when you consider it's the biggest thing anyone's ever tried. In this fiscal year alone, there are $2\frac{1}{2}$ billion dollars going into ballistic missiles. This program has mastered extraordinarily difficult technical problems—very basic technical problems, in guidance, propulsion, re-entry, and let's not forget the design and engineering of operational facilities which turned out to be the toughest job of all. These problems have been solved by a team of 30 prime contractors, 200 major subcontractors and 200,000 suppliers.

Now this is a very quick review of the ballistic missile program and I think it is an essential background for the things I want to say about our space program because it was in this ballistic missile program that the Air Force laid the base for space. As far back as 1955 we began exploratory development which we hoped would lead to satellite systems which would complement our ballistic systems.

Today, space is being investigated from two points of view. From one point of view, space is being looked at in terms of pure scientific investigation. From the other point of view, we are looking at it in terms of military application. From the scientific point of view, space is a gigantic laboratory just loaded with question marks. And from a military point of view, space is an observation post, a communications center, and an arena for future deterrence.

Many of you know the name Midas—I suppose all of you do. Midas is a name which stands for Missile Defense Alarm System. And Midas takes advantage of space as an observation point. Midas carries infrared sensing devices and orbits hundreds of miles above the earth looking for the telltale exhausts of ballistic missiles. This satellite can nearly double the warning time available to us from any other system. There are two of these 5000-lb Midases in orbit this morning as I speak to you. We have radio communication with both of them and they are both prototypes of the complete Midas system. Midas helps us to counteract the Soviet advantage of surprise in a time period when it only takes 30 minutes for an ICBM to get from the Eurasian land mass to the United States. The very fact that we have the capability to be warned of an attack will serve as a deterrent to that attack.

There's a rather new name coming into the space vocabulary now and it's Saint. Saint is just beginning to cut its first hardware; it's a co-orbital inspector system and it will be our first satellite that has a capability of rendezvousing with another spacecraft. The purpose of the rendezvous will be to assess the missions of potentially hostile satellites.

There are 27 American spacecraft in orbit this morning as we sit here. If you caught the early morning news, there may be 28. People aren't going to be sure, they say, until tomorrow. But, there are 27 sure birds in orbit. Every one of these birds owes a great deal to the Discoverer program. Discoverer is our testbed satellite in the Air Force and it's proving out many of the critical components that have to go into Midas and Samos long before they go into the birds. Comparatively it's a very inexpensive way of testing these components. The key problems are things like temperature, stabilization, control, communication, separation, auxiliary power, and the really big problem, recovery. These are the areas in which Discoverer is doing a lot of pioneering for us and feeding into our more long-term program.

The 11th of August 1960, just a little over a year ago, is the date that is rather historic in the space business, because on that day the first Discoverer capsule was recovered from orbit. This has been repeated five times; four of these times the pickups have been in midair by specially instrumented C-119 aircraft and one of these times, recently, a pickup has been made, as a matter of fact it was in June, by a crew of Air Force paratroopers literally jumping right into the middle of the Pacific Ocean. Their comment on this is, "It's fun."

The Air Force has a very deep personal interest in every satellite that goes into orbit from the United States these days because in September 1959, Mr. McElroy, then Secretary of Defense, assigned to the Air Force the responsibility for the development, production, procurement, system integration, and launching of space booster-stages for all of the services.

This assignment covers one of the most difficult parts of the space business. It's an assignment which we in the Air Force are taking very seriously because the stature of the national space program depends quite largely on how well we do our part on this job.

We have launched a wide variety of spacecraft and I've selected just a few of these to give you an indication of how the nation is moving into space.

Last year we launched NASA's Pioneer V. By the way, we launched the vast majority of NASA flights. Pioneer V is the paddlewheel satellite and it holds the long-distance transmission record of $22\frac{1}{2}$ million miles. It is now in its 311-day solar orbit and it has an expected lifetime of several million years. Now I recognize that I'm reaching back pretty far into the archives for Pioneer V, but I wanted to mention it this morning because it was highlighted a few months ago by the failure of the competitive Soviet Venus probe.

One of our Thors is used to boost Tiros into orbit. Tiros goes into equatorial orbit and I think you are familiar with the performances of Tiros I and Tiros III. Over a period of 78 days, Tiros I collected over 23,000 pictures of cloud cover around the world. Tiros is serving today to move the expression "neph analysis" back into the meteorologist's vocabulary.

An Air Force Thor-Able-Star combination boosts Transit every time a Transit flies. Transit is the Navy's navigational satellite and repeated tests have shown that Transit is a very feasible and a potentially excellent navigational system.

Every report to the stockholders should contain some pictures and I have a very short film here which highlights some of our activities during the past 12 months and I would like to inject it at this point in my speech.

Now today, as we review these events, there are new satellites and new space probes being readied in the assembly buildings and on the launch pads at Patrick Air Force Base and at Vandenberg Air Force Base.

Students of history know that some of the most stirring and momentous occasions have had no torchlight parades and no ruffles and flourishes and no signposts saying, "This way to the excitement." There is sort of a normal human propensity for ignoring great events. Breughel was so impressed by this human trait that he painted a famous canvas in which he shows a peasant stolidly plowing his field in the foreground, while in the background Icarus is ending his legendary flight into the Aegean.

We are moving out of the Kitty Hawk of the space age, and events such as those you have been listening to are beginning to blur in our minds. They are becoming frequent, almost commonplace. Only the so-called catastrophes are on page one of the morning newspapers; completely successful ballistic missile and satellite launchings are usually tucked away around page 26, well below the fold. And perhaps, in a sense, this in itself is a sign of our growing maturity in the space age; however, we must always remember, through this increasing blandness, that Icarus may have just flashed by.

How are we doing in space? This is a vital question, and I think one of the best ways to help you answer it is to call the roll of some of the key space events of the last few years. Here are some, and not nearly all, of the significant U.S. "firsts" in space.

The first scientific flight above the atmosphere.
The first photograph of a complete tropical storm.
The first detection of X rays in the high atmosphere.
The first exposure of animal life above the atmosphere.
The first motion pictures from space.
The discovery of the Van Allen radiation belt.
The first use of solar cells for electrical power.
The first precise geodetic use of a satellite.
The first communication satellite.
The first satellite on a polar orbit.
The first still television picture from space.
The first computer operating in space.
The first weather satellite.

The first restart of a large rocket engine in space.
The first successful dual-satellite launching.
The first mapping of total ionizing flux.
The first measurement of interplanetary magnetic fields.
The first micrometeorite observations from a space platform.
The first radio communication from a position beyond the moon.
I have selected these items pretty much at random from a list of over 75 United States "firsts in space."

Let's take a look at another indicator of our progress: two numbers which those of us in the Space Systems Division call the space boxscore. There have been 58, and maybe as of this morning 59, successful earth satellite launchings. Thirteen of these have been by the USSR, 45, maybe 46, have been made by the United States of America.

Now these comparisons are very heartwarming and they represent a magnificent effort by a very dedicated group of Americans. However, I would be very distressed if you went away from this room feeling that the Air Force, Department of Defense, or the nation is overconfident or complacent or so amateurish as to think that perhaps anyone can put space in its place. Certainly, you'll never find us setting up the equation that you can see on the second floor of the Hayden Planetarium in New York. There, as you approach the stairwell, is a very elegant, illuminated sign which says, "This way to the solar system and restrooms."

To me there is a consideration that is very much more important than lists of "firsts," more important than these international boxscores, no matter how gratifying they may be. And this consideration has to do with the motives, the corporative objectives which guide the planning of our space program and furnish the spiritual dynamics for the progress of these programs.

In 1957, our corporate objective was very, very simple. It was simply to get a satellite—any satellite—into orbit, and it was at that time that the concept of the space race began. That concept is still with us. Fortunately, as this nation has grown in experience and in skill in space technology, it has recognized the necessity for choosing its own course and its own goals. I have seen this change. And this recognition has brought a gradual end to the very defensive objective of awaiting the other contestant's moves and then attempting to outdo him one move at a time. Such a race is an intramural contest. The big league in this business demands that we set our own goals, that we point toward them with broad, logical programs, and then follow our courses with energy, determination, and confidence.

This big-league approach calls for us to measure our progress against the most demanding standard of all—the standard of our own capabilities, our own hopes and, I would add, our own dreams. The first energetic steps in this direction are being taken; you and I are watching them. Why are we building a Tiros? Why is this nation building a Transit, a Pioneer, a Samos, a Midas, a Saint? Each of these is first in its field. Each of these is alone in its field. Each of these is our national response to our national needs. Tiros is keeping a weather eye on the earth and it will enhance weather forecastings for both military and civil purposes. Transit demonstrates what satellites can do to give us new global navigational devices. Pioneer is proving the way for reliable, uninterrupted, instantaneous global communications. Saint opens the door to space rendezvous and inspection. Midas sees the world in infrared, guarding us from surprise ballistic missile attack. These are the first sure steps in a new direction.

This shift in motivation—this measurement of progress against our own needs and aspirations—may well be the most important space event of the last two years. We do not know all of our needs as yet; we have not fulfilled all of those which we know; but we have found our major goals and we are moving toward them with energy and confidence, perhaps recalling Emerson's observation that "Power... resides in the moment of transition from the past to a new state... in shooting the gulf... in the darting to an aim."

Well this is our report to the stockholders. I have described the key developments in our ballistic missile and space programs and I have spent some time commenting on the curious paradoxes of how we are doing, pointing out that in the context of narrow intramural goals we can regard the national space product with pride, while in a broad big-league sense these comparisons are interesting, but not nearly as important as the aggressive fulfillment of national space objectives.

I think we can all be proud of the fact that in the past 10 years man's knowledge of his universe has increased more than it did in the centuries between Galileo and Einstein, and I think we will always remain humble in the knowledge that no matter how dramatic we make the account of how we moved into space, grandchildren will always ask, with that wide-eyed, diabolical innocence, "Why did it take you so long?"

ADVANCES IN
ELECTRONIC CIRCUIT PACKAGING

MATERIALS FOR ELECTRONIC PACKAGING

D. A. Beck

The Bendix Corporation, Research Laboratories Division

Electronics, by its name, implies electrical circuitry and manifestation of electrical phenomena. If this could take place without the need to provide the material to surround the "tunnel" of a tunnel diode, like the hole in a doughnut, or to separate the plates of a capacitor, or to make this or that functioning circuit transportable from here to there, as in the case of a missile, we would have the electronic engineers' Utopia and there would be no such thing as an Electronic Circuit Packaging Symposium.

The electronic engineer is faced with one prime objective, to design a circuit that will perform some specific task. It is the packaging engineer's responsibility to confine, hold together, separate, heat, cool, and otherwise assemble these electrical manifestations in a "black box" such that the final product will fulfill its intended mission. It is not the purpose of this paper to consider the materials that are used in the electronic components, such as resistors, capacitors, tubes, transistors, and the like. I do want to point out the advantages and disadvantages of some materials that are used to contain or support these components and to promote confidence in the use of a few materials which have not had as wide an acceptance as they deserve. This lack of acceptance has been due primarily to too much lethargy among the packaging engineers. Engineers have been too prone to save weight until they are "blue in the face" — as long as they do it with steel, aluminum, or some standard "garden variety" material with which they are familiar.

Principal materials to be covered in this paper are: 1. beryllium, 2. magnesium, 3. aluminum, and 4. titanium.

Had all of the aluminum now in orbit been replaced by beryllium or magnesium, there could have been a saving of many pounds — pounds that could have meant an increased payload or reduced fuel requirements. Add to this the additional weight that could have been saved by use of these two materials in the booster stages, and one would have a pretty significant contribution to the space race. Since it takes from 15 to 30 lb of thrust to put 1 lb of payload in orbit, or putting it another way, since one pound saved in a military aircraft is worth anywhere from $500 to $2000, the necessity for using the lightest and best materials should be self-evident.

Figure 1 shows the relative weights of the various materials. Note that aluminum is 54% heavier than magnesium or beryllium, while titanium is 64% heavier than aluminum and over 154% heavier than magnesium or beryllium, and steel — but perish the thought!

Throughout this discourse I want to point out that the charts and curves are shown principally to indicate a comparison of materials. While the data are accurate for the particular alloy, temper, or method of production, they will change with different alloys, tempers, etc. This is not meant to be a designer's handbook; there are plenty of those already, plus a wealth of producers' literature from which one can get specific values on specific alloys to do a job best. The

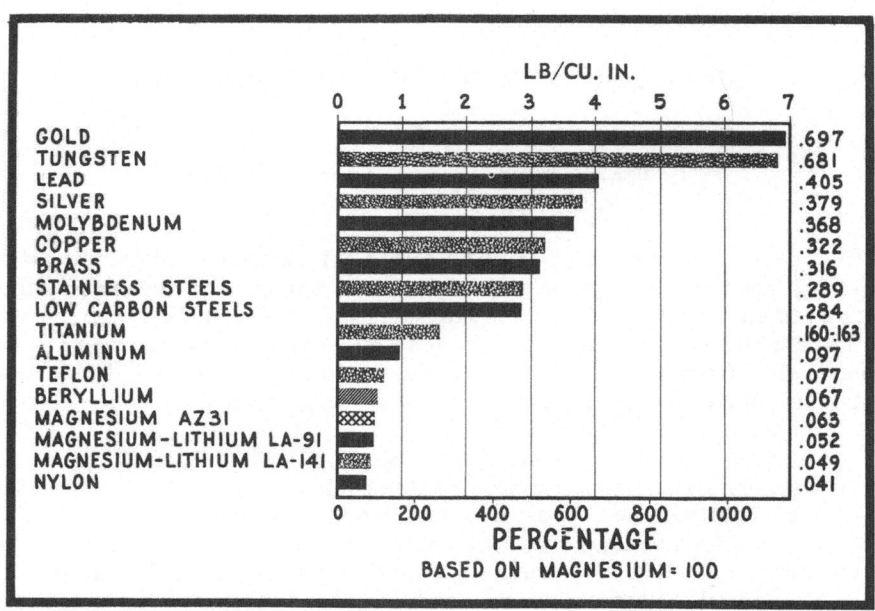

Fig. 1. Density comparison.

comparisons made here will not markedly change with different alloys, tempers, etc.

Figure 1 also shows the percentage weights of these materials compared with magnesium. With the exception of beryllium, the less dense the material the less its tensile strength, and theoretically the more material you have to use to do a given job. Practically, however, other requirements operate to negate this factor. More often than not the packaging engineer is faced with minimum usable gauges. Stiffness, vibration, fabricating difficulty, personnel handling, and many other requirements make it necessary to use greater thicknesses than are required by the actual tensile or compressive stresses present. This is where the low density materials pay off. For instance, Table I shows the relative stiffness of various materials when used in a beam of constant cross section. Now look at the last tabulation on this figure and see what happens to stiffness when the same weight of material is put into a beam of the same width, but of varying depth to accommodate the less dense material. The beryllium beam is 1400 times as stiff as the aluminum beam. Obviously, this improvement is due to the increased moment of inertia and higher modulus of elasticity. Figure 2 shows a comparison of the E of various materials and Fig. 3 shows the effect of temperature on E. The most expedient design will actually lie somewhere between these extremes with a gain in stiffness and yet a reduction in weight. There are cases where one just doesn't have the room to increase the depth and in these cases use of the high-strength materials is best. However, don't forget our new friend beryllium. This material, if it weren't for the almighty dollar sign, is the answer to many a designer's prayer. It certainly has enough good features to make anyone think several times before not using it. In fact, this is the thing that irks me most — most designers do not consider beryllium or magnesium first. They have too often considered steel or aluminum only because it is what

TABLE I. Relative Beam Bending Strength and Stiffness. (Rectangular beams of constant width using aluminum alloy 6061-T4 and minimum yield strengths as comparison basis.)

Comparison	Material		Depth	Strength	Stiffness	Weight
Beams of equal depth	Magnesium	AZ31 - 0	100	98	63	66
		AZ31 - H24	100	166	63	66
		AZ80 - T6	100	200	63	66
	Beryllium		100	318	435	69
	Aluminum	3003 - 1/2H	100	100	100	100
		6061 - T4	100	100	100	100
		2024 - T4	100	252	100	100
		7075 - T6	100	416	100	100
	Titanium	A - 55	100	318	149	163
		C110M	100	625	149	163
	Steel	SAE 1025	100	227	286	286
		SAE 4130	100	852	286	286
Beams of equal strength	Beryllium		57	100	78	38
	Magnesium	AZ80 - T6	71	100	23	48
	Aluminum	7075 - T6	49	100	12	50
	Magnesium	AZ31 - H24	78	100	30	52
	Aluminum	2024 - T4	63	100	25	63
	Titanium	C110M	40	100	10	65
	Magnesium	AZ31 -0	102	100	66	67
	Titanium	A - 55	57	100	27	93
	Steel	SAE 4130	35	100	12	100
	Aluminum	3003 -1/2H	100	100	100	100
		6061 - T4	100	100	100	100
	Steel	SAE - 1025	67	100	83	190
Beams of equal stiffness	Beryllium		82	207	100	55
	Magnesium	AZ31 -0	117	133	100	78
		AZ31 - H24	117	227	100	78
		AZ80 - T6	117	270	100	78
	Aluminum	3003 - 1/2H	100	100	100	100
		6061 - T4	100	100	100	100
		2024 - T4	100	251	100	100
		7075 - T6	100	414	100	100
	Titanium	A - 55	89	234	100	145
		C110M	89	490	100	145
	Steel	SAE - 1025	71	114	100	204
		SAE - 4130	71	426	100	204
Beams of equal weight	Beryllium		149	683	1400	100
	Magnesium	AZ31 -0	154	232	230	100
		AZ31 - H24	154	400	230	100
		AZ80 - T6	154	485	230	100
	Aluminum	3003 -1/2H	100	100	100	100
		6061 - T4	100	100	100	100
		2024 - T4	100	252	100	100
		7075 - T6	100	416	100	100
	Titanium	A -55	61	117	35	100
		C110M	61	234	35	100
	Steel	SAE - 1025	350	28	12	100
		SAE -4130	350	105	12	100

they are accustomed to using. They haven't used the lighter metals because there is always that "first" time, and because they are not sure of them, they fall back on the "tried and true" materials of grandfather's day. It also too often happens that, because the designers had not previously specified it, the model or prototype shop doesn't stock magnesium or may not be equipped to machine beryllium so the pressure is on to get that first unit out, be it "breadboard" or prototype, and the quickest and easiest material at hand is aluminum, or brass, or steel. So, the first system gets out and it weighs X^2 pounds, but by then the "heat" is on to get the production design released, so the engineer's plea is, "we made it of ferromanurium and it held together OK. We haven't got time to change it now. Get the drawings out!" So it stays made of ferromanurium.

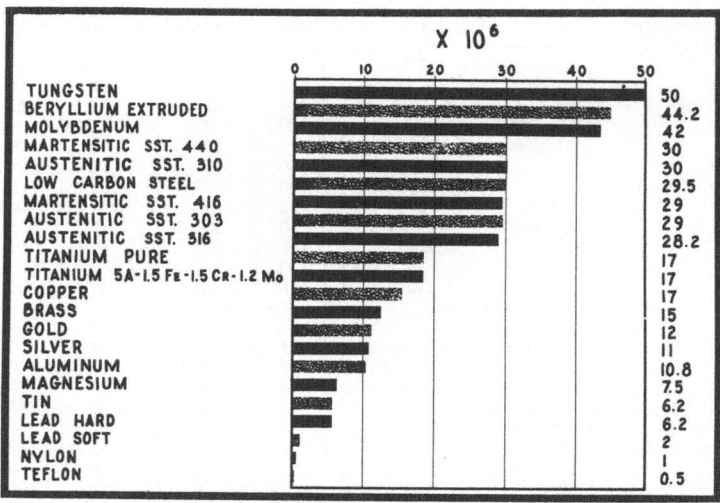

Fig. 2. Modulus of elasticity (in tension).

Getting back to beryllium — its modulus and high-temperature characteristics are enough to make it first choice for many applications. (Only some of the cermets and a couple of exotic metals have a higher modulus.) Its E of 44 million makes it a logical choice for seeker antennas, gyros, or any system using gimbals or subjected to high accelerations but yet requiring the lowest possible deflections. Figure 4 shows its high strength at medium-high temperatures to be far better than aluminum or magnesium, the two materials now commonly used for such applications. Electronic chassis subjected to high accelerations and deflections could well be made of beryllium. Beams and other structural members could use this material at no increase in weight over magnesium, but with a terrific gain in stiffness. However, beryllium is not without its drawbacks. "Block" beryllium, not extruded, has very low elongation. Therefore, you must use care to insure that your design does not have sharp interior corners that may cause a notch effect. "Warm extruded" beryllium has much improved elongation, which is somewhat a function of the extrusion ratio. The mechanical properties of "extruded" beryllium are also considerably better than the "block" material, but there is a somewhat abnormally wide spread between the longitudinal and transverse properties. "Tensile yield" stress and "proportional limit" are common "check-point" values used in design. However, these do not guarantee that there will be no "yield" if they are stressed below the proportional limit; in fact, one must go to an absurdly low stress to achieve absolute preclusion of any "yield" or strain, which is the desirable situation in inertial guidance equipment. In this area of use the usual "yield" stress based on 0.2 offset strain is not used, but instead the "precision elastic limit" is used which is strain at the rate of 1μ in./in. However, at least one major beryllium supplier, the Brush Beryllium Company, is in the last stages of developing material which has a precision elastic limit of 10,000 psi ±1000. This material is called instrument

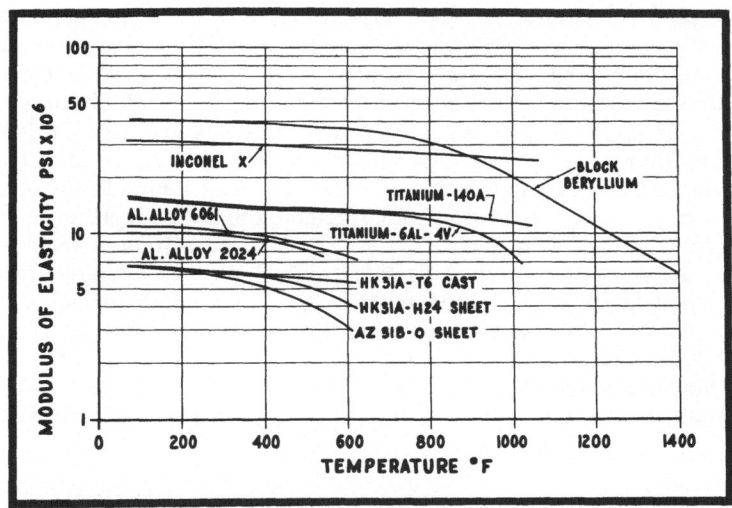

Fig. 3. Modulus of elasticity (psi x 10^6).

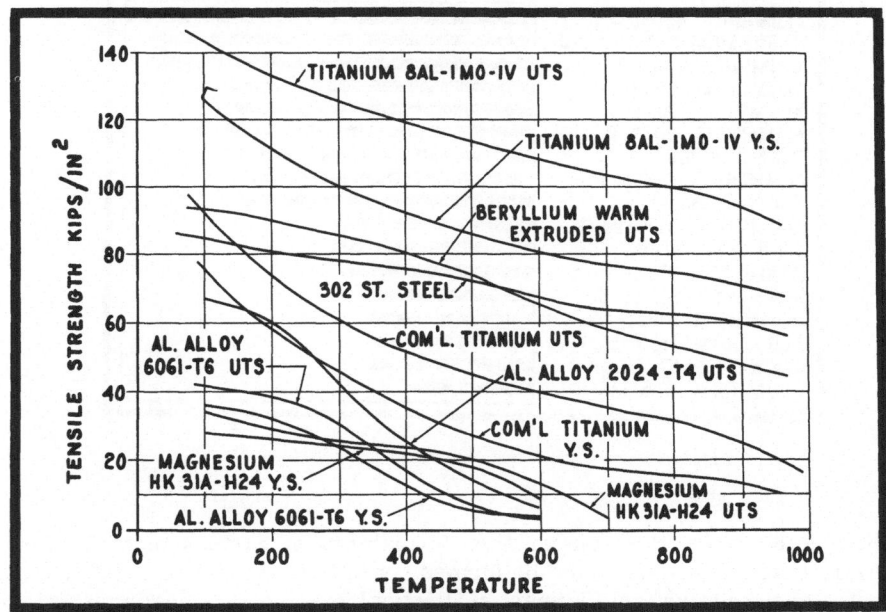

Fig. 4. Comparative tensile strengths vs temperature.

grade and presumably is produced by use of subsieve powders causing a subsequently high oxide content.

Beryllium dust and fumes are toxic and proper fabrication and ventilation methods must be observed, and these precautions will cost money. But an initial outlay is necessary for that lathe or mill in the first place in order to machine the part out of any material. The mill was an improvement over the file, but was paid for because the improvement was worthwhile. So it is with beryllium over magnesium or aluminum. Aside from this ventilating problem, machining requires only standard toolroom equipment. Beryllium may be plated with practically any standard material, including electroless nickel. It may also be soldered directly, using proprietary soldering compounds which include the flux and solders. It may be brazed in an inert atmosphere, using silver lithium solder. Beryllium is high in the electromotive series, in fact, higher than aluminum or magnesium, and therefore attention must be paid to proper insulation and protection from corrosion. General machining of beryllium can be compared fairly closely with cast iron. Drill and tap life are short, but other operations can be accomplished satisfactorily by using carbide-tipped tools. In ordinary bar stock form it currently costs in the neighborhood of $75.00 to $80.00 per pound, but producers are hopeful of cutting this price in half in about two years. It is available in all standard forms, sheet, plate, tube, bar, etc. It does cost more, but one gets more, and the only way the price will come down is to increase production, which merely means that packaging designers must specify it.

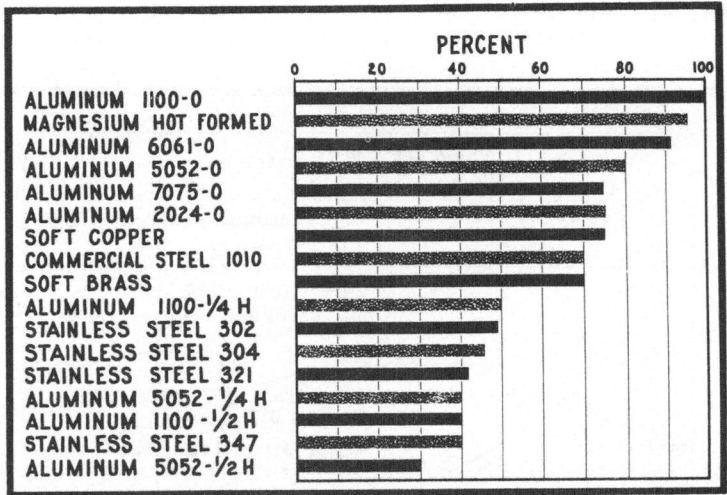

Fig. 5. Comparative formability of some sheet materials.

Figure 5 shows the comparative formability of various materials. Note, however, that magnesium must be formed hot at between 350 and 650°F. Figure 6 shows the relative coefficients of thermal expansion. These must be considered in dissimilar metal assemblies and are equally important in welded assemblies. Weld shrinkage is a common property of many materials that is generally forgotten by many designers. Thermal conductivity, compared in Fig. 7, must be considered not only in cooling electronic components; it is equally important in welding and brazing operations. Use of both thick and thin sections in brazed aluminum parts will cause your shop no end of grief due to warped chassis caused by unequal cooling rates of the large and small masses of material. Emissivity, which must be considered in heat dissipation of electronic components, is compared in Fig. 8. Figure 9 shows the average power required to machine some of the materials used in electronic packaging.

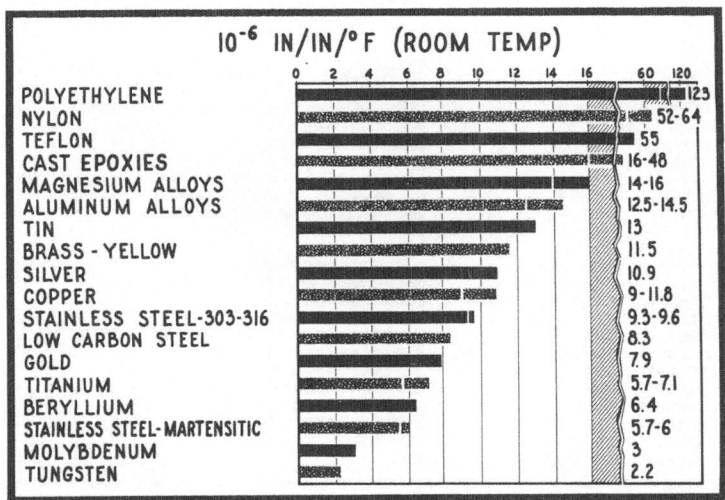

Fig. 6. Average thermal expansion coefficients.

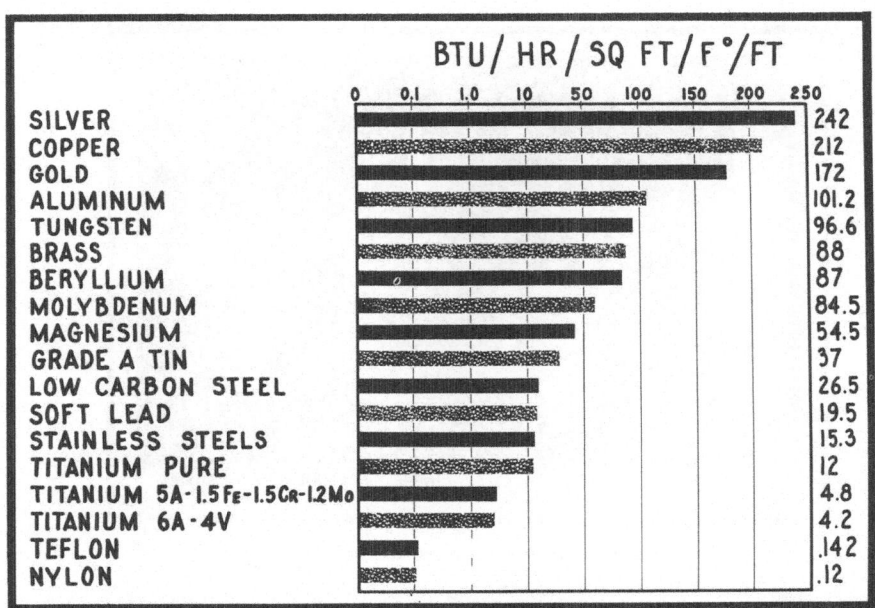

Fig. 7. Comparison of thermal conductivity at room temperature.

Behind beryllium, but not too far, is magnesium. This metal has probably been unjustly maligned more than any other structural material. The material itself is certainly not to blame — only the designer. Like any material, there are places where it is the best material to use and there are places where it is not the best. To use magnesium and not consider corrosion is folly, but it is done every day. Properly protected or insulated, it can be used to do a fine job structurally and still save up to $\frac{1}{3}$ the weight of its aluminum counterpart. Standard protective finishes are Dow 1 or 7, and Iridite 15 which serve as a good

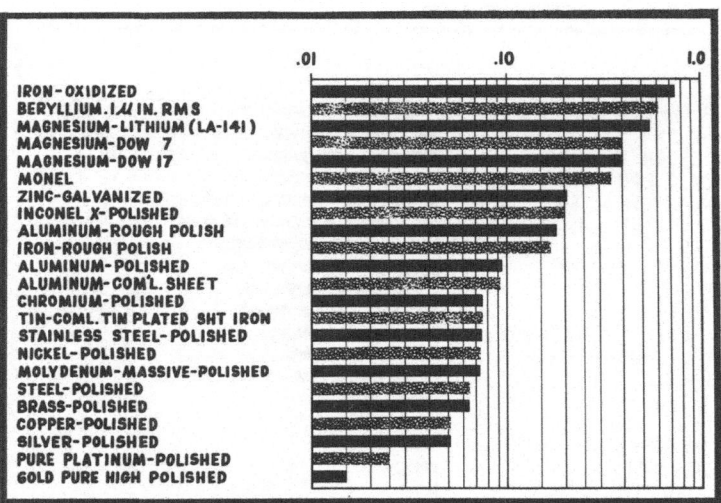

Fig. 8. Emissivity coefficient (212°F).

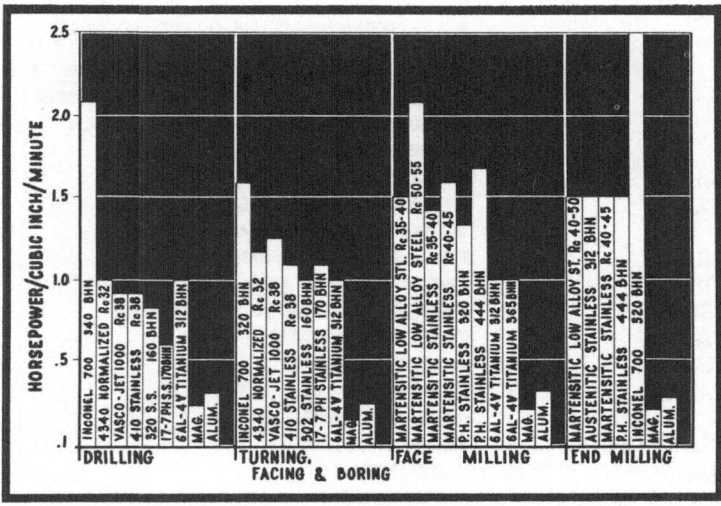

Fig. 9. Average unit power for machining.

paint base, but for more severe corrosion protection, use Dow 17 or HAE, ceramic-type anodic finishes that, when sealed with epoxy resin, provide hundreds of hours of protection in salt spray tests. It is pretty much a toss-up as to whether Dow 17 or HAE is better. I have found that HAE test panels may show isolated pits a little earlier, but Dow 17 reached failure stage first. Other tests might show the reverse. I like the green color of Dow 17 better, and maybe that is a good method of choosing between them. HAE is harder and I think it may have some good potential if used in instrument gearing — servos and the like where you must have low inertia. The surface would have to be filled and ground to eliminate the original abrasiveness of the HAE. Steel or brass ring gears may be shrunk on magnesium hubs and wheels with improvement in the inertia

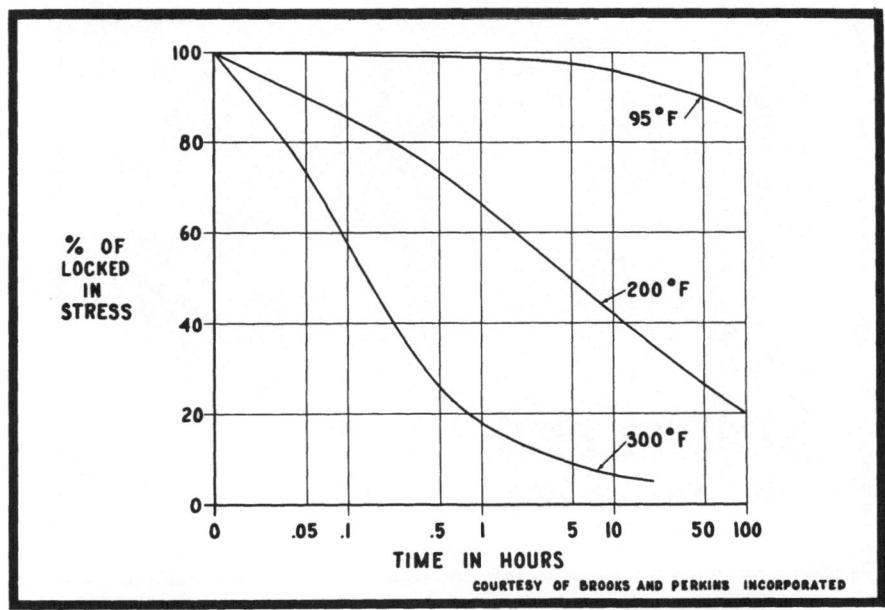

Fig. 10. Relaxation of AZ31X sheet.

problem. In fact one major electronics company has run gear simulation tests with bare magnesium. The principal problem was to get past the first fifty hours of operation, but once that was done the material seemed to run on and on. However, I feel that their answer to this problem is proprietary and cannot be given here.

Fairly recently Dr. W.F. Higgins in England developed a Fluoride Anodize for magnesium. This necessitates an extremely simple setup — requiring only an ac supply, a voltage regulator such as a Variac, some cheap chemicals, and a license from Dr. Higgins. It gives very good corrosion protection, and is an excellent paint base. If anyone attends the fall meeting of the Magnesium Association in New York, and I would recommend it, he will see a Fluoride Anodize system in operation.

Dow 17, HAE, and Fluoride Anodize each present problems in rf grounding since they are all good insulators. If the metal is sufficiently thick, holes may be tapped and grounding screws used with a sealer to prevent moisture entry. Tape or bolted-washer masking for grounding terminal spots may also be used with HAE, Dow 17, or Fluoride Anodize. Dissimilar metals must be insulated from magnesium at their faying edges to prevent setting up an electrolytic cell. To neglect this point is to ask for trouble. It must be remembered that it is the designer who fails when magnesium does not do the job properly. The electrolyte is the critical ingredient in any electrolytic corrosion system. As dry-charged storage batteries have tremendous shelf life, the magnesium-dissimilar metal joint will have tremendous life if the electrolyte is kept away from the critical interface, in this case usually moisture.

Magnesium can be easily welded by the heli-arc tungsten electrode process or by the heli-arc consumable-electrode process on heavier sections. Gas welding can be done, but it is not recommended. AZ31 alloy parts should be stress relieved after welding to reduce the possibility of cracking. Figure 10

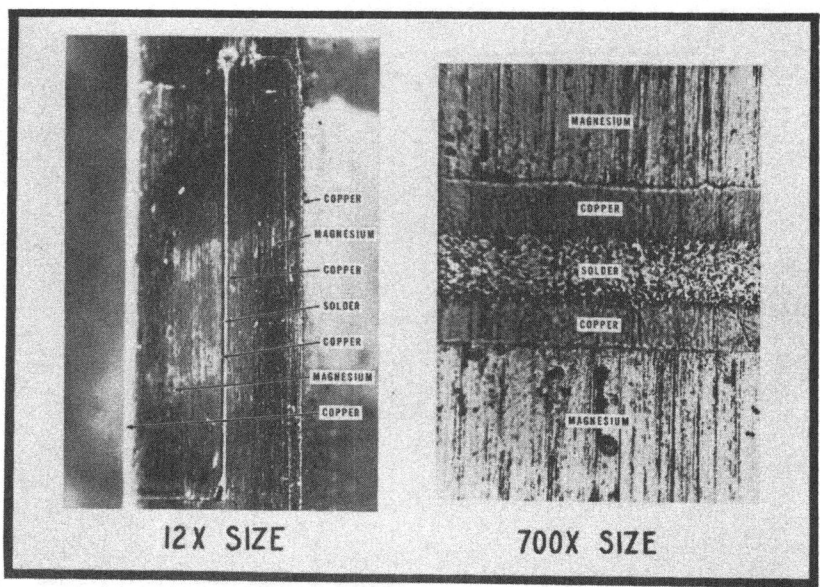

Fig. 11. Magnesium tube lap joint (copper-plated, soft-soldered).

shows the effect of stress relieving AZ31 alloy. Other alloys follow the same general pattern but with different values. The time at temperature is, of course, the time after the complete mass has come up to temperature. As the percent of "locked in" stress goes to zero, the allowable stress of the material is that of the annealed state. For all practical purposes, however, stress relief down to within 50 or 25% of the annealed state is satisfactory in most applications. The newer ZE10A alloy does not need stress relieving and yet has most of the other desirable characteristics of the more common AZ31. Spot welding is accomplished on any spot welder capable of producing acceptable spots on aluminum but requires less current per gauge than aluminum. It will require cleaning of the electrodes slightly more often. Brazing is not satisfactory on magnesium alloys except the old M1 alloy which is rarely used any more except in waveguides. The trouble with brazing is that the brazing alloy melting temperature is too close to that of the parent metal and requires excessively close control for satisfactory shop use. Magnesium may be soldered by the ultrasonic process or may be soldered with a soldering iron using indium with no flux. Once the part is tinned with indium then soft solder may be applied to the indium. However, soldering is not a process that I recommend unless there is no other way of doing the job at hand. A magnesium-soldered joint looks about like that of aluminum and certainly leaves much to be desired from the esthetic viewpoint. Magnesium cannot be silver soldered with the silver solders in general use. Magnesium may be plated by first immersing the part, after cleaning, in a zincate bath, then applying a copper flash, then a copper plate, and from this copper base any other material may be plated — i.e., tin, gold, silver, etc. Electroless nickel plating has been done on magnesium but from what I can gather its performance has been spotty. Figure 11 shows sections of a magnesium tube with a lap joint. The tube was copper plated with less than 1 mil of copper and then soft soldered. One of the proprietary "hard" soft solders with a plastic range of from 640 to 740°F was used to solder a copper-plated magnesium flange to a copper-plated waveguide. The plated parts were baked at 450°F for two hours and then soldered. There was no evidence of blistering and adhesion was excellent. Figure 12 shows a small magnesium case that was selectively plated about six years ago. These cans were butt-lapped face-to-face and tinned; then a copper "peel" strip was sweated on the joint. This peel strip was then peeled off much as you open a sardine can. The can was then resoldered-and-peeled seven additional times and at no time did the copper plate blister or pull away from the magnesium. Incidentally, this soldering was done with an acetylene torch and obviously the temperature was not accurately controlled. Figure 13 shows another hermetically sealed can with the "peel" key in place. Figure 14 shows a round glass face soldered into a magnesium simulated instrument case. Figure 15 shows glass insulated feed-throughs soldered into a plated magnesium case. These are instances to show that magnesium cases can be used to provide true hermetically sealed enclosures. Obviously if such cases are made, selectively plated, and soldered, a protective finish should be applied over the dissimilar metal interfaces to prevent corrosion.

Fig. 12. Magnesium hermetic-seal case. Selectively plated peel strip not shown.

AZ31 alloy may be bent cold providing the bend radius is five times the thickness or greater; however, when worked hot at between 350 and 650°, the bend radius may be reduced to as low as one times the thickness or less, provided some care is used.

Magnesium is a wonderful deep-drawing material and can be drawn far deeper than either aluminum or steel on one pass. Figure 16 shows a single-stroke stamping with top and side stiffening beads incorporated to enable the case to withstand 30 psi external pressure. This stamping, after trimming, was welded to a shell-molded magnesium casting that included an O-ring pressure groove as well as an rf choke groove. Figure 17 shows a magnesium hemisphere made in one draw — one of the most difficult types of drawing because of the buckling of the large unsupported area. In drawing, both the metal being drawn and the die and punch must be heated and while it may frighten one to see a magnesium hot press in operation for the first time, because of the flames playing around the dies and oil dripping from the press, nevertheless I am not aware

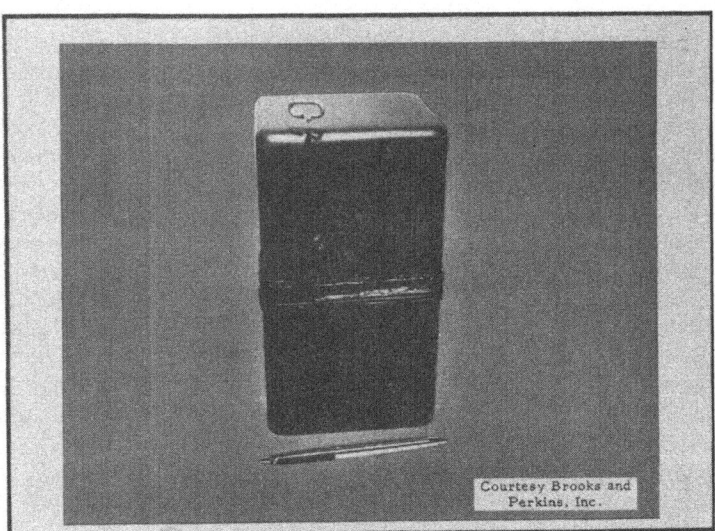

Fig. 13. Deep-drawn magnesium hermetically sealed container. Soldered peel
strip shown.

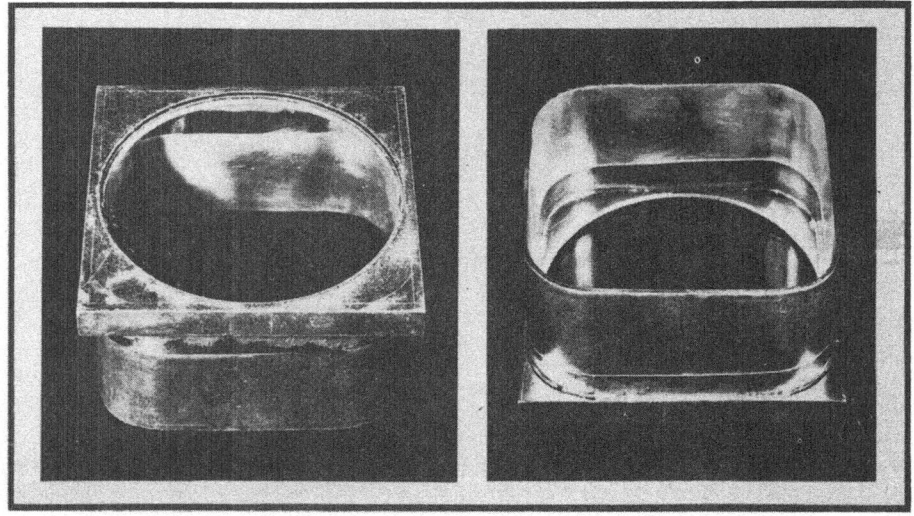

Fig. 14. Soldered magnesium. Glass window soldered in simulated instrument case.

of any case where a fire has caused any damage. Things like this become accepted only by familiarity with use of the material and that is my prime purpose
in this paper — to promote confidence in specifying magnesium and beryllium
in designs.

The chief limitation of magnesium being milled or turned in the lathe is not
the metal, but in most cases is the machine operator or the machine. Speeds and
feeds may be used on most magnesium parts that exceed the present capability
of most milling machines and operators. It just doesn't come naturally to the
average machinist to do a "hogging" job in rough machining a part and be forced

Fig. 15. Soldered magnesium. Glass insulated feed-throughs selectively copper plated.

Fig. 16. Single-stroke drawn magnesium pressurized case.

to have a good strong air blast to keep the chips away so he can even see the part being machined. This again comes under the category of proper education about magnesium. Fine finishes down to 4 to 7 μin. have been achieved on a single cut with a properly designed end mill. Figure 18 shows cavity plates from a microwave printed circuit assembly on which this finish was achieved.

Assembly of magnesium parts may be by any of the standard methods, screws, adhesive bonding, or riveting. However, when riveting, 56S aluminum alloy rivets should be used, with no dissimilar-metal corrosion protection required.

The fire hazard has been exaggerated to the point where many people are afraid to use magnesium. Magnesium will burn, provided enough oxygen is supplied to maintain the material above ignition temperature; however, the usual case is that, except in very thin gauges, magnesium will not support combustion due to the rapid conduction of heat away from the point of burning. If one insists on using dull cutting tools or tools not sharpened with proper clearance angles, then one will run the risk of starting a fire on fine chips. Also, if good house-keeping is not practiced and chips are not swept up at regular intervals, one may expect trouble should a fire start. However, because people are running the machines and are subject to all humans ills, it still makes good sense to have a bucket of G1 powder or other proprietary magnesium extinguishers adjacent to the machining operation. Once a magnesium fire has started and is in material fine enough to support combustion, the results can be quite spectacular if some-one sprays a stream of water on the burning mass. Again, these things only happen because of carelessness or ignorance and since the proper equipment must be used to put out an electrical fire, other proper equipment to put out a gasoline fire, and still other proper equipment to put out a wood fire, there is no reason to expect that one will not use the proper equipment to put out a mag-nesium fire. It must be borne in mind that magnesium was used in incendiary bombs, but early in World War II this material had to be replaced by aluminum powder since magnesium would not burn with sufficient intensity. Magnesium also was used in photographic flash bulbs; now aluminum is used because magnesium doesn't burn brightly enough. The people who produce commercial metal powders fear aluminum powders as a fire and explosion hazard far more than magnesium. Gasoline is used in automobile engines because it is, currently, the best material to use and one accepts the fire hazards that go along with gasoline. By the same token it is unrealistic to refuse to use magnesium because there is a remote possibility of fire. It is also unrealistic to refuse to use beryllium because of the health hazard mentioned previously.

Courtesy Brooks and Perkins, Inc.

Fig. 17. Single-stroke deep-drawn magnesium hemisphere.

A magnesium-lithium alloy is currently in the later stages of development by Battelle Memorial Institute and Dow Metal Products Company. For operation at room temperature and below, it has some interesting properties. Its density, in one alloy, is 0.049 lb/in.³, which is only 75% of the common magnesium alloys. Considering its stiffness in a beam such as is described in Table I we find that in beams of equal stiffness the magnesium-lithium beam will weigh 20% less

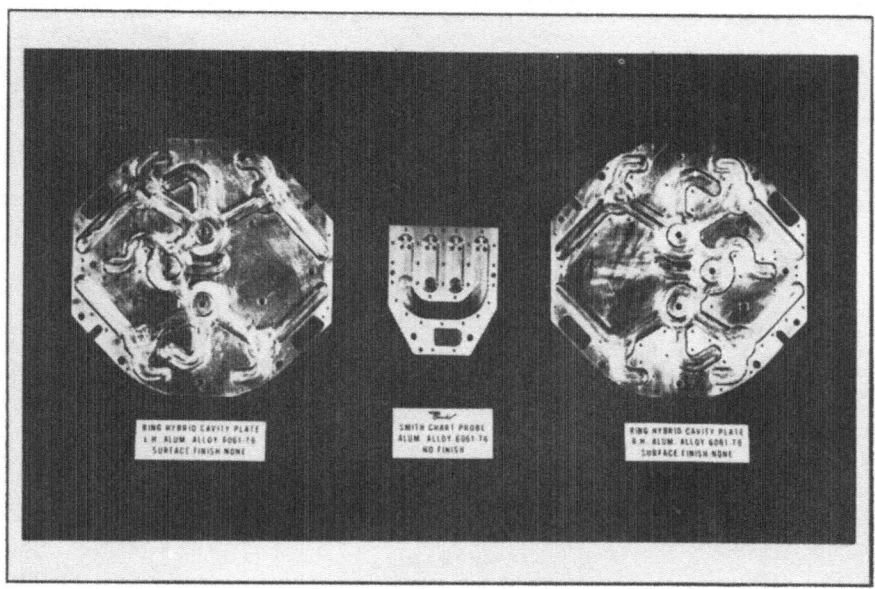

Fig. 18. Microwave printed circuit plates machined on Bendix numerical-control milling machine.

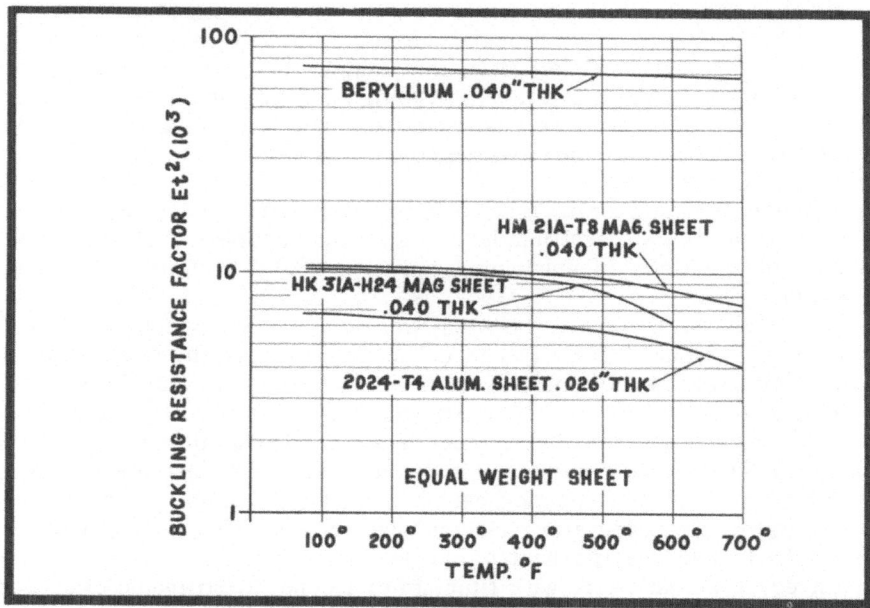

Fig. 19. Sheet buckling resistance comparison.

than in conventional magnesium alloys and will be 40% less than an aluminum alloy beam. For equal weight, the magnesium-lithium beam will be about twice as stiff as one made of the common magnesium alloy and almost five times as stiff as an aluminum-alloy beam. It can be welded, using AZ92A rod, and stress relief is required after welding. It takes anodic coatings such as Dow 17. It

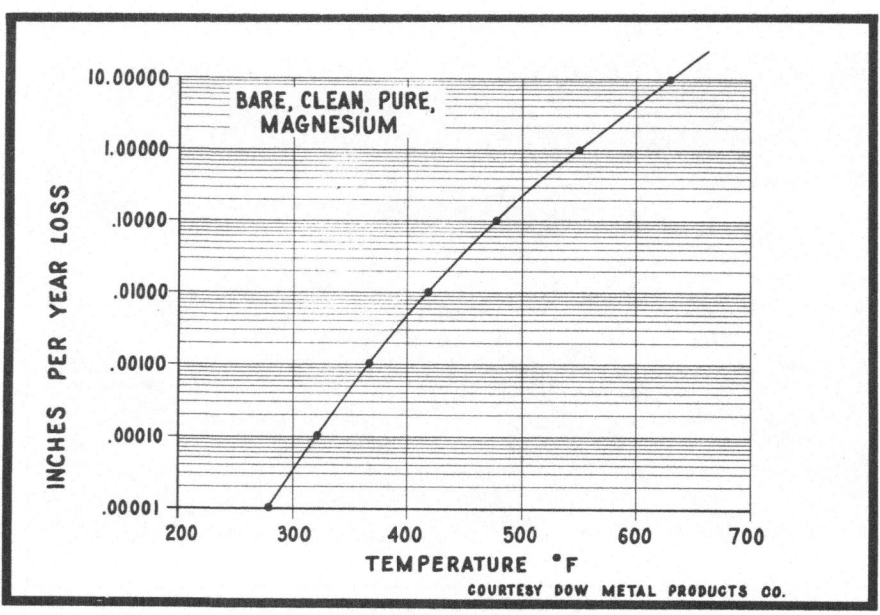

Fig. 20. Sublimation rate in absolute vacuum.

requires somewhat more precautionary measures in welding, machining, and handling since there is some evidence that fumes, dust, and skin contact will cause respiratory and skin irritation. Due to the high lithium content, the magnesium-lithium probably is more easily ignited than other magnesium alloys. It is being evaluated by several selected companies and we will, no doubt, hear more of it in the future.

For use at elevated temperatures there is no better light metal than beryllium. As can be seen in Fig. 4, beryllium holds its strength. Figure 3 illustrates that its modulus of elasticity is much more suitable than that of either aluminum or magnesium.

Figure 19 shows the sheet buckling resistance comparison at elevated temperatures of various materials and one can readily see the reduced weight possibilities by using beryllium. However, if that dollar sign blurs your vision too much, take a look at the magnesium-thorium alloys — both HK31 and HM21. These two materials maintain usable tensile properties up in the 600 to 800°F range whereas 6061 and 7075 aluminum and magnesium AZ31 lose their strength rapidly after 300 to 400°. A further gain is made when thicker gauges of magnesium are used; the temperature rise per second of elapsed time will be less because of the higher specific heat of magnesium alloys. This holds true when used on an equal weight basis with titanium or Inconel X. Stiffness, of course, will be greatly increased because of the greater material thickness.

One drawback to the use of bare magnesium as a space material is its sublimation rate. Figure 20 shows the rate of sublimation of magnesium with temperature as a variable. However, bear in mind that this curve is for pure magnesium. Very little, if any, data are available on sublimation rates of metal alloys. One would presume, however, that alloys containing constituents with higher vapor pressures would decrease the sublimation rate of the alloy. Also bear in mind that in space applications the magnesium probably will be surface

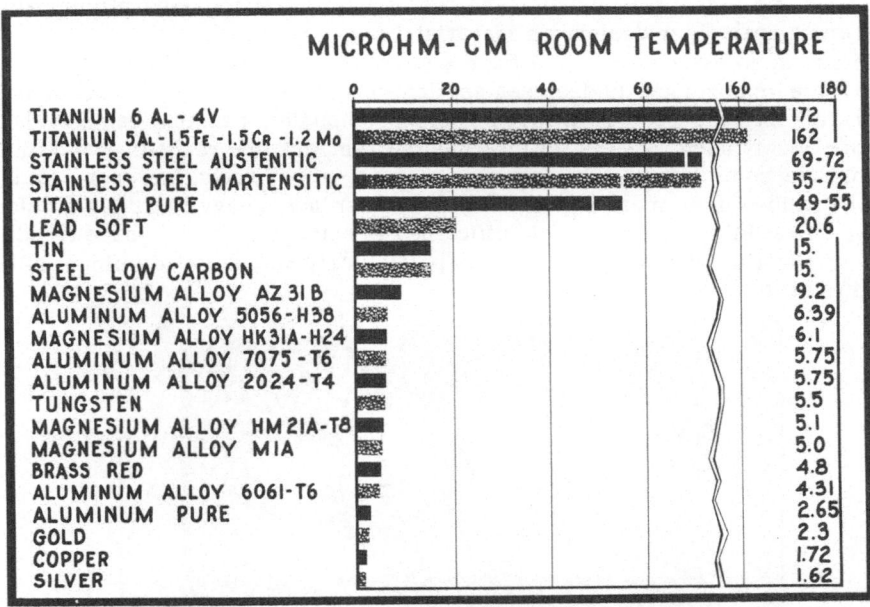

Fig. 21. Electrical resistivity comparison.

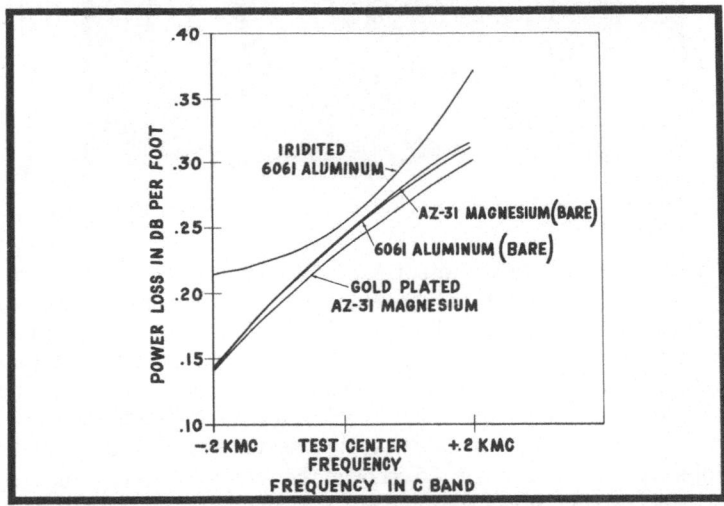

Fig. 22. Power loss in microwave printed circuit.

treated with electrolytic coatings such as Dow 17, HAE, Fluoride Anodize, or they may be plated. Copper, tin, gold, aluminum, and beryllium all have vapor pressures high enough that sublimation at normally encountered temperatures is not a problem. Very little has been published on vapor pressures of plastic materials.

Aluminum, while it suffers from excessive density, does have some features that merit its use on occasion. These are greater tensile and compressive strength at lower temperatures and its relative ease of brazing, and under a few conditions these features make it the best material to use. Generally speaking, however, it is the lazy designer's material.

Titanium is extremely useful from a corrosion standpoint if this attribute is of extreme importance. In deep sea applications, where stainless steel or other materials are not good from a corrosion standpoint, titanium can and should be more widely used. Also in the intermediate temperature range of 700 to 800°F it can have some useful applications. As most of you know, however, titanium must be welded in an inert atmosphere or at least have heavy inert gas shielding during the welding process. Weld efficiency is extremely high in pure titanium. Some of the titanium alloys, however, particularly vanadium-aluminum alloys, are difficult to weld.

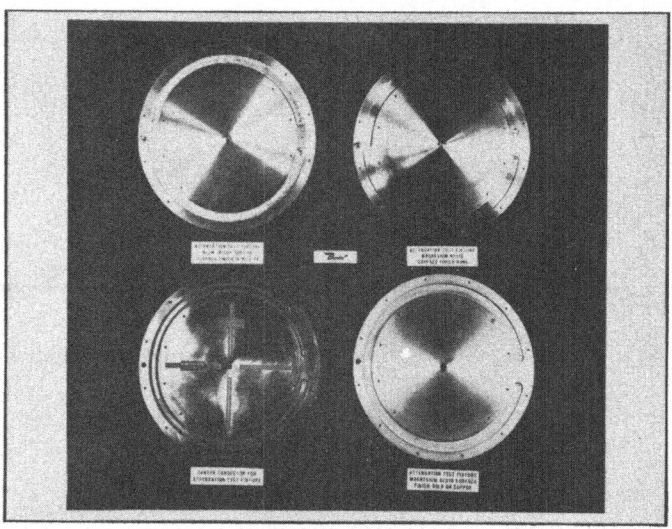

Fig. 23. Microwave printed circuit plates used in attenuation tests.

Magnesium has a higher electrical resistivity than aluminum, brass, or silver — common microwave component materials. Figure 21 compares the resistivity of various materials. While I don't recommend magnesium, yet, as a hook-up wire material (although new resistance welding techniques could change my mind), it is a good material for microwave plumbing and microwave printed circuits. Figure 22 shows the results of some tests made at Bendix Research Laboratories on printed circuit cavity plates made of aluminum and magnesium with various finishes. Figure 23 shows the circular cavity plates and center conductor used in this test. As can be seen, the differences are pretty small — but in favor of magnesium. Maybe at lower frequencies the differences would be greater, but at "C" band the penetration of electrical energy into the material is so small that the resistivity of the material has little practical effect. One company, using magnesium waveguides for years, has found that anodizing aluminum has little effect on attenuation. Adding Dow 17 to magnesium increases attenuation from 10 to 20%. Adding Dow 7 increases attenuation 5 to 10%. I do not know of any tests on magnesium with Fluoride Anodize, but think it would be quite low since the coating appears to be only a few molecules thick. HAE coating had practically no effect on attenuation with some samples, +10 to 15% with others.

Incidentally, the cavity plates in Fig. 18 with their highly "compacted" pattern and consequent saving in volume and weight were made practical by the use of a Bendix numerically controlled milling machine. The numerical control equipment for this machine was developed by the Bendix Research Laboratories and produced by the Bendix Industrial Controls Section. The wandering centerlines of the cavity slots were held to within 0.001 in., the depth to within 0.001 to 0.0015 in., and the surface finish finally varied from 4 to 7 μ in. Normal milling time on these cavity plates, including a bucketful of small holes, was only one hour and six minutes; this included rough and finishing cuts. Setup time was approximately 50 minutes. Complete tape preparation time was 31 hr. The total time, including tape preparation, paid for itself in making just one plate and considerably greater accuracy was achieved than could have resulted if a hand-operated milling machine had been used. The center conductor was a double-copper photo etching on epoxy filled glass laminate.

The foregoing is merely a splash in the ocean of evidence showing the desirability, if not downright necessity, of using these lighter materials in this space age electronic packaging. And as a parting shot, this must be borne in mind — one may resist using these better materials, but there are some competitors who a r e using them.

DISCUSSION

Question: Jake Rubin, General Electric, Philadelphia. I would like to compliment Mr. Beck on his excellent paper and ask if he would care to comment on dip brazing of magnesium and aluminum in widely different sections? Has he had any success in this?

Answer: Well, I wouldn't compare dip brazing in magnesium and aluminum. I don't think there is any comparison; it's a contrast. The only alloy that is dip brazable in magnesium, is the old M alloy. I think extrusions are about the only way you can buy that now unless you have a special mill run. You can buy it in waveguide form which can be and is being dip brazed. It is not as good a brazing material as aluminum because of the small temperature differential between the brazing alloy and the part being brazed. The temperatures are too close together for reasonable control. Except on the M alloy waveguides, I don't recommend brazing. Maybe, someday, there will be a good brazing alloy that will be better than we have now. I hope so. (In editing this discussion, I see I did not answer Mr. Rubin's basic question. In dip brazing magnesium or aluminum in widely differing sections, it is easy to get into considerable trouble. First, in preheating the part when the temperature is brought up too quickly and, in extracting the part from the brazing bath too quickly. Aluminum as well as magnesium knows its own coefficient of expansion far better than any designer and if it is not treated with due regard to this characteristic, you will get all kinds of perverse distortion. This is a trap that many designers spring when attaching bosses, etc., to sheet metal chassis.)

Question: Doug Connors, Boeing. I did not see any figures giving the comparative formability for beryllium. Do you have one?

Answer: Well, the reason that isn't shown is that at the time the slides were made there wasn't any information on it. Formability comes into the picture with sheet but, since there isn't a large quantity of sheet being used, I wasn't able to get comparative figures on it. We don't have any data at Bendix and I don't believe Brush or the Beryllium Corporation have any figures on the relative formability either. Due to its high strength and high E, it is going to be fairly rugged from a forming standpoint, but I can't give you any comparative figures.

Question: Mr. Beckman, Sperry Gyroscope, Great Neck, New York. I have a question in regard to joining beryllium-to-beryllium either by riveting, welding, or adhesion. Do you use any, and what do you prefer?

Answer: Well, it can be welded; it can be spot welded; it can be riveted, of course. For riveting it may be in the same category as magnesium. You use 5056 or some of the similar types of aluminum alloys for riveting magnesium, and I think that you will run into the same type of thing with beryllium. But it can be welded; it can be brazed; and it can be soldered. There are some proprietary solders wherein the solder is contained in the flux. So I don't think joining beryllium is a real problem. As with any other material, the specific parts to be joined will determine the process.

Question: You talked about the fact that the greater the use of beryllium, the lower its price. There is another factor you mentioned, its toxicity. To properly ventilate say, a lathe, drill press, milling machine, or shaper, and run it into a ventilating system and recover it by a wet process like cycloning would run well into five figures. Wouldn't that more than offset advantages of using these materials? Especially these days when the armed forces are getting tighter and tighter with money, and you are competing against many other companies, and you have to throw in the cost of beryllium at $70.00 a pound or more plus the cost of a ventilating system and hooding of machine tools, you would never get a contract from any of the armed forces.

Answer: I think that the armed forces probably looks at beryllium somewhat like they looked at magnesium a few years ago. They were willing to pay a little extra to get the extra performance out of the particular parts that were in question. If you can get extra performance out of the parts by the use of beryllium, I don't think that you would have too much trouble. If somebody else is smarter than you are and can do the same thing in a cheaper material, then you have rough competition. You have to assume that they are not going to perform as well as you are.

Question: With magnesium, for instance, you would not need the precautions you need with beryllium because beryllium, besides being toxic, can be fatal.

Answer: With magnesium you are probably going to be faced with hot forming. In beryllium you are going to be faced with ventilation, and in sheet forming probably hot too, and with higher temperature than magnesium.

Question: Haven't you found in your ventilation problems you actually have to have an enclosed system and take it out wet?

Answer: We don't have a ventilating system.

Question: You don't? Well, we had a case of beryllium pneumosis once, and it could be fatal.

Answer: I think you will find that beryllium parts, for probably sometime to come, unless you are willing to go into it on a pretty extensive scale, will be better handled by custom machine shops—custom fabricators—like Speedring in Detroit. A large share of their business is in machining beryllium and I think you will find that right now it is cheaper to have somebody else make the ventilation setup. They have done that and found that it's no real problem as long as they have enough volume to justify it. If you have one piece to make, obviously you're not going to set it up. If you're going to go into it, if you admit the possibility that it has a good future, then it's going to cost some money. But as I said in the paper, it cost you money for the lathes and other machine tools in the first place when you started forming steel or aluminum. You have to make some initial investment to realize a gain from it.

Question: Bob Rooney, RCA, Burlington, Massachusetts. Do magnesium castings require special surface treatment or finish in order to maintain a pressure differential under space vacuum conditions?

Answer: Any casting can be made porous and quite often it is inadvertently made porous. Under those conditions, where you don't have sufficient foundry control to make sound castings, you have to impregnate them. That will be true in magnesium or aluminum. It will be more true of aluminum than it will be of magnesium because, generally, the magensium casting alloys have lower microporosity than aluminum. You can get into a bad foundry in any part of the country and you can have trouble. But castings can be made tight and if they are not made tight, they can be impregnated to maintain pressure tightness. On top of that, if you are talking about space applications, you are very likely going to plate it or paint it or treat it in some way. The surface treatment will fill up these leaks or at least it will stop any sublimation. It is not normal, except in some cases where you know your temperatures are going to be below 350°, to use bare magnesium. In a good many satellite applications even aluminum is plated for emissivity, or for any number of reasons, and you do the same thing with magnesium.

CONTROLLING THE MECHANICAL RESPONSE OF PRINTED CIRCUIT BOARDS

R. P. Thorn

Lord Manufacturing Company, Erie, Pennsylvania

The methods for packaging electronic equipment circuits have evolved to a high level of efficiency. The use of solid state devices and miniaturized components results in high component density and many new and different electrical and mechanical design problems. It seems the design tendency today is to produce a circuit configuration that will function electrically for the proposed application and to accept whatever mechanical arrangement and response happen to result.

One typical method of constructing electronic equipment is to build modular or functional elements where the various electrical components are supported on a circuit board with a printed or deposited interconnection matrix. Where such equipment is used for high-performance aircraft, missiles, and ground support applications, the system must not only perform satisfactorily as an electronic device, but must operate reliably in operational environments, that include vibration, shock, noise, and other conditions. In addition to living through the expected environment, most equipment must also pass an accelerated life or qualification test. These tests are generally covered by broad military specifications often calling for continuous system operation under high levels of vibration and shock, which are chosen to simulate the potential degradation due to mechanical disturbances expected in actual operation.

These specifications have tended to become more severe and demanding over the years. It is becoming more and more apparent that the design engineer cannot just toss together a group of circuit components to meet an electrical requirement and expect the system to pass accelerated life tests or to perform satisfactorily in an operational system. The mechanical response characteristics of the complete package must be evaluated and controlled as carefully as the electrical performance to provide a suitable system capable of meeting both electrical and mechanical performance. As described in the paper entitled, "Controlling Structural Response of Electronic Chassis" [9], presented at the first Packaging Symposium, this can be a complicated problem.

Consider a typical modern airborne electronic unit including isolators for the control of shock and vibration, with associated tie-down and mounting base structure (see Fig. 1). When the base is mounted in a vehicle with the equipment fastened in place, the system may be schematically described as shown in Fig. 2. The top spring-mass damper system represents a fragile or the most fragile component mounted on a structural support. Mechanically, this component may be described as having some elasticity, some damping, and some mass. The mass is relatively easy to determine while the elasticity and damping are functions of the internal construction of the part and the method of connection to the remainder of the system. The next spring-mass damper system represents the circuit board which supports the components. This is a distributed structure which may be approximated by the simple lumped elements. The next spring-mass damper system represents the equipment frame or main chassis of the unit, while the one below that represents the distributed mass of the top tray of the mounting base, and the vibration and shock control elements or isolators. The next spring-mass damper system represents the distributed characteristics of the vehicle structure which is, again, approximated by simple lumped components. Finally, somewhere along the line, there is an acceleration, velocity, or motion

Fig. 1. Typical airborne electronic unit with isolator.

input due to a mechanical disturbance or excitation of the vehicle structure. A typical frequency response or transmissibility curve for such a simple system would have a series of peaks as shown in Fig. 3. Note the frequency ranges for elements in the complex system. No effort has been made to consider coupled effects, or the multidegrees of freedom expected in distributed mass systems. The gross complexity of the design problem soon becomes overwhelming.

Typical design practices in the industry today suggest that the various natural frequencies of the composite unit be separated as far as possible. A good figure is one to two octaves. This involves choosing the spring and mass combinations of the system to tune the resonant frequencies so they are non-coincident or occur in noncritical regions or outside of the frequencies included in the band of excitation. Unfortunately, good design practices, reasonable equipment weight, satisfactory stress levels, and component density or equipment size usually result in typical natural frequency ranges for various system elements (see Fig. 3). This means the design engineer has little real control over the many possible structural resonances and must concern himself with the few critical frequencies likely to cause trouble. The only other solution available is to control the amplitude of the response through damping. This may be provided in many ways, either by chance or by design.

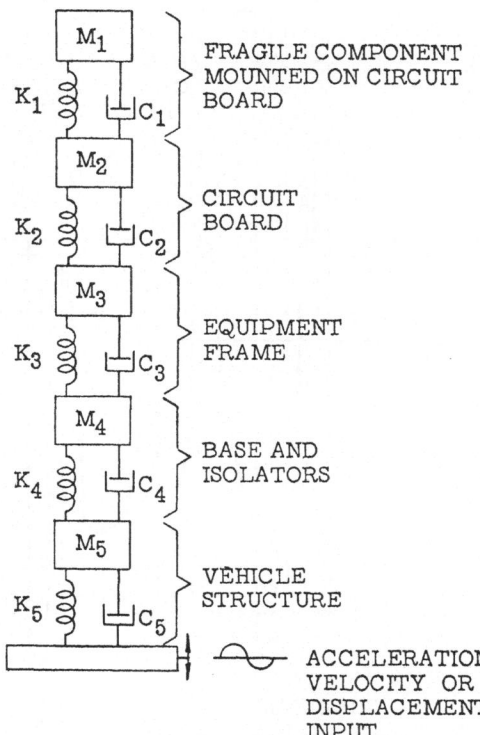

M₁ — FRAGILE COMPONENT MOUNTED ON CIRCUIT BOARD

M₂ — CIRCUIT BOARD

M₃ — EQUIPMENT FRAME

M₄ — BASE AND ISOLATORS

M₅ — VEHICLE STRUCTURE

ACCELERATION VELOCITY OR DISPLACEMENT INPUT

Fig. 2. Lumped constant schematic of typical complex electronic unit.

To appreciate the effect of damping in solving such a problem, it is necessary to consider the response characteristics of a simple single-degree-of-freedom system, that is, one spring-mass damper combination from the complicated system. The equivalent diagram and response characteristics are shown in Fig. 4. The natural frequency of the suspension system and the magnification at resonance can be predicted. The damped natural frequency equals $(1/2\pi)\sqrt{k/m - (c/2m)^2}$, the resonant amplification equals $\sqrt{1 + 1(2c/c_c)^2}$. Note that the natural frequency depends primarily upon the spring (k) to mass (m) ratio, and the magnification at resonance depends entirely upon the amount of damping or fraction of critical damping c/c_c in the system.

From this simple response curve, it is easy to see that any spring-mass damper combination has a response that is frequency sensitive. Location of the natural frequency and amplification at resonance can be controlled by a proper choice of the spring-mass damper arrangement. Generally, the safest design procedure to pursue is to shoot for a minimum number of structural responses, each with the lowest possible magnification at resonance. The mass associated with any circuit configuration is generally pretty well fixed by the function of the circuit itself and the location or interconnection of the various parts. Spring rate or structural stiffness is usually limited by upper package weight and size. The lone remaining element to be considered is damping for the control of structural response.

As discussed in [9], it is definitely possible to add a significant amount of structural damping to a system to adequately control the mechanical response

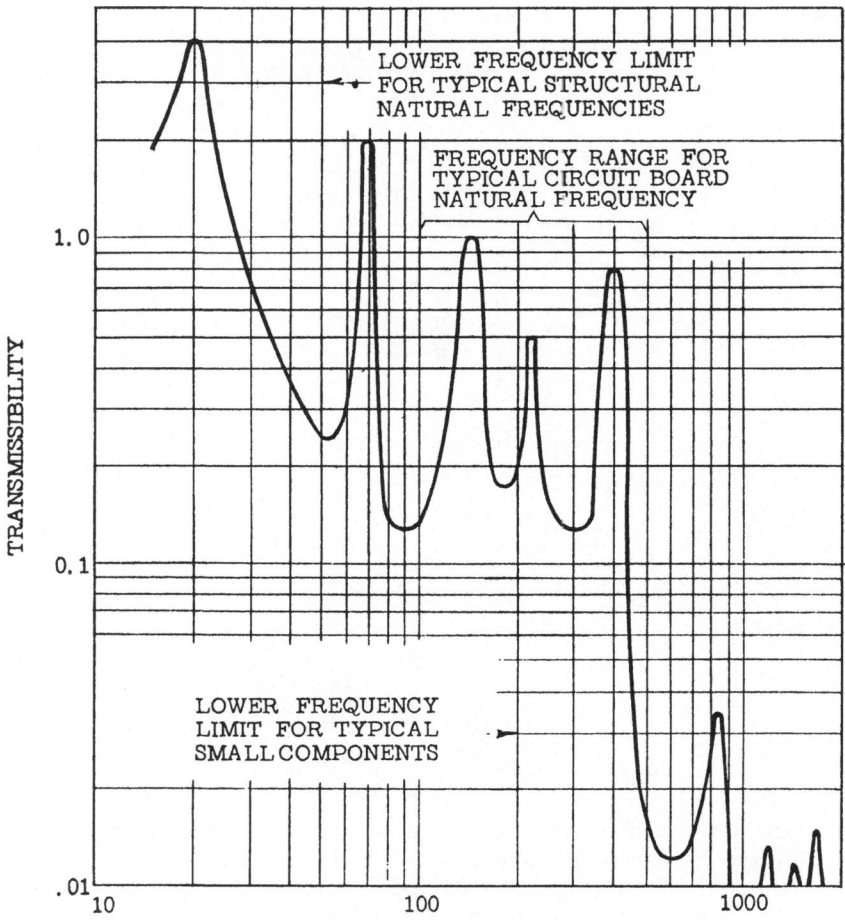

Fig. 3. Response curve for typical electronic unit.

at resonance. However, experience has shown that controlled damping added to the external frame members and structures only may not be sufficient to provide satisfactory levels of attenuation. A typical example of this is shown in Fig. 5. This is a simple U-shaped frame or structural module supporting a 3 in. by 4 in. printed circuit board at four corners. Under 10 g vibratory sweep, components were popping off the board. Several attempts were made to introduce damping into the sheet metal structure. None of these efforts proved to be particularly successful due to the responsiveness of the circuit board itself. This led to the development of a printed circuit board with controlled internal damping. The addition of predictable amounts of damping into the circuit board may be compared to the results obtained by adding damping to a single-degree-of-freedom system (see Fig. 4). There is a reduction in the amplification at resonance which results in a reduction in total dynamic strain in the system or motion across the component and a reduction in the dynamic stress level. Typical damped circuit board construction is shown on Fig. 6. These are basically extensions of the simple three-ply dyna-damped panels available in many types of metal skins

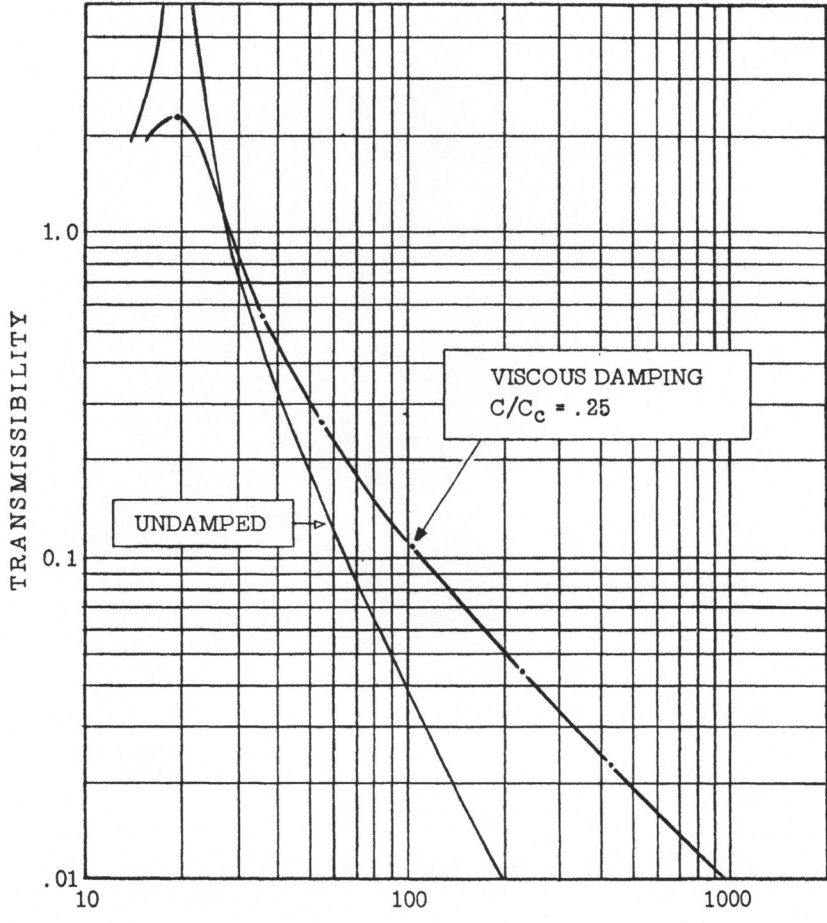

FREQUENCY IN CPS

Fig. 4. Response curve for simple single-degree-of-freedom system.

and thicknesses for primary structural members. The successful application of damping to pointed circuit boards was primarily the result of the development of special processes to bond the elastomeric core to the fiber-reinforced epoxy board.

The core material is BTR, a special highly-damped elastomeric material compounded by Lord Manufacturing Company for use in both isolators and damped structural configurations where high strength and high damping over a temperature range of −65 to + 300° is essential. The amplification at resonance of the basic material itself is shown in Fig. 7 as a function of temperature. Note the near constant characteristics over the entire temperature range. In addition, BTR is also amplitude and frequency sensitive, as shown in Fig. 8. This means that the larger the strain across the material, the more highly damped it appears. Other special highly-damped elastomeric materials are also available for limited temperature ranges, depending upon the application.

Three basic thicknesses are available as shown in Table I. All are simple three-ply configurations shown in Fig. 6. In each case, the effective damping is

Fig. 5. Printed circuit board with U-shaped structural suspension.

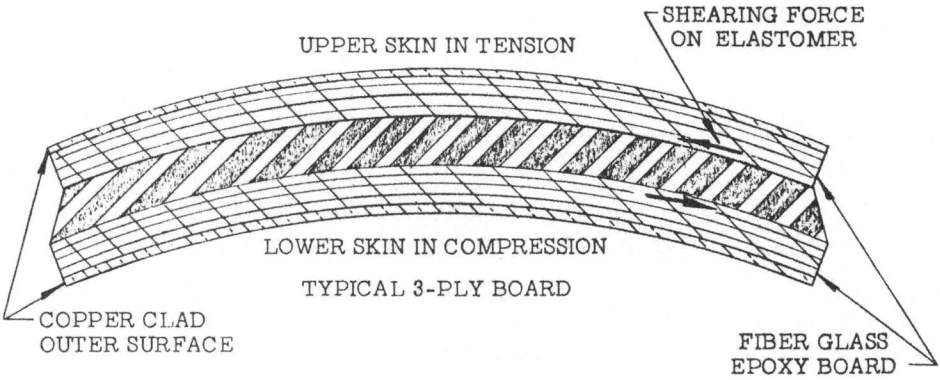

Fig. 6. Resultant forces due to deflection of damped circuit board.

obtained by shearing forces in the fully bonded elastomeric layer. Effective resonant control is obtained by dissipating a portion of the energy exchanged by the distributed mass and spring through hysteresis damping in the elastomeric layer.

Several methods are available for evaluating the performance characteristics of these boards. Logarithmic-decay characteristics of simple cantilever beams made of damped and undamped glass epoxy materials are shown in Fig. 9. The obvious advantage of damping is quickly seen.

Vibratory performance characteristics of identical boards were also compared using a simple four-point support. This is shown in Fig. 10. In all cases,

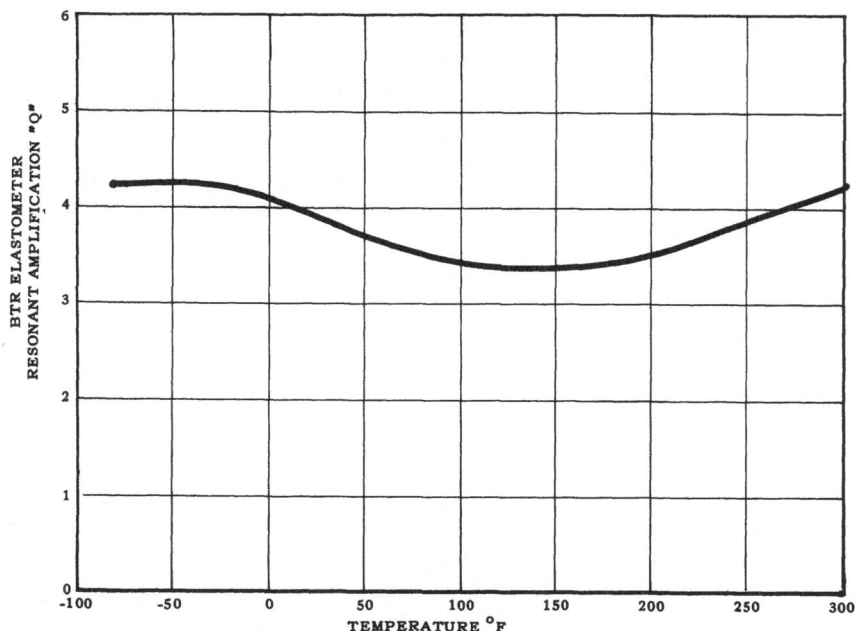

Fig. 7. Effect of temperature on dynamic characteristics of BTR elastomer.

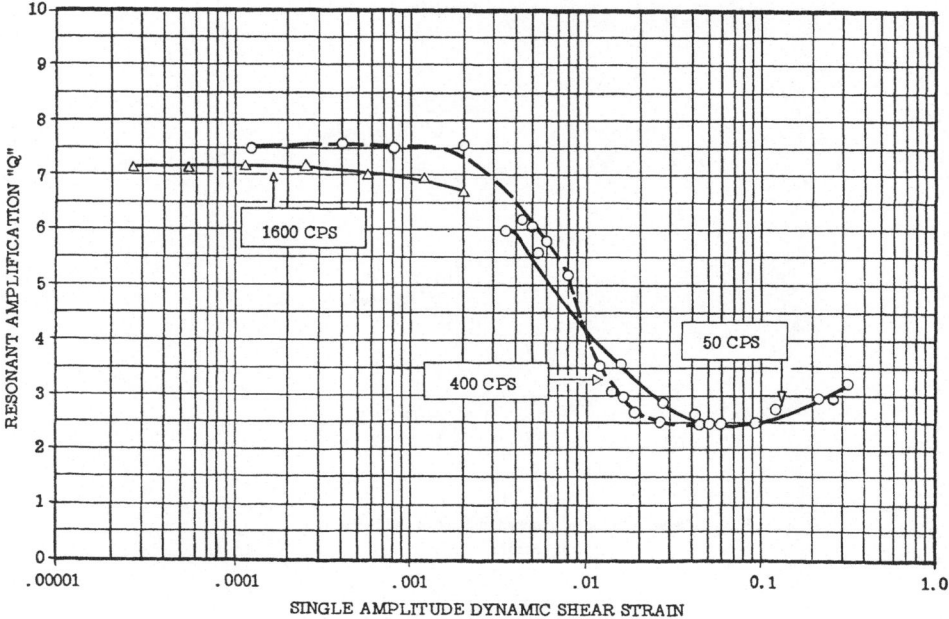

Fig. 8. Single amplitude dynamic shear strain.

TABLE I. Circuit Board Thicknesses

Total thickness	Tolerance on thickness	Skin thickness	BTR core thickness
0.062	± 0.0075	0.020	0.022
0.093	± 0.009	0.030	0.033
0.125	± 0.012	0.046	0.033

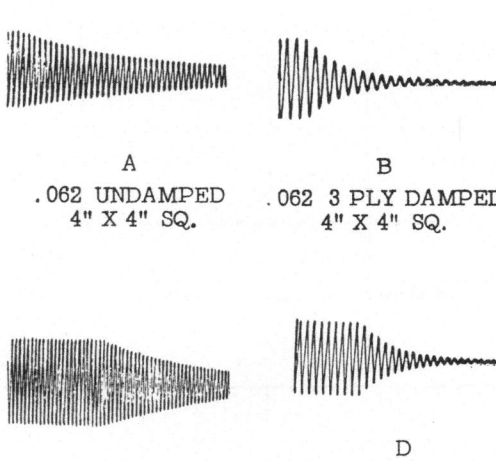

A
.062 UNDAMPED
4" X 4" SQ.

B
.062 3 PLY DAMPED
4" X 4" SQ.

C
.093 UNDAMPED
5" X 5" SQ.

D
.093 3 PLY DAMPED
5" X 5" SQ.

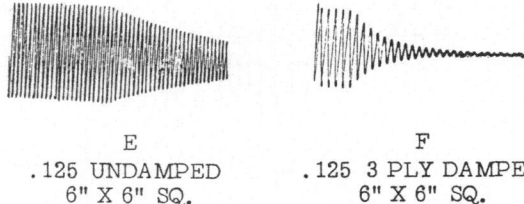

E
.125 UNDAMPED
6" X 6" SQ.

F
.125 3 PLY DAMPED
6" X 6" SQ.

Fig. 9. Decay curves for damped and undamped printed
circuit boards.

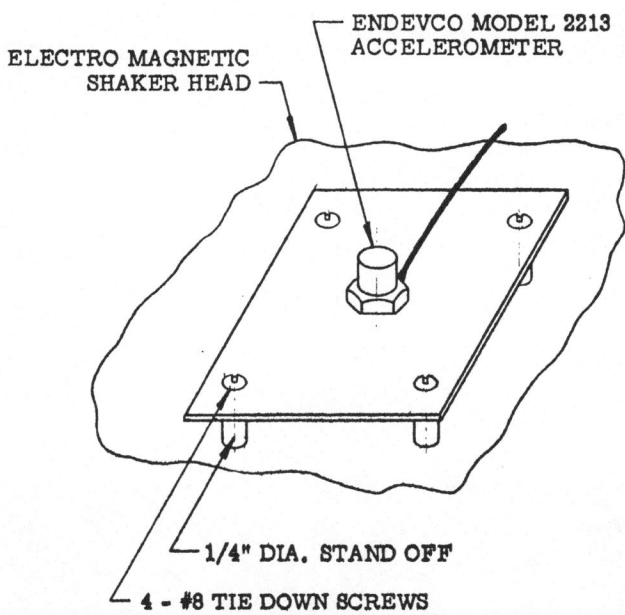

ELECTRO MAGNETIC
SHAKER HEAD

ENDEVCO MODEL 2213
ACCELEROMETER

1/4" DIA. STAND OFF

4 - #8 TIE DOWN SCREWS

Fig. 10. Test setup for measuring response of circuit boards.

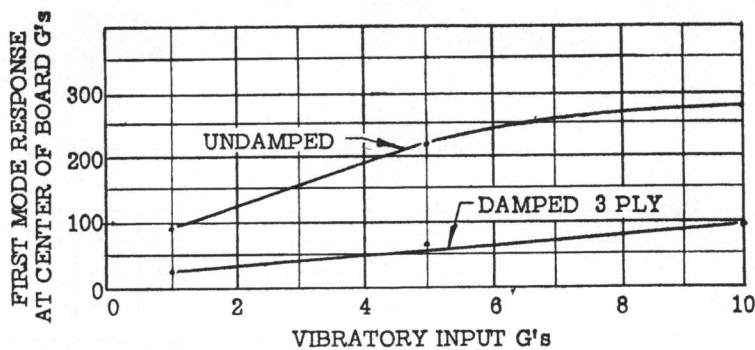

Fig. 11. Response of 4 in. by 4 in. 0.062 glass epoxy circuit board, fixed at 4 corners.

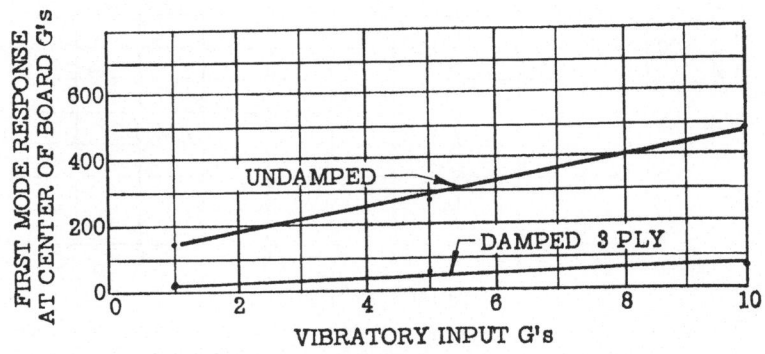

Fig. 12. Response of 5 in. by 5 in. 0.093 glass epoxy circuit board, fixed at 4 corners.

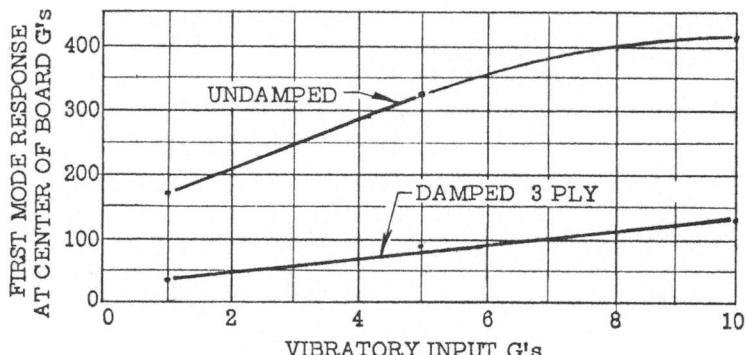

Fig. 13. Response of 6 in. by 6 in. 0.125 glass epoxy circuit board, fixed at 4 corners.

Fig. 14. Production configuration of printed circuit board. Left: damped; right: undamped.

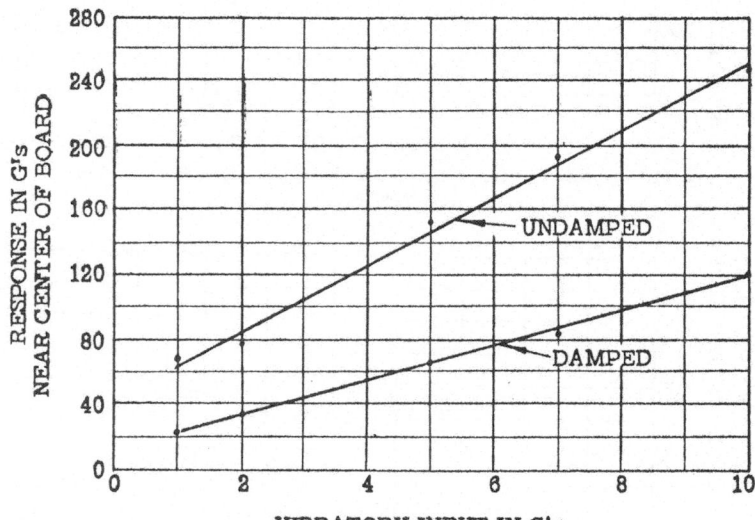

Fig. 15. First-mode response complete module using damped and undamped circuit boards.

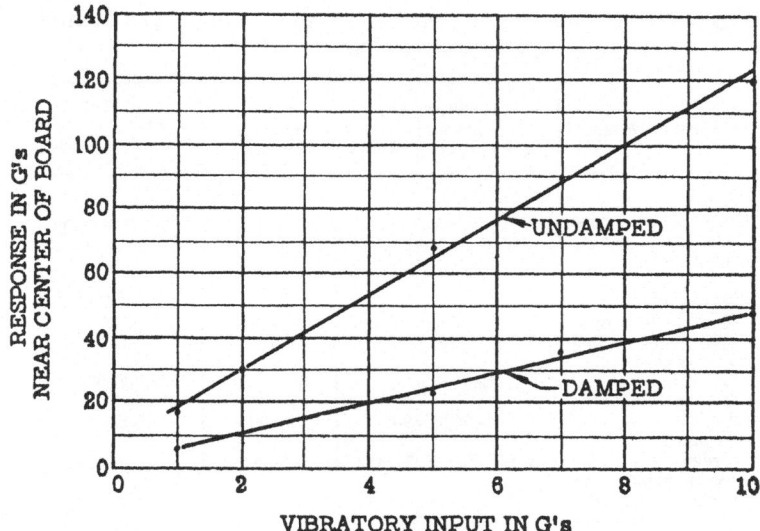

Fig. 16. Second-mode response complete module using damped and undamped circuit boards.

the boards were loaded only by an accelerometer fastened directly through the board. Typical response curves for the undamped boards and the damped boards at various inputs are shown in Figs. 11, 12, and 13. These indicate a surprising linear trend and suggest a method of extrapolating expected system response for high-level test inputs based upon low-level vibration runs. Knowing the minimum vibration tolerance of any component in an electronic network, it would be possible to project the maximum input level that a particular board could withstand based upon low-level tests. This approach could also be used to estimate the minimum clearances needed between boards for a particular level of vibratory input. The nonlinear response of the undamped board at high-input g levels is due to friction damping at the tie-down points.

This circuit board material was designed to be a direct replacement for existing Nema G10 and Nema G11 boards. The material can be handled in exactly the same manner as the base glass epoxy board. It can be sheared, punched, drilled, sawed, milled, and processed easily with present-day etching, cleaning, and soldering methods. The standard board is available in 18 in. by 18 in. nominal sheets and is supplied with Nema G11 skins unclad or clad with 2 oz copper. However, clad or unclad Nema G10 paper base epoxy, glass-reinforced Melamine, and paper-base Melamine can also be supplied.

The physical characteristics of the damped composite boards are compatible with the requirements of MIL-P-13949B. A typical comparison of test results for these boards is shown in Table II.

It is quite reasonable to conclude that no degradation of the capabilities of the board has resulted.

A typical application of the damped circuit board to a production configuration is shown in Fig. 14. This shows an actual circuit board constructed of undamped and damped material. The original test module was used to support the boards during test. Vibratory input was normal to the plane of the boards. The comparative performance characteristics of the damped and undamped construction are shown in Figs. 15 and 16 for both first- and second-mode

TABLE II. Properties of 0.062 Damped Circuit Board Tested
to MIL-P-13949B

Property to be tested	Conditioning procedure (See MIL-P-13949B)	Acceptable property values	Average property values damped circuit boards
Solder dip:			
Unetched specimens	A	No blistering or delamination	
Etched specimens	A	No interlaminar blistering	
Peel strength, min., lb/in. width 2 oz copper:			
After solder dip	A	9	12
After tem. cycling	(3)	9	12
Volume resistivity, min., meg-cm.	C-96/35/90	100,000	7,000,000
Surface resistance, min., meg-cm.	C-96/35/90	1,000	200,000
Water absorption, average, max., %	E-1/105+des+D-24/23	0.035	0.018
Dielectric breakdown (parallel to laminations), average, min. kv:			
Step-by-step test	D-48/50+D-1/2/23	30	40
Dielectric constant average max., at 1 Mc	D-48/50+D-1/2/23	5.8	3.9
Dissipation factor, average, max., at 1 Mc	D-48/50+D-1/2/23	0.045	0.007
Flexural strength, average, min., psi:			
Lengthwise	A	55,000	21,800
Crosswise	A	45,000	22,000
Arc resistance, average, min., sec	D-48/50+D-1/2/23	60	97

resonances. Typical reductions in magnification at resonance run about 3 to 1 for most usual board constructions. This means that a printed circuit board capable of surviving 3.5 g of peak input when undamped could take 10.5 g peak input if damped. In many cases, this is more than sufficient to allow the system to pass qualification tests without any change in design.

It is now possible, through the use of controlled damping in printed circuit boards, to extend the control of structural response all the way down to the level of the individual component. This approach leads to several attractive results:

1. Improved reliability due to the reduction of resonant response.
2. Increased package density due to the reduction of motions at resonance between adjacent boards.
3. The possibility of using less expensive components with lower vibration tolerance.
4. Reduction or elimination of redesign during the qualification-test phase because of higher vibration tolerance due to the reduced resonant peaks.
5. Predictable performance based upon a fully bonded part cured under high temperature and high pressure to provide a high-temperature time-stable part.

Through good design methods and the use of proper materials, the design engineer c a n successfully cope with vibration problems.

REFERENCES

[1] J. P. Den Hartog, "Mechanical Vibration" (McGraw-Hill, New York, 1956).
[2] G. W. Painter, "The Measurement of the Dynamic Modulus of Elastomers by a Vector Subtraction Method," ASTM Bull. (Oct., 1951), pp. 45-57.
[3] G. W. Painter, "Dynamic Characteristics of Silicone Rubber," Trans. ASME (Oct., 1954), pp. 1131-1135.
[4] G. W. Painter, "Dynamic Properties of B. T. R. Elastomer." SAE National Aeronautic Meeting, Los Angeles, California (Sept. 29 - Oct. 4, 1958) Paper No. 83B.
[5] R. P. Thorn, "The Mobility Method," Machine Design (December 10, 1959), pp. 144-157 (Dec. 24, 1959) pp. 104-111.
[6] B. W. Campbell and R. P. Thorn, Engineering Report No. 325 on Damped Structural Sections and Laminated Panels (Lord Manufacturing Company, Erie, Pennsylvania).
[7] R. Plunkett, Editor, "Colloquium on Mechanical Impedance Methods for Mechanical Vibrations," Trans. ASME (Dec. 2, 1958).
[8] G. W. Painter and B. W. Campbell, "Evaluation of Damped Three-Ply Laminates and Structural Elements," In Plant Engineering Report No. 3, Project 01783.
[9] R. P. Thorn, "Controlling Structural Response of Electronic Chassis," First Intern. Electronic Circuit Packaging Symposium Papers.
[10] B. W. Campbell, "Elastomers Applied to Structural Damping," ASME Paper No. 60-RP-19 (October 9-12, 1960).

DISCUSSION

Question: Herb Flatto, Hughes, Fullerton. Have you made any investigation into controlling the response of welded modules with increased component density and potting compounds which in themselves result in damping of the structure?

Answer: This is a much more complicated problem because of the three-dimensional configuration. The introduction of the potting compounds virtually eliminates any approach which might use a material such as this in a constrained layer damping configuration. It is entirely possible to introduce a potting material that will have sufficient damping properties to control, successfully, the amplitude of response of the components within the welded three-dimensional configuration. The unfortunate thing is that these materials must be used as an extensional damping medium. That is, only the surface of the part is coated, and there is no continuous constraining layer to force the material into shear. This means that the modulus of the potting compound (this is the dynamic elastic modulus) and the dynamic damping modulus of the material must be quite high. To get these properties it is necessary to use a class of materials that are rather limited in their temperature range. If a limited temperature range of operation is satisfactory for the particular device I would say yes, it is possible to have a material that will work satisfactorily in this manner.

Question: Does the increased component density decrease the problem of dynamic response as oposed to the flat configuration of your circuit board?

Answer: I would say that it does. Certainly, if you can decrease length. For a simple beam equation the natural frequency is proportional to $(1/ml^4)^{1/2}$ where m is the mass per unit length. Any time l can be reduced there is a significant increase in the natural frequency of the structure. For a constant "G" input a higher natural frequency means lower strains and possibly less trouble. This increase in natural frequency still produces a resonant response which can only be controlled by damping.

Question: Ken Plant, Minneapolis-Honeywell, Minneapolis. Because at least part of this BTR damping medium must be in immediate contact with the circuitry, I would be interested in knowing how this damping material reacts to environment, such as humidity and changes in temperature—high temperature. Do the electrical properties of this medium change as it absorbs moisture or is it necessary to seal the edges of boards to completely seal the components, and so forth?

Answer: If you will look on page five of the paper you will find some of the properties of the damped circuit board when tested to MIL-T-13949-B., including water absorption, dielectric constant, and dissipation factor. These will give you some of the characteristics of the board in accordance with the MIL specification tests. We have found that the boards themselves, the composite damped circuit boards, do not significantly change their properties after the dip soldering process. This is probably as severe a temperature change as you'll ever find in any application. The dissipation factor and most of the electrical properties also remain virtually unchanged. We know that the material which we are using is both time and temperature stable. It is cured under high pressure and high temperature and, as such, from a chemical standpoint has reached its stabilized condition. It is cured under very high temperature, far higher than would normally be encountered in use. The temperature range which we recommend is up to 250° ambient. The boards were replaced in an oven at 350° for about four hours, just to see what the effect would be. The boards were unchanged; so I think our 250° upper limit is quite conservative.

Question: Tom Mahoney, Boeing Compnay, Seattle. I would like to ask two questions. First, does the damping material exhibit any fatigue breakdown after an extended period of vibration? And the other question is—most of our circuit board and electrical connections are subject to fatigue failures. According to your Fig. 6A, you indicate that due to the shear forces you would expect to see a considerable shearing or bending motion on any connections which were connected to the circuit board. Have you experienced any failure of this kind?

Answer: Well, let me give you a practical example. Here is a circuit board; this one happens to be the damped configuration NEMA-G-11. This circuit board is undamped NEMA-G-10. Both boards were tested in exactly the same manner. This board, without damping, has a vibration tolerance of about $3g$. At $3g$, after just one or two sweeps, circuit components were popping off the board. The damped board has been subjected to innumerable (over 10) extended sweeps, from 5 to 2000 cps, at a $10\text{-}g$ level in the resonant range of the board itself without component failure. Now this particular board is made with eyelets. We gave actual test data from users who have successfully through-plated these boards. There is no failure experienced in properly plated through holes. We have found that all of our trouble has been breaking the component leads near the attaching point and not failure of the connection through the circuit board itself. In all cases the addition of damping has significantly reduced the amplitudes at resonance.

Question: You indicated there were sweeps involved. Over what period of time were these involved—how many hours, minutes?

Answer: The sweep rate was approximately $\frac{1}{2}$ octave per minute. I'm ashamed to admit it, but there were so many tests on this board that I can't give you an accurate number as to how many times it has been exposed. It has really been given a pretty rough test. The original qualification test on this particular part is $5g$ at a sweep rate of 1 octave per minute.

Question: Joe Vasilik, RCA, Van Nuys. You implied that the number of components does not change the response of the board. Yet this kind of damping depends on relative motion between the constraining layers. Are the components attached through both constraining layers or just one?

Answer: They are connected through both layers. It is a typical through-eyelet application where the leads are inserted through the eyelet, bent over, and then the part is dip soldered in the normal fashion. As a matter of fact, take this board and use it just like you use the regular board. Now the surprising thing is, and I honestly don't have a good theoretical explanation for this, that the through connection does not grossly modify the response characteristic of the board. The number of components also does not strongly affect the response. I think you can see that by looking at the response curve of the simple laminate with a single accelerometer connected to the middle of the board and comparing it to the response of the more complicated structural configurations we have here, where the acceleration was measured at a point fairly close to the center of the board. Fortunately, this is one of the gifts of nature. I can't explain it, but it does work.

A LOW-DENSITY POTTING COMPOUND

A. J. Quant

Sandia Corporation, Albuquerque, New Mexico

INTRODUCTION

The development of certain electronic packages for missile and airborne applications has created a need for a lightweight potting compound that offers greater environmental protection than that provided by the available polyurethane foams. Preliminary efforts indicated that this objective could be best achieved by incorporating low-density fillers into conventional epoxy resins.

Several such fillers * were investigated during the course of this effort, but none possessed processing characteristics and end properties equal to those of glass microballoons. An exhaustive review of possible resin systems was required as this type filler proved more difficult to handle than conventional solid fillers. It is not interchangeable with other fillers and cannot be freely substituted into existing formulations since it must be processed over rather limited ranges of time, temperature, and viscosity. Failure to maintain these conditions yields nonhomogeneous castings or mixes which are too viscous to process.

DEVELOPMENT

It should be borne in mind that the formulation developed is intended for use in the encapsulation of complex, expensive, electronic circuitry designed to withstand extremely severe environments. Therefore, a difficult-to-process material is not objectionable when it yields an end product of the high confidence level required.

The system finally selected parallels a previous Sandia system using the standard filler, mica, in that diethanolamine-cured Epon 828 was utilized for many of the same reasons that made it attractive earlier. Several of these reasons are:

1. good pot life
2. fluidity at the processing temperature
3. low exotherm
4. almost complete absence of toxicity.

A processing study of the microballoon-filled mix showed that other resins of a similar epoxide equivalent are not all interchangeable since differences in viscosity and reactivity result in less homogeneous castings.

Other variables, in addition to the choice of resin and hardener, were necessarily fixed to ensure homogeneous mixes. The concentration of filler was fixed at its maximum processable limit; anything less results in thinner mixes which permit flotation of the hollow spheres. The cure temperature was firmly fixed at 150°F; lower temperatures enhance flotation because the mix is in a fluid state for a longer time. Increased temperatures also enhance flotation since the viscosity of the mix is sharply reduced. Batch size was also restricted as

---

*Urea-formaldehyde, phenolic, and silica microballoons as well as polystyrene and aluminum silicate hollow spheres of a much larger particle size were investigated.

35

small batches lose heat readily and become too viscous to process. Large batches exhibit unusually high exotherms because of the thermal insulation imparted by the hollow spheres; this results in a reduced pot life. Timing is very important if the mix is to be successfully evacuated during periods of optimum viscosity. The type of mixer and time of mixing must also be defined to avoid excessive crushing of the fragile microballoons.

These investigations of the properties of the several raw materials and their interaction within a mix resulted in the recommendation of a single formulation and a tightly defined process.* This single formula is the one which yielded the maximum advantages of the glass microballoon filler consistent with a minimum of processing limitations.

From the foregoing, it can be seen that the processing of a microballoon-filled epoxy resin is critical. If such a critical process is to remain manageable, reasonable control must be exercised over the individual constituents of the formulation. Fortunately, the resin and hardener give little difficulty. However, the glass microballoons sometimes exhibit serious batch-to-batch variations that lead to processing difficulties or a reduction in end properties of the cured mix.

When these batch variations became evident, a program was set up to make several properties measurements on the batches on hand and to continue the measurements on all future batches.†

The property of greatest importance is average particle density. Since the formulation is based on parts by weight, any change in particle density is reflected by an adverse change in viscosity due to the variation in the volume of filler incorporated. Particle size distribution is also important since it controls the packing of individual particles within the resin matrix. Moisture content, the last requirement considered mandatory, is controlled because excessive moisture results in undesired agglomeration of the filler particles. Additional properties presently being determined on an "information only" basis are alkalinity, water solubility, and flotation ratio.

Cold-shock tests are being run by Sandia on all lots of microballoons received. This test is not part of the microballoon specification, since it must be evaluated from the final mix rather than from the microballons alone. The tests are primarily intended to ensure that a processable mix which will pass thermal and mechanical shock testing can be guaranteed. A continuing investigation is under way to determine what properties and what limits of these microballoon properties best define this objective.

The study of the properties of early batches of glass microballoons led to the issuance of a tentative specification.‡ From the data listed in Table I, it can be seen that the more recent batches generally meet the limits established in the specification and yield processable mixes that pass thermal shock testing.

An observer studying Table I may question the broad limits on the sieve analysis. However, all attempts to narrow this range and to include information on further sieve sizes have resulted in a mass of confused data.

The data for average particle density have proven to be more valuable. It can be seen that Batch 21-S, dated December, 1959, exhibited a particle density of 0.35 g/cc. Because of this low density, the formulation which is based on parts by weight was too viscous to pour, since it contained an abnormally large volume

*See Appendix 1.
†Standard Oil of Ohio developed and originally produced this material. It is now being produced under exclusive license from Sohio by Emerson and Cuming, Inc., as Eccospheres-R.
‡See Appendix 2.

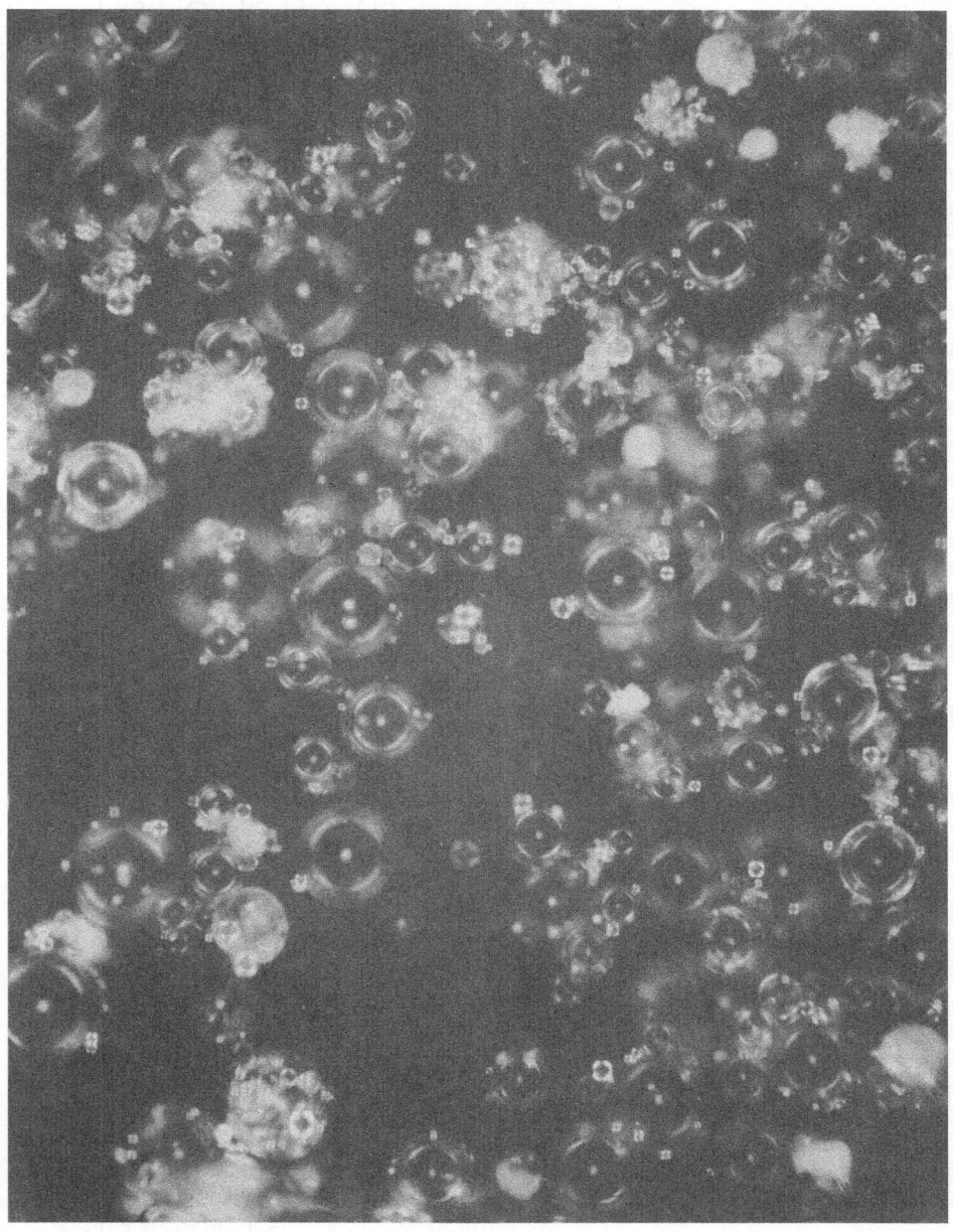

Fig. 1. Eccospheres-R. 100x.

TABLE I. Eccospheres-R Glass Microballoons Properties Chart

	9/59 20-S	10/59 20-S	11/59 20-S	11/59 21-S	12/59 21-S	1/60 23-S		3/60 24-S	8/60 27-S	1/61 30-S	Spec.
	RT	RT	RT	RT	RT	RT	CM	CM	CM	CM	
Sieve analysis, wt. %											
+60 mesh (250 microns)	0.1	0.0	0.0	0.1	0.0	0.0	0.1	0.2	0.1	0.3	0.5 max.
−200 mesh (47 microns)	47.8	24.3	61.2	37.2	25.4	25.0	29.9	34.1	41.1	38.4	20–50
Avg. particle density, g/cc	0.48	0.48	0.44	0.49	0.35	0.45		0.46	0.45	0.45	0.39–0.46
Moisture content, wt. %	<0.05	<0.05	<0.05	<0.05	0.09	0.02		0.04	0.10	0.06	0.20 max.
Alkalinity, meq/g	5.3	5.0	5.1	5.0	5.0	4.8		5.1	4.9	4.75	
Water solubility, wt. %	41	32	42	35	37	32		41	40	37	
Flotation ratio*	9:1	7:1	11:1	6:1	9:1	6:1		9:1	9:1	9:1	
Thermal shock†	2P 1F	2P 1F	2P 1F	0P 3F	‡	5P 0F		5P 0F	4P 0F	5P 0F	

P = Passed
F = Failed
RT = Ro-Tap shaker
CM = Cenco-Meinzer shaker

*Volume ratio of floters to sinkers in water.
†Surface of test specimen improved and number of test specimens increased for greater reliability starting January, 1960.
‡Too viscous to pour.

of filler. Conversely, Batch 21-S, dated November, 1959, exhibited an average particle density of 0.49 g/cc, which is above that allowed by the specification. This material was rejected; it yielded castings that completely failed the thermal shock test since the mix did not contain a sufficient volume of filler. This discrepancy within one batch has led to a better system of defining and identifying batches.

Batches 27-S and 30-S both exhibited peculiar characteristics that emphasized the need for further studies attempting to define acceptable microballoons. The average particle density is determined by measuring the density of the cured casting. Knowing the density of the cured resin, it is a simple calculation to determine the average particle density of the included balloons. These measurements are made on hand-stirred mixes to eliminate variations due to the crushing of the balloons by mechanical mixing. Batches 27-S and 30-S, when incorporated into the usual 70% resin - 30% microballoon weight ratio, proved to be too viscous to pour, indicating too-low average particle density. However, when measured, they were both found to be 0.45 g/cc, near the high end of the acceptable range. When mixed mechanically (yielding an average particle density of 0.51 g/cc) the microballoons proved to be quite processable and gave acceptable end properties. Earlier batches had been carefully mixed mechanically and showed an average density increase of only 0.02 g/cc. This larger increase for Batches 27-S and 30-S may indicate that these are more fragile balloons. Why they exhibit a high particle density and still yield viscous mixes has yet to be explained.

To date nothing significant has been noted in the data accumulated for those properties measured for "information only." The alkalinity of each batch has been observed because it is well known that the reaction of the epoxy cure is base catalyzed. The values for water solubility appear alarmingly high, but the electrical properties of the cured mix , owing to the protection offered by the resin matrix, are relatively unaffected after exposure to high humidity. The ratio of particles that float in water to those that sink does not seem to affect processability or thermal shock resistance within the limits observed.

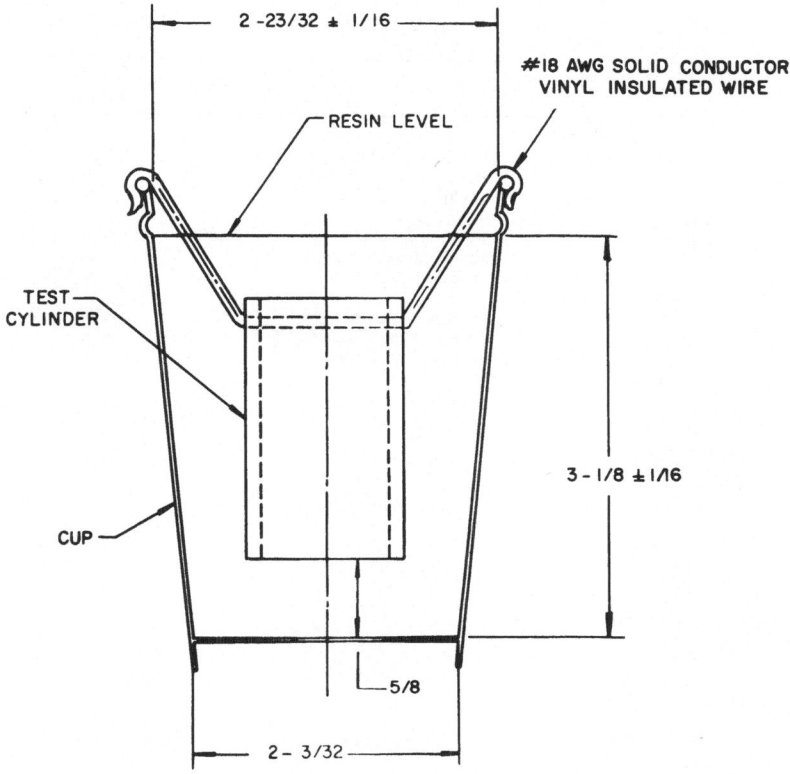

Fig. 2. Cold-shock test; assembly and test conditions. Notes: 1) Maintain concentricity of cup and cylinder within 1/32 in. 2) Maintain other tolerances to ±1/32 in. except as shown. 3) A container which fits these requirements is the 8-oz. Dixie No. 2108 or the 8-oz Dixie "Mira-Glaze," No. 4338. The container shall be removed following cure. 4) Subject the embedded specimen three times to the following cycle: 4 hr at room temperature (75±10°F) immediately followed by 4 hr at -65°F, then immediately return to room temperature. Visually inspect for cracks following each cycle. (All dimensions in inches.)

In addition to Eccospheres-R,* Emerson and Cuming also produces refined grades of microballoons with improved chemical, temperature, and moisture resistance. Unfortunately, the refining process reduces the range of the particle distribution and eliminates all high-density "sinkers." The greater range of particle distribution for Eccospheres-R, although admittedly quite variable, results in castings that are superior in cold-shock resistance to those containing refined microballoons. Furthermore, the elimination of "sinkers" in castings processed and cured according to the procedure outlined in Appendix 1 results in an area at the bottom that is devoid of filler. Therefore, it is desirable to have a certain concentration of high-density particles to offset the flotation of the low-density particles. In view of these shortcomings, it is felt that Eccospheres-R are better suited to the needs of this program.

Of the other low-density fillers examined, all proved inferior to glass microballoons in one or more respects, including undesirable particle size or distribution, lack of uniformity, fragility, high viscosity of the resulting mixes, and poor cold-shock resistance. Phenolic microballoons are being re-examined and, because of the processing experience that has been gained, they now appear

* The "R" indicates raw product.

TABLE II. Low-Density Potting Compound Properties

		−65°F	Room temp.	+165°F
Tensile ASTM D638	Maximum stress, psi	5,480	4,660	2,200
	Strain at max. stress, %	1.1	1.1	2.6
	Modulus, × 10^5 psi	5.3	4.4	1.8
Compressive ASTM D695	Maximum stress, psi	14,600	10,600	3,480
	Strain at max. stress, %	3.9	3.3	10.6
	Modulus, × 10^5 psi	4.4	4.2	0.9
Flexural ASTM D790	Maximum stress, psi	7,230	5,850	2,920
	Strain at max. stress, %	1.6	1.4	4.6
	Modulus, × 10^5 psi	4.7	4.4	2.0

Thermal coefficient of expansion	Temp. range (°C)	Expansion rate × 10^{-6} in./in./°C
ASTM D696	−60 to 0	35
	0 to 40	39
	40 to 70	42
Thermal conductivity Btu-in./hr/ft^2/°F	1.1	
Specific gravity ASTM D792	0.85	
Heat distortion temp. 264 psi ASTM D648	170°F	
Heat resistance, wt. loss, % MIL-I-16923	0.04	
Moisture absorption, wt. % MIL-I-16923	0.37	
Fungus resistance	Nonnutrient	
Dielectric constant ASTM D150	10^3 cps 3.02	
	10^6 cps 2.72	
Dissipation factor, % ASTM D150	10^3 cps 3.96	
	10^6 cps 2.55	
Dielectric strength, vpm ASTM D149	430	
Volume resistivity, ohm-cm ASTM D257	Room temp.	$2.05 \cdot 10^{14}$
	150°F	$1.97 \cdot 10^{13}$
	200°F	$9.30 \cdot 10^{10}$
	250°F	$2.43 \cdot 10^9$
	300°F	$3.68 \cdot 10^8$

TABLE III. Comparative Properties of Potting Compounds of Various Density

	Polyurethane foam*	Glass-microballoon-filled epoxy	Mica-filled epoxy†
Density, lb/ft^3	10	53	102
Tensile strength, psi	350	4,660	7,700
Tensile modulus, × 10^5 psi	0.1	4.4	12
Compressive strength, psi	325	10,650	17,970
Compressive modulus, × 10^5 psi	0.1	4.2	10
Flexural strength, psi	450	5,850	11,300
Flexural modulus, × 10^5 psi	0.2	4.4	12

*Average properties for several types of 10 lb/ft^3 foam, (the density most commonly used by Sandia).
†50 pbw Epon 828, 50 pbw 1000 mesh mica, 6 pbw diethanolamine.

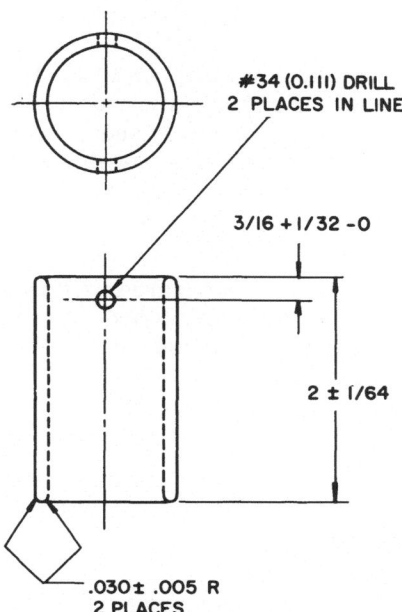

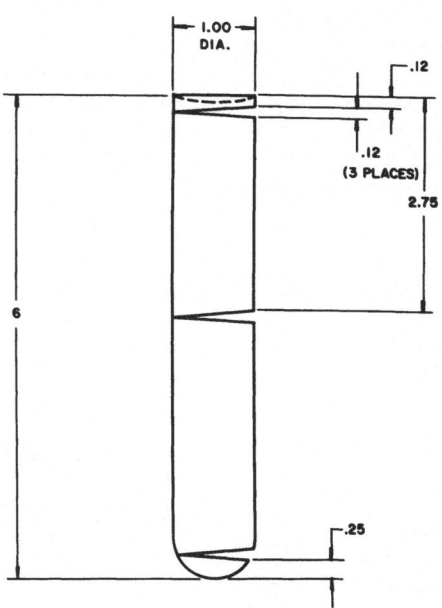

Fig. 3. Cold-shock test; test cylinder. Notes: 1) Scale: full size. 2) Material: seamless steel tubing 1-1/4 in. od 0.120 in. wall, low-carbon, MIL-S-11486, cold-drawn. 3) Finish: machine or polish to obtain a bright surface of 50 μin. rms inside the outside. (All dimensions in inches.)

Fig. 4. Specific gravity specimen locations. (All dimensions are approximate and are given in inches.)

much more promising. The phenolic balloons have a considerable price advantage, but when one considers the confidence and reliability that have been established for the glass microballoon system in various critical applications, then raw materials costs become relatively insignificant.

CHARACTERISTICS

Capabilities

1. The use of this microballoon-filled epoxy resin can result in considerable weight saving where the weight of the encapsulating material is an appreciable part of the total weight of a package. The specific gravity of the cured system averages 0.85 as opposed to 1.64 for a general-purpose system.

2. There is presently no dynamic test available to permit a quantitative evaluation of the high-level shock resistance of this material. However, components encapsulated with it have survived an estimated 5000 g, with a 2-msec rise time, whereas the use of conventional materials has led to complete failure under these conditions.

3. In addition to its excellent mechanical shock resistance, this material also possesses superior thermal shock resistance. To date, it is the only rigid system to pass Sandia's most severe thermal shock test.

Limitations

1. To achieve maximum properties, the system has been loaded to the point where its viscosity imposes a processing problem. Therefore, the electronic

package designer, taking this factor into account, should assign work only to experienced potting organizations.

2. Because of the thermal insulating nature of the microballoon filler, the cure exotherm frequently reaches a peak of 190°F. Furthermore, processing characteristics presently preclude the possibility of a reasonably short-time, low-temperature cure for thermally sensitive components, as is possible with general-purpose systems.

APPLICATIONS

With careful design consideration, this resin system can be incorporated into electronic packages of the type that are currently being encapsulated with the more familiar filled resins. However, the advantages in a particular application must outweigh the disadvantages listed above.

APPENDIX 1

PROCESS FOR ENCAPSULATION OF ELECTRICAL COMPONENTS

1. SCOPE

1.1 This process designates the requirements for encapsulating electrical components in a low-density filled epoxy resin.

2. REQUIREMENTS

2.1 Formulation — The encapsulating compound shall be formulated from the materials listed below, taken in their exact weights. The tolerance on weights of materials shall be ± 1%. The approved materials satisfactory for use in this process are listed in Section 3.1.

Epoxy resin	700 g
Filler	300 g
Hardener	85 g

2.1.1 Master Batches — Master batches of filled epoxy resin may be prepared by combining 70 parts, by weight, of epoxy resin with 30 parts of filler. The final formulation shall then be prepared by combining 1000 g of this mix with 85 g of hardener.

2.1.2 Filler Material — Since the filler material tends to segregate and stratify in its shipping container during shipping and handling, it shall be blended, but not sieved or screened, before use to ensure a uniform particle distribution throughout the material. The filler shall be a free-flowing powder at time of use.

2.1.2.1 When any filler is removed from a new drum of material, a sufficient quantity of bagged activated desiccant shall be inserted in the inner polyethylene bag to maintain dryness, and the polyethylene bag shall be tightly resealed each time it is opened.

2.2 Mold and Component Preparation — The mold (or assembly housing if the mold remains as a part of the finished unit) and electrical components shall be preheated in a forced-convection oven to a temperature of 175 ± 5°F at the time of pouring the encapsulating compound. When not an integral part of the assembly, the mold shall be coated with a thin even film of a suitable mold-release agent and allowed to dry (see 3.1.5).

2.3 Mixing of Encapsulating Compound — For best results during the subsequent deaeration step, it is recommended that the container have a diameter approximately equal to its height and a volume at least four times the volume of the mixed compound. Combine the epoxy resin and filler and mix thoroughly in a planetary-action mixer with a flat beater until the filler has been wetted and thoroughly dispersed throughout the resin. The mixing operation shall be completed in approximately 10 min to avoid excessive crushing of the filler. (A Hobart mixer has been found satisfactory.)

2.3.1 Heat the resin-filler mixture of 2.3 above to 150 ± 5°F in a forced-convection oven and add the hardener. Mix thoroughly in equipment described in 2.3 above until the mixture is uniform throughout (approximately 3 min).

2.4 Deaeration of Encapsulating Compound — Immediately upon completion of mixing, the mixture shall be returned to the 150°F oven. Eight minutes ± 30 sec after addition of hardener, the mix shall be evacuated at a pressure of 1-3 mm Hg absolute. Continue the evacuation for 1 min after the initial foam rise collapses. The entire evacuation process shall take a maximum of 3 min. The deaerated resin mix shall be used in the subsequent steps as soon as practicable.

2.5 Pouring of Encapsulating Compound — At the time of pouring, the encapsulating compound shall be at a temperature of 180 to 190°F, preferably not lower than 185°F. The preheated components and mold shall be partially filled with the warm, deaerated encapsulating compound. Degas the mold and components by placing them in a vacuum chamber and evacuating at a pressure of 1-3 mm Hg absolute and maintaining this reduced pressure for at least 3 min. Allow pressure to return to atmospheric. In no instance shall this step be completed later than 30 min after the addition of hardener (2.3.1).

2.5.1 In some units the configuration and spacing of components may be such that in order to remove all entrapped air it will be necessary to introduce the encapsulating compound in two steps, evacuating as above after each step.

2.5.2 Add additional encapsulating compound to bring level up to height indicated on the product drawing or to $\frac{1}{4}$ to $\frac{1}{8}$ in. of indicated level if unit is to be topped as specified in 2.5.3. For this last step, the encapsulating compound may be utilized for as long as 40 min after addition of hardener. No further evacuation is necessary. Any bubbles which appear at the surface of the filled mold prior to gelation of the encapsulating compound shall be broken by spraying the surface with resin defoamer (3.1.4).

2.5.3 Topping — If shrinkage of the potting compound is such as to leave an uneven surface, it may be desirable to apply a topping. When a topping is to be applied, the mold or housing shall be filled to within $\frac{1}{4}$ to $\frac{1}{8}$ in from the top. The compound shall be permitted to gel by heating in a forced convection oven for a minimum of 3 hr and not more than 6 hr at a temperature of 150 ± 5°F. After the compound has gelled, the mold shall be filled to the required level with encapsulating compound. For the topping operation the encapsulating compound may be utilized for as long as 40 min after addition of hardener (Section 2.3.1).

2.6 Cure — The filled housing or mold shall be placed in a forced convection oven maintained at a temperature of $150 \pm 5°F$ and shall be cured at that temperature for a minimum of 24 hr.

3. NOTES

3.1 Sources of Supply

3.1.1 Epoxy Resin — The only epoxy resin approved for use in this process is
      Epon 828   Shell Chemical Corporation
                                        500 Fifth Avenue
                                        New York 18, New York

3.1.2 Filler Material — The filler used in this process consists of expanded glass beads. The only approved source is

      Eccospheres-R   Emerson and Cuming, Inc.
      glass microballoons   869 Washington Street
                                 Canton, Massachusetts

3.1.3 Hardener — The only hardener approved for use in this process is di-ethanolamine. It is available from the following sources:

                          Primary sources
      Diethanolamine   Union Carbide Chemicals Co.
                                2770 Leonis Boulevard
                                Los Angeles, California

      Diethanolamine   Dow Chemical Company
                                Midland, Michigan

                          Secondary sources
      Diethanolamine   Distillation Products Industries
      Cat. No. 1598   Division Eastman Kodak Company
                                Rochester 3, New York

      Diethanolamine   Fisher Scientific Company
      Cat. No. D-45   2850 S. Jefferson Street
                                St. Louis 18, Missouri

3.1.4 Resin Defoamer — A satisfactory resin defoamer is

      Resin defoamer   Par Industries, Inc.
      No. 1   2193 E. 14th Street
                                Los Angeles 21, California

3.1.5 Mold Release — A mold release agent which has been found satisfactory when casting in a separate mold is

      Garan 225   Ram Chemical Company
                                Gardena, California

3.2 The ingredients of this encapsulating compound may be toxic. Hence, adequate ventilation should be provided in the handling of these materials to prevent undue exposure. Ingestion or skin contact with these materials shall be avoided. If accidental skin contact should occur, the exposed areas should be washed immediately with soap and water.

APPENDIX 2

SPECIFICATION FOR
GLASS MICROBALLOON FILLER MATERIAL

1. SCOPE

1.1 Scope — This specification designates the requirements for glass micro-balloons used as a filler for encapsulating resins.

2. REQUIREMENTS

2.1 Form — The filler shall be a white, free-flowing powder and shall be free from any visible contaminants or foreign materials.

2.2 Sieve Analysis — When tested, as specified in Section 3.3.1, the sieve analysis of the filler material shall be as follows:
   a. No more than 0.5% by weight of filler shall be retained on a 60 mesh sieve.
   b. No less than 20% nor more than 50% by weight of filler shall pass through a 200 mesh sieve.

2.3 Water Content — The water content of the microballoon filler shall not be greater than 0.20% by weight when determined as specified in Section 3.3.2.

2.4 Particle Density — The average particle density of the filler material shall be within the range of 0.39 to 0.46 g/cc when determined as specified in Section 3.3.3.

3. QUALITY ASSURANCE PROVISIONS

3.1 Lot — A lot of microballoon filler material shall be: (1) that quantity made during one production run from one batch of ingredients and offered for delivery at one time; or (2) a homogeneous blend of two or more lots described in (1) above. Each lot shall be identified to distinguish it from all other lots.

3.2 Sampling — At least one entire shipping container of microballoon filler material shall be randomly selected from each lot.

3.2.1 Blending — All the filler material from the sample container shall be placed in a tumbler and tumbled until the particles are uniformly blended (maximum of 20 min). After blending, a representative sample of filler material shall be taken for testing, and the remaining filler material shall be immediately returned to its original package. A sufficient quantity of activated bagged desiccant with a suitable humidity indicator shall be placed with the filler material in the polyethylene bag, and the container shall be tightly closed. The test sample shall be stored in a dry, airtight container.

3.3 Test Procedures

3.3.1 Sieve Analysis

3.3.1.1 Apparatus
   a. Weigh out a 100 ±0.1 g sample of filler.
   b. Assemble the clean sieves with the 60 mesh on top, followed by the 200 mesh, and with the pan at the bottom of the series. Transfer the sample to the 60 mesh sieve.

c. Place the covered sieve series on the shaker and shake for 20 min at shaker setting of "6."

d. Separate the nested sieves and pan. Weigh each sieve and pan and its contained filler material to the nearest 0.1 g. Carefully remove the residue and clean each sieve. Reweigh the cleaned sieves and pan. The difference in weights will equal the amount of residue on each sieve or pan.

Note: Sample recovery from the sieves and pan shall total at least 98 g.

3.3.2 Water Content

3.3.2.1 Apparatus
a. Drying oven, capable of maintaining a temperature of $250 \pm 5°F$.
b. Analytical balance.
c. Platinum dish, 100 ml capacity.
d. Desiccator.

3.3.2.2 Procedure
a. Heat a clean 100-ml platinum dish in a drying oven for 2 hr at $250 \pm 5°F$. Remove the dish and cool in a desiccator. Tare the dish.
b. To the platinum dish add an approximate 10 g microballoon sample, accurately weighed, to the nearest milligram.
c. Place sample and dish in the drying oven for 2 hr at $250 \pm 5°F$. Remove from oven, cool to room temperature in a desiccator, and then weigh.

3.3.2.3 Calculation
Percent water content $= (W-B) \times 100/W$
where $W =$ original weight of sample
$\quad\quad B =$ sample weight after drying.

3.3.3 Particle Density

3.3.3.1 Apparatus
a. 25×150 mm test tubes
b. Fisher chain gravitometer

3.3.3.2 Sample Preparation

3.3.3.2.1 Formulation of Casting Compound — Tolerance on weights of materials shall be $\pm 1\%$.

Epoxy resin	70 g
Microballoon filler	30 g
Diethanolamine	8.5 g

3.3.3.2.2 Mold Preparation — Clean the test tubes and coat the inside surface with a thin film of mold release. Allow to dry.

3.3.3.2.3 Mixing and Deaeration of Casting Compound
a. Combine the epoxy resin and filler and mix thoroughly by hand until the filler has been wetted and uniformly dispersed throughout the resin.
b. Heat the resin-filler mixture to $150 \pm 5°F$ in a forced-convection oven and then add the hardener. Mix thoroughly until the resultant mix is uniform (approximately 3 min).
c. Immediately upon completion of mixing, return the mixture to the 150°F oven. After 10 min have elapsed from the addition of hardener (step b, above), remove the mix from the oven, place it in a suitable vacuum chamber, and evacuate to a pressure of 1 to 3 mm Hg absolute. Continue

the evacuation for 1 min after the initial foam rise collapses. The entire evacuation process shall be completed within 3 min.

3.3.3.2.4 Pouring of Casting Compound

 a. Immediately after evacuation pour the mix into the prepared test tube mold, filling to a level approximately $1\frac{1}{2}$ in from the top.
 b. Degas the mold and casting compound by placing it in a vacuum chamber and evacuating to a pressure of 1 to 3 mm Hg absolute and maintaining this reduced pressure for at least 3 min.
 c. Add additional casting compound to bring level up to the top of the test tube. No further evacuation is necessary.

3.3.3.2.5 Cure — Place the filled mold in a forced-convection oven maintained at a temperature of $150 \pm 5°F$ and cure at that temperature for at least 24 hr.

3.3.3.3 Test Specimens

 a. Remove the casting from the mold.
 b. Take three wedge-shaped specific gravity test specimens from the casting as shown in Fig. 4. Specimens shall be free of entrapped bubbles.
 c. Using a fine abrasive, sand or grind the flat surfaces smooth. Remove dust by brushing with a stiff-bristled fiber brush.
 d. Determine density of each specimen by means of the gravitometer. The specimen shall be held by wedging it in the underside of one of the slots of the gravitometer basket. A small quantity of wetting agent shall be added to the water to facilitate release of bubbles.

3.3.3.4 Particle Density Calculation — The average particle density shall be calculated as follows:

$$\text{Average particle density} = \frac{30}{108.5/D - 65.42}$$

where D = average density of three specimens.

3.3.3.5 Note — If the nature of the filler material is such that a 70-30 resin-filler mix is not pourable, then a 75-25 resin-filler mix shall be prepared, and the following changes shall be made in the determination.

 a. Section 3.3.3.2.1   Epoxy resin   75 g
                                Microballoon filler   25 g
                                Diethanolamine   9 g
 b. Section 3.3.3.4   Average particle density $= \dfrac{25}{109/D - 70}$

3.4 Acceptance Tests — Each lot of microballoon filler shall be tested for average particle density and water content and shall be subjected to a sieve analysis. If any specimen fails to meet requirements of this specification, then the lot represented by that specimen shall be subject to rejection.

3.5 Certification — Each lot of filler material shall be certified as to water content, sieve analysis, and average particle density.

4. PREPARATION FOR DELIVERY

4.1 Packaging — The microballoon filler material shall be packaged in 5-lb and in 25-lb-size fiber containers having polyethylene-bag liners.

4.1.1 <u>Marking</u> — The container shall be labeled with the manufacturer's name, product designation, lot number, and date of manufacture of individual lots, as defined in 3.1.

5. NOTES

5.1 <u>Sources of Supply</u> — The only microballoon filler material qualified under this specification is

Eccospheres-R glass microballoons   Emerson and Cuming, Inc.
869 Washington Street
Canton, Massachusetts

EVALUATION OF MICROBALLON FILLER MATERIAL
"FOR INFORMATION ONLY"

1. GENERAL

1.1 <u>Intended Use</u> — This document is intended for use in conjunction with "Specification for Glass Microballoon Filler Material," for the gathering of additional engineering information. It is not used for procurement.

2. REQUIREMENTS

2.1 <u>Properties to be Evaluated</u> — Each lot sample obtained in accordance with Section 3.2 of the above specification shall be analyzed or tested for the following properties:
Leachable alkalinity
Water solubility
Flotation ratio
At least three specimens shall be taken and analyzed for the determination of each property.

3. TEST METHODS

3.1 <u>Alkalinity</u>

3.1.1 <u>Apparatus</u>
    a. Electric heater.
    b. Erlenmeyer flask, 250 ml with 24/40 ground-glass top.
    c. Reflux condenser, water-cooled, with 24/40 ground-glass joint.
    d. Burette, 50 ml, graduated.

3.1.2 <u>Reagents</u>
    a. Hydrochloric acid, standardized, 0.1N solution.
    b. Phenolphthalein indicator, 0.1% solution in methyl alcohol.

3.1.3 <u>Procedure</u>
    a. Place an approximate 1-g sample of microballoon filler, accurately weighed to the nearest milligram, in a 250-ml Erlenmeyer flask. Add 100 ml distilled water and connect the flask to a water-cooled reflux condenser.
    b. Heat the water to boiling and allow to reflux for 2 hr. Allow flask to cool until it can be handled. Disconnect flask from reflux condenser.
    c. Add several drops of phenolphthalein indicator solution and titrate to the pink end point with standard 0.1N HCl. Record the titer obtained.

  d. Connect the flask to the condenser again and continue to reflux for an
     additional hour. Remove flask from the condenser and titrate to the end
     point again. Record the titer.
  e. Continue refluxing the sample for 1-hr periods, titrating after each period
     until the titer for an individual titration is less than 1 ml of 0.1N HCl.
     (This point is usually reached at the end of the fourth or fifth reflux
     period.) Record the total quantity of HCl used for all the titrations.

     IMPORTANT! Save the flask and sample for the determination of solu-
                 bility of microballoons in water.

3.1.4 Calculation

Alkalinity in milliequivalents of HCl per gram = $A \times N/W$
where A = total milliliters of HCl used in titrations
      N = normality of HCl
      W = weight of sample in grams.

3.2 Solubility

3.2.1 Apparatus
  a. Filter flask, 500 ml, connected to water aspirator.
  b. Filtering crucible, low form, fritted glass, medium porosity, 30 ml.
  c. Walter crucible holder.
  d. Analytical balance.
  e. Drying oven, capable of maintaining a temperature of $250 \pm 5°F$.

3.2.2 Reagents — Methyl alcohol, reagent grade.

3.2.3 Sample — The sample for this determination is contained in the flask re-
tained from the determination of alkalinity.

3.2.4 Procedure
  a. Place a tared, filtering crucible in a crucible holder of the filtering flask
     and start the water aspirator.
  b. Transfer the contents of the flask to the filtering crucible. Wash the flask
     several times with methyl alcohol. (The fine particles of filler are more
     easily transferred by alcohol than by water.)
  c. Wash the filter well with water, followed by several washes with alcohol.
  d. Place the crucible and insoluble residue in an oven and dry at $250 \pm 5°F$ for
     2 hr. Remove the crucible and allow to cool to room temperature in a
     desiccator. Weigh the crucible when cool.

3.2.5 Calculation

Solubility, percent = $(W - A) \times 100/W$
where W = original weight of microballoon sample
      A = weight of insolubles filtered.

3.3 Flotation Ratio

3.3.1 Apparatus
  a. 250-cc graduated cylinder.

3.3.2 Procedure
  a. Place 100 cc of microballoon filler in the 250-cc graduate.
  b. Fill the graduate to 250 cc with water and shake well to wet the micro-
     balloons.

c. Allow the mixture to stand for 2 hr.
d. Determine the ratio of the volume of floating particles to that of sunken particles.

DISCUSSION

Question: How does the thermal conductivity of this compound compare with other potting compounds when the heat has to be conducted and dissipated?

Answer: I have the data on that in my preprint. Thermal conductivity is not as low as you might think. It is 1.1 Btu-in./hr-ft^2-°F and compares favorably with unfilled systems. It is about one-third that of the solid-filled system.

Question: You were speaking about compatibility of this potting compound with the structure that it's going into. Have you had any experience with a molded shell of diallyl phthalate and the compatibility of this potting compound with it?

Answer: That is the principal moding material that we use; generally, compatibility has been good. We have had some separation where the potting compound separates from the housing, but it has been rare and we haven't been able to conclusively isolate the cause. We think there may be variations in the molding compound. Generally, the bond to the diallyl phthalate is excellent, and remains excellent throughout the intended environment.

Question: Charles Ross, Martin, Denver. You talked about the shrinkage problems. I am wondering if you have any problem with glass diodes and this type of component breaking because of shrinkage.

Answer: No, generally not. We sometimes put a mold release on the metal-to-glass seal of the diode to prevent adhesion of the potting compound to that surface so that shear stresses are not imposed in thermal cycling. If that doesn't do it, we generally use a cushioning compound such as sponged silastic on our pressure- or shear-sensitive components.

Question: Hugh L. Uglione, Martin, Orlando. I noticed you talked about the glass microballoons. How about the phenolics?

Answer: They are probably our second choice. Generally, they do not exhibit the cold-shock resistance of the glass microballoons. Their resistance is somewhat reduced. However, cost is quite a factor. They cost about $1.00 per pound as opposed to $5.00 to $8.00 per pound for the glass microballoons. We are reinvestigating the phenolic microballoons since we have only a single source for the glass microballoons. We would like to have another material to fall back on.

Question: Milton Ross, Milton Ross Company. Have you done any experimenting with the plating of the microballoons? We have experimented with a flash silver and gold to increase their heat dissipation. Have you found this of any advantage?

Answer: That is an area we have not investigated.

Comment: (Mr. Ross). You spoke of the epoxy pulling away from the diallyl phthalate shell. We have found that by sand blasting the shell and also by ultrasonically cleaning the shell we get a very good adhesion under most conditions.

Answer: We also use sand blasting. We have found that ultrasonic cleaning is unnecessary, if the shell is properly sand blasted. This is a standard technique with us.

Question: Carl Holzbauer, Space Technology. Have you considered the use of dilluents with the use of your resins to increase the amount of filler?

Answer: Yes, but as this requires quite a lengthy answer, I would rather discuss it with you later.

Question: Galen Maurer, General Electric, Cincinnati. What sort of a procedure do you use to eliminate voids when you are potting, and what sort of gauge do you use to know that you have no voids?

Answer: Proper adherence to a very restrictive process is our best control. This ensures that the material is at its maximum fluidity when being processed. However, we do radiograph (X ray) potted units and we do section a certain number. Voids are sometimes a problem but with experienced operators and good designs we can keep voids to a minimum.

Question: Did you attempt to fill high-density welded modules with these foams?

Answer: No.

Question: Are you aware that there is a material called Helic which has just about the same properties as described here? I don't know who the manufacturer is.

Answer: The same physical properties?

Question: Right.

Answer: Particle size, particle distribution, and particle density are extremely important properties. We have found that we get an optimum balance of these properties in this material; we have looked at many other similar materials but none seem to approach it.

Question: Now how about the flow of this material? Can you get it to go in between components?

Answer: Yes, it flows reasonably well if properly handled.

Question: Do you find separation between the potted components and this material during the curing process?

Answer: No, not if the components are clean and if they have surfaces to which we can bond (not teflon or polyethylene, etc.).

Question: Charles Vannamen, Lear, Grand Rapids. I have two questions. Is this material considered repairable and can components be removed from it and replaced?

Answer: Yes, we have done it on prototypes. We do not consider it a normal practice. It is probably more repairable than some of the solid materials and less repairable than polyurethane foams.

Question: Also, is it considered a humidity-proof material when it is completely encapsulated?

Answer: Yes.

AN EFFECTIVE USE OF CASTINGS IN A
LIGHTWEIGHT ELECTRONIC PACKAGE

Harold Ferris and John B. Willbanks

Bell Helicopter Company, Fort Worth, Texas

INTRODUCTION

Many electronic packaging engineers may have overlooked the advantages that might be gained by using castings instead of the conventional sheet-metal housings. This paper will describe an airborne electronic stabilization kit for helicopters, and point out the advantages of using castings.

The unit has the following components:

1. Two printed circuit boards using solid-state circuitry (eight transistors, thirty-nine resistors, thirteen capacitors, three transformers, and four trimpots).
2. A rate gyro.
3. A size eight servomotor driving a size eight synchro control transformer through a 2000:1 gear train.
4. One 15-w special transformer.
5. An inductor.
6. One four-pole double-throw microminiature relay.
7. One 10-μf 200-v Mylar capacitor.

The unit has two functions — directional dynamic stabilization and heading hold. One or both functions are to be installed in the aircraft, and the unit must conform to MIL-E-5400 Class 1A specifications. This means that two separate packages were designed which may be combined into a single package. The problem resolved itself into what was the best way to package these components into the smallest possible space, with a minimum weight, at a reasonable cost.

FACTORS AFFECTING PACKAGING PHILOSOPHY

Vibration. Vibration problems in helicopters differ from those in other aircraft in that rotor-aerodynamic induced oscillatory loads reflect into the structure at low frequencies and at relatively high amplitudes. Bell helicopters will excite electronic packages at 5 and 10 cps (1 and 2 per revolution) and may under extremes reach 0.3 g peak. Avionics gear equipped with vibration isolators to satisfy MIL-T-5272A vibration requirements typically will resonate at these frequencies or their first harmonics; consequently, we rigidly mount electronic packages to the helicopter structure whenever possible. This means that any heavy internal components must be rigidly mounted in the package so that all resonant frequencies are 25 cps or higher. For instance, a transformer, gyro, or servomotor cannot be mounted on a thin-gauge sheet of aluminum. A casting design with its inherently good damping properties and rigid structure will more than adequately meet these requirements. Mounting bosses are easily arranged to properly tie down all components and modules. Shock loading will not be a factor when rigid structure is used. The rate gyro, however, is shock mounted inside the housing in such fashion as to provide a natural frequency along all axes of approximately 25 cps.

Fig. 1

Fig. 2

Fig. 3

Fig. 4

Components. The components in this electronic unit are quite different from those used in computers in that a wide variety of electronic parts is used. In addition to widely varying components, the unit also requires a precision gear train and gyro mount. This necessitated numerous mounting bosses and some accurate machining for the servomotor, synchro, and gear train. It was found that much space can be saved if the gear box is made part of the housing, with the electronic components mounted around the mechanical parts. (See Figs. 2 and 4.) Since machining was required, it was very economical to drill and tap the numerous bosses needed for component mounting while set up for gear box machining;

Fig. 5

several facing cuts needed for sealing could be made at the same time. The decision to use castings for this set of requirements was based on past experience
with castings and sheet metal housings. Rigidity and dimensional stability and accuracy are very important for proper operation of the gear trains; this rules out
sheet metal chassis construction. If more than one part is desired, milling out
a billet is a costly operation; consequently, a casting was chosen.

Environment. The environment under which this equipment must operate is
peculiar to helicopter operation. High altitudes, extreme temperatures, and high
g loadings are not the problem they are in airplanes and space vehicles; but since
a helicopter spends much of its airborne time close to the ground or water level,
its operating regime requires protection against dust, humidity, and salt-water
spray. Therefore, sealing is very important and sheet metal housings would
require extensive operations to seal corners, joints, and covers. Potting is not
feasible because of the precision gear train and gyro. The best answer to this
problem is the use of machined castings separated with gaskets, which allows a
completely dustproof and moistureproof package. A desiccant bag keeps internal
humidity low. The reliability of the electronic components thus protected is
greatly improved.

Logistics and Maintenance. The military services are becoming very concerned
about the number of spare parts needed in the pipeline to keep systems in operation. Closely allied to this is the problem of maintenance or servicing with
personnel trained for only the basic technical requirements. Designers are solving
these problems by using modular-type construction whereby subassemblies
rather than individual components can be stocked. Bell's stabilization package
uses this concept (see Figs. 2 and 3). Printed circuit board-wiring bundle subassemblies and separable subsystems are used. The castings, when machined,
provide mounting surfaces and bosses for mounting all components and modules,

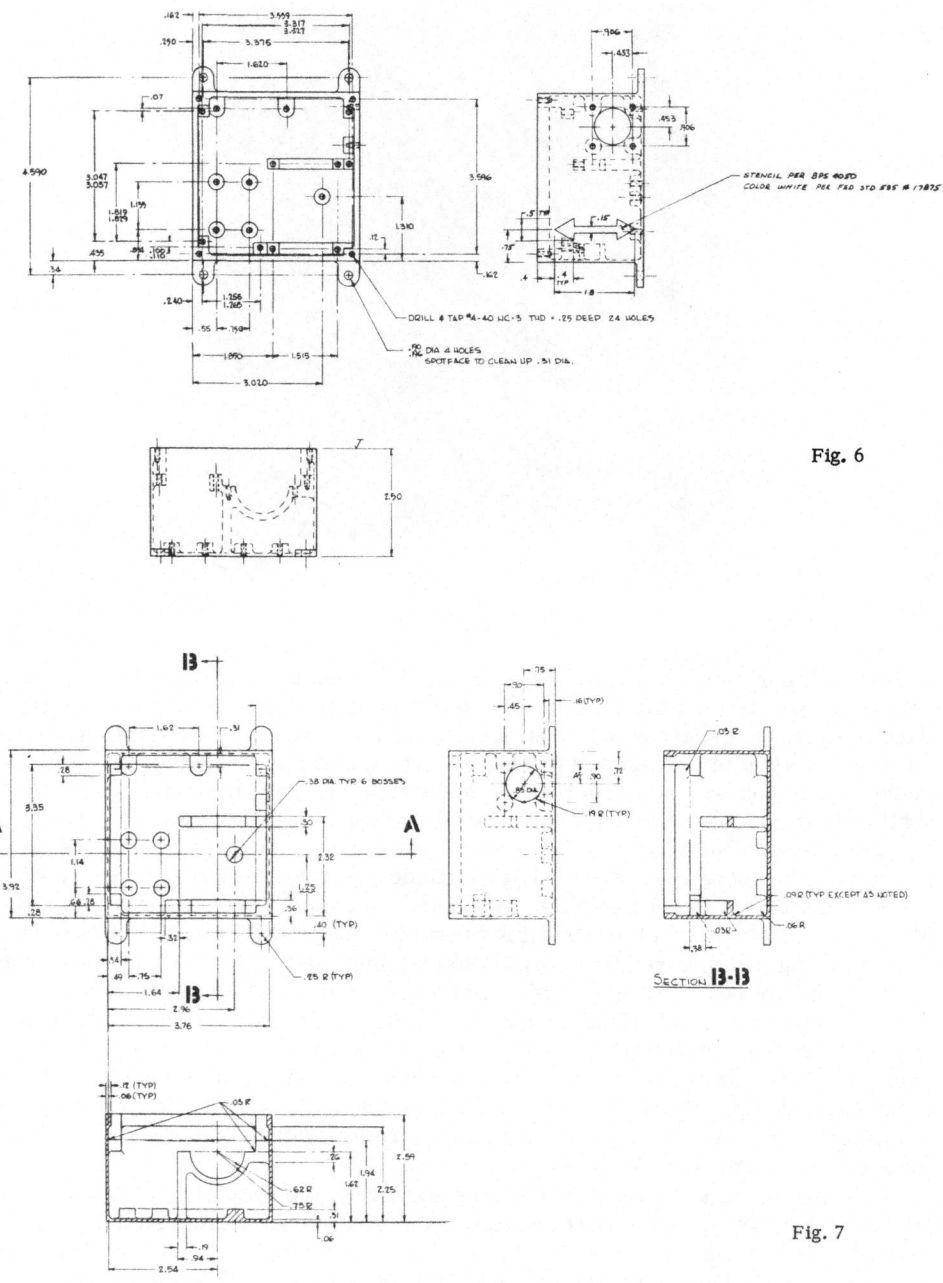

Fig. 6

Fig. 7

thus eliminating numerous extra parts, which would be needed for a sheet metal design. Also, external mounting lugs are easily cast integral with the housing. Accessibility and ease of maintenance are assured by using a split housing that can be opened like a book, exposing the parts in a convenient manner (see Fig. 3). The wiring bundles are routed to form a service loop, and arranged so both halves can be laid open and the printed circuit cards folded out.

Heat Rejection. The servomotor—generator is the largest heat producer in the package and this heat must be removed to allow for proper operation of the package. By using a casting in a manner such as shown in Fig. 4, heat can efficiently be removed by radiation and conduction. The motor is placed close to a black-wall area that is ideal for radiation absorption. The mounting surface is integral with the housing and will effectively carry heat away by conduction. The unit is mounted in the helicopter so that housing heat is removed by conduction, convection, and radiation.

Reliability. Most of the previously mentioned factors affect reliability and it has been shown that, for this package, the use of castings is the best route. In addition, we can say briefly that internal reliability of 1000 hours mean time to failure is desired. Components conforming to MIL specifications were used and were derated as much as possible. Printed circuit boards do not require spring connectors. Funnel-flared tubelets with positive fastening were used and each tubelet was solder-filleted on both sides. Components were then hand soldered to the board and appropriate holders were used where required. Service testing will provide for a reliability-test program and the results, we believe, will show that careful attention to reliability factors early in the design phase will assure a 1000-hour package.

HOUSING DESIGN

Prototype Units. For the first units, simple sand castings were designed without cores, close tolerances, thin walls, or fine surface finish, in an effort to obtain housings quickly and at low cost. The circuits had been breadboarded and the components fixed before the package arrangement was laid out. Efforts were made in component arrangement to allow a coreless casting design, i.e., the mahogany pattern could be drawn directly from the sand without loose pieces or core problems. This policy paid off when we received our prototypes ten days after drawing release. The cost, with patterns prorated, was $17.50 each for four top castings and $19.50 each for four lower housings. Time and cost factors dictated the choice of aluminum alloy No. 356 QQ-A-601 Temper T6 over A291C magnesium. Wall thicknesses were 0.10 ±0.020 in.; surface finish, 250 rms; draft angle, 0.5°; corner radii, 0.03 in.; and fillets, 0.05 in. These dimensional tolerances can be obtained economically with normal foundry techniques. X-ray inspection or impregnation was not required.

The castings, when received, were checked with a height gauge to determine the exact location of each boss center relative to certain other bosses and surfaces. A machined casting drawing had been prepared without these dimensions. The advantage of using this technique is that the bosses can be much smaller and still have No. 4-40 tapped hole edge distance throughout the tolerance stack-up of dimensions. This is possible because the pattern and casting tolerances have been all but eliminated, leaving only the machining tolerances. It should be noted that the draftsman's dimensioning for hole centerline locations and mating parts must follow the most direct route (see Figs. 6 and 8). For a small electronic casting, the success or failure of the unit may very well depend

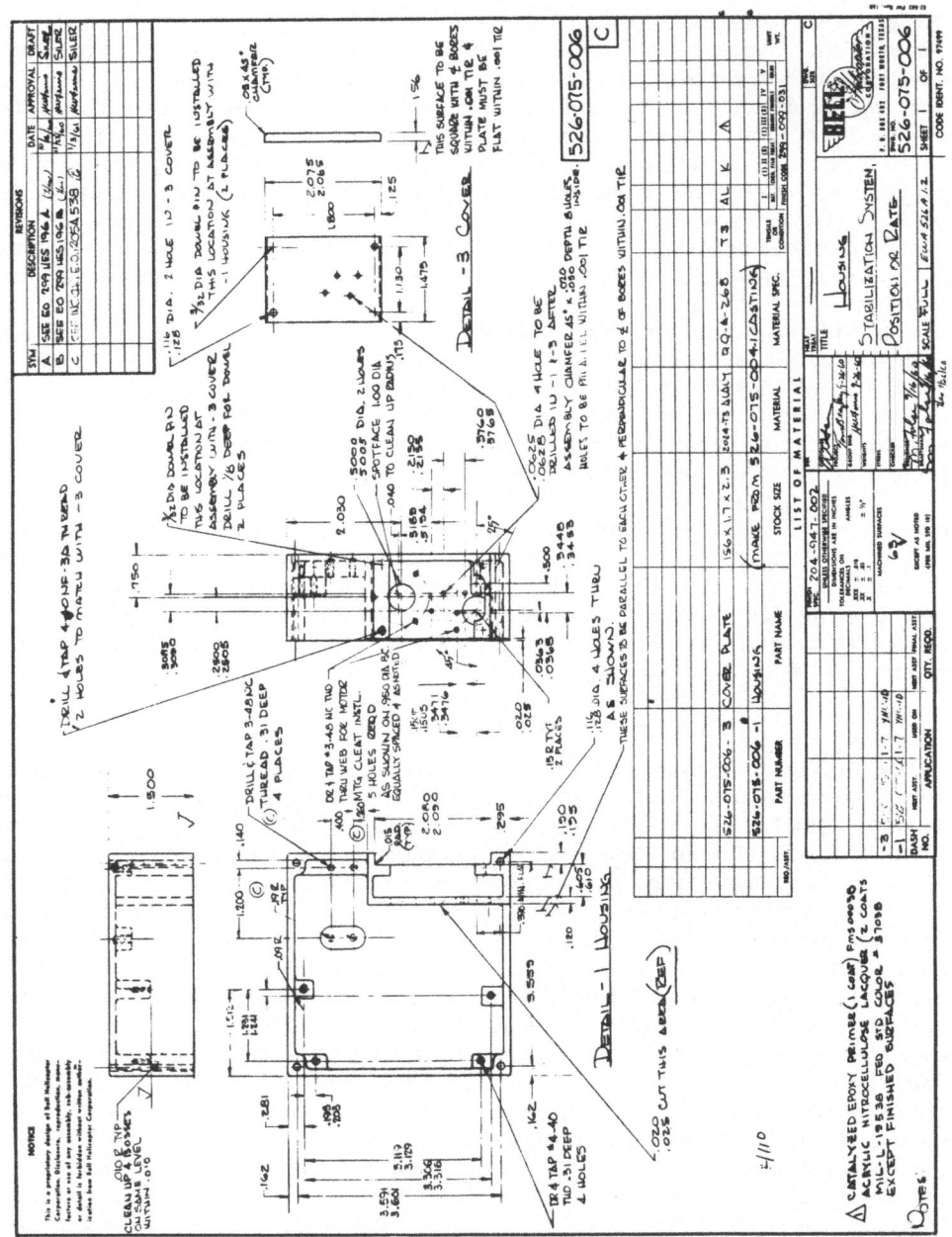

Fig. 8

upon holding supports and bosses to a minimum in size, since otherwise a larger and heavier housing will be needed with a resultant loss of the advantages gained over sheet metal housings.

When the operations called for on the machined casting drawing have been completed, the housing is ready for buildup. The assembly-of-components phase will reveal changes that are required for optimum spacing of parts and for correction of errors. A system checkout will probably require other changes that must be incorporated on the drawings before a production release is possible.

Production Units. The prototype phase has allowed a package to be checked out in a short time and at a low cost. The housings can now be changed for a different casting technique that will reduce weight and improve the esthetic features. Four production techniques were investigated: metal patterns with CO_2 cores inside and out, investment castings, plaster patterns, and die castings. Die castings were ruled out because of uneconomical production runs (less than 1000 units). Plaster pattern foundries are few and far between, with desired delivery and quality a question. Bids were received for investment and CO_2 cored processes using A291C magnesium. Typical results were:

1. Metal pattern and CO_2 core boxes $500.00
       51 pieces and up: $5.00 each
2. Investment castings — molds: $1500.00
       51 pieces and up: $19.00 each

The first method was chosen mainly because of cost but also because most good local nonferrous foundries can deliver acceptable housings with this technique and in a shorter time period.

The CO_2 cored process allows a 125 rms finish because a much finer-grained sand can be used without a moisture — gas problem. Wall thicknesses can be held between 0.06 and 0.09 in., with 0.5° draft angles, and dimensional stability is excellent. The production castings, due to the coring process, are much lighter than the prototypes since all excess material is eliminated.

Magnesium is 30% lighter than aluminum while retaining comparable strength. Costs are about 40% higher than for aluminum castings due to more extensive melting control, molding techniques, and heat-treatment. Care must be used to prevent fire during machining and the surface must be especially protected against salt spray. The material is softer and may require inserts for frequently removed components. In spite of these disadvantages, the weight advantage warrants using magnesium whenever weight is a critical factor.

SUMMARY

Electronic circuit packaging techniques and system design have changed considerably in the last decade with the advent of subminiature components and solid-state circuitry. In many instances, the cross-over point has been reached where castings are a better housing medium than the conventional sheet metal chassis. The weight of the housing described is comparable to the weight of an equivalent sheet metal package. When this is possible, the casting may often be the best packaging medium and should be considered.

DISCUSSION

Question: Bob Rooney, RCA, Burlington, Mass. Could you give me the name of a foundry that could produce 0.060 wall castings? The best we can do, and we have searched high and low, is about 0.160, and most of them want ± 0.030 or better. Can you recommend anyone that is around 0.060?

Answer: I was quite surprised to find, when I was talking to Mr. Beck, that in L. A. there is an outfit (Osbrink Mfg. Co.) that can cast to 0.030. I thought 0.060 was as good as they could do. Now we have two foundries in the Ft. Worth—Dallas area (Texas Bronze, Ft. Worth) and I am sure that most metropolitan areas of any size should have foundries that can handle it. Now let's be practical about this. We give them 0.060 but they take 0.020 tolerance so you may end up with 0.080 in some places and 0.040 in others. But this doesn't hurt because this casting is not subjected to high stresses. We are more interested in rigidity, especially for that very close tolerance gear train required in the servo system.

Question: Gordon Short, IT&T Industrial Product Division. I would like to know what you do about the prevention of galvanic corrosion between your casting and your steel inserts, if you use them, or any other attaching hardware in your magnesium casting. Also, I want to get the name of the place in L.A. that can make the 0.030 walls.

Answer: The firm in Los Angeles is reported to be Osbrink Mfg. Co. We have this problem on our magnesium transmission cases. We usually use an aluminum washer against the Dow 7 protected magnesium face surface. It is 5356 alloy. At any rate, this galvanic protection for magnesium is all pretty well covered in MIL-F-7179A. The painting process, and the Dow 7 treatment, and using this special aluminum alloy washer is a pretty safe route. Inserts should be installed using zinc chromate primer.

Question: Obviously, you are gasketing the cover plate on the side of the gear box. I was wondering what that gasket was. Also, you are not using bearings. How do you prevent wear on the casting from the play if you have a gasket with possibly varying thickness?

Answer: Our space between the two gear shaft surfaces is about 0.005. Now there is very little side load on these gears, being spur gears and driving a synchro transformer at a practically no-load condition. We have enough face width to take care of a little wear. So far we have been able to get by without using instrument bearings. No gasket is used between the plate and housing since both surfaces are machined and close tolerances are required. A sealant similar to MIL-S-8802 is inserted around the edges and a sheet metal cover is bonded over the plate. This procedure prevents moisture and dirt from entering the gear train.

Question: Ben Owens, Aeronutronic Division, Newport Beach, Calif. What operations do you go through on your jig boring operation versus your finish to come up with bearing tolerances in your finished casting?

Answer: If you had time to notice the tolerances on the machine drawing, they are ±0.00025. This is no problem with the tape-driven jig borer which can hold them within ±0.0001. We don't really care where the hole pattern is on the casting but we do care where each hole is in relation to the next hole.

Question: Do you do this jig boring before you apply your Dow finish, or after, in order to come up with bearing fits on your shaft?

Answer: We bore the holes, apply the Dow 7 treatment which does not change the dimension, and install the gear journals with Dow Corning XC145 grease. This is an adequate protection, providing we control the moisture level as we do.

Question: Did you mention the casting alloy?

Answer: Yes, that was in the paper. AZ-91 for the magnesium. Actually, you should go to your local foundry and find out what they are best suited for pouring. In the aluminum field there are probably a dozen good alloys. We use the one that the local foundry is most familiar with.

Question: Johnston, Ball Brother Research, Boulder, Colo. What inspection technique did you use to assure yourself of a sound casting?

Answer: When you say a sound casting, just what do you mean?

Comment: Well, one that was free of cracks either externally or internally, and free of voids.

Answer: This is not a high-stress application. So we do not make an X-ray inspection. We do zyglow inspection after all of our machining operations, which will show up any cracks that you can't see with your eye. We don't impregnate either, because so far we have not been troubled with porosity.

Question: Beck, Bendix. The question was brought up about inserts in the magnesium casting. I don't know if you are using them or not; it did not look as though you were. Is that the case?

Answer: Those were the aluminum prototypes that you saw.

Comment: Inserts in mag castings are not any problem anyway, because to get corrosion you must have an electrolyte to complete the electric circuit. And as long as you keep the electrolyte away from the similar metal interface, you won't have corrosion. So it's really no problem because any time you use an insert there is a mating part or a bolt head or some such thing covering up the interface. You can put an assembly through salt and you won't be able to find the corrosion between the helicoil and the magnesium.

Answer: That's right. In the aluminum casting we use no inserts. If you watch your carburetors being torn down sometime you will see that there are no inserts. There are none in my Chevrolet carburetor. They should take a carburetor apart a lot more often than the really reliable electronics package.

HEAT SINKS AND ENCAPSULANTS FOR VOLUMETRIC PACKAGING

Lawrence V. Gallacher

ARMA Division, American Bosch ARMA Corporation

The growth of the volumetric packaging concept in the past few years has been truly phenomenal. This technique, originally conceived as an interim device to last until the perfection of true microminiature circuitry, now promises to surpass printed circuit packaging in stature in the military field.

Volumetric packaging is characterized by the utilization of two basic techniques. First, the components are mounted three-dimensionally instead of two-dimensionally as in conventional printed circuit packaging. Components may be assembled in bundles with high volumetric efficiency this way. Second, component interconnections are made by welding. Welded connections can be made close to the body of a component without fear of component overheating. The net result is an electronic assembly with very high component density.

The volumetric or welded electronic packaging concept promises the ultimate in volumetric efficiency using "off the shelf" components. Functioning circuits containing 25 or more components per cubic inch have been produced. Part of the over-all scheme is to break circuits down into several discrete modules, each module representing a logical circuit division. The modules are assembled, encapsulated, and finally incorporated into a larger unit. The encapsulants can be color coded according to the function of the particular module.

Attendant with the high volumetric efficiency achieved in 3-D modular packaging is high heat dissipation per unit volume. It is possible to encounter wattage densities approaching 4 w/in.^{3} in transistor output circuits, while 0.5 w/in.^{3} can be reached in resistor-capacitor networks. Unfilled epoxy resins are poor heat conductors, with thermal conductivity coefficients around 1.5 Btu-in./ft^2-°F-hr compared to 1400 for aluminum. Filled resins can be brought up to a conductivity coefficient of 9.0. On the surface, then, it appears unlikely that epoxy resin encapsulants can dissipate much heat from operating components.

The proximity of components in volumetric packaging makes it extremely difficult to achieve complete resin penetration unless the resin has a low working viscosity. This ordinarily eliminates highly filled materials.

In addition we are usually faced with the problem of heat sensitive components (germanium diodes and transistors) which limit the exotherm and curing temperature of the encapsulant to approximately 85°C.

There are additional resin requirements not peculiar to volumetric packaging, namely, good electrical insulating characteristics before and after environmental testing, low curing shrinkage, and low coefficient of expansion.

In order to solve these and other problems connected with welded electronic packaging, the Development Services Department at Arma has been carrying on an extensive investigation of basic parameters. Some very early work in volumetric packaging has demonstrated the need for complete coordination of all efforts in the development of functioning packages; as a result there is a day-to-day tie-in among groups investigating encapsulants, heat sinks, welding variables, components, circuitry, and physical design.

It was decided very early in the encapsulant evaluation that the problems of encapsulation and heat sinking should be explored concurrently. Actually an earlier investigation had proved two things which led us to this decision:

1. To design heat-sinking and heat-transfer means into a package without considering the encapsulant usually results in excess heat-transfer capacity and thermal-shock problems after encapsulation.

2. Encapsulants do play a significant role in heat transfer. In a test where resistors were encapsulated in identically sized blocks of various modified epoxy resins, and resistor temperature rise versus wattage was plotted for each resin, it was found that temperature rise per watt was higher in air than in unfilled epoxies. Furthermore, in the best of the resins tested, the temperature rise per watt was less than one third the rise in air, indicating considerable heat dissipation capacity.

We felt strongly that by viewing the problems of heat transfer and encapsulation together we could eliminate the occurrence of situations where no encapsulant would meet all of the requirements.

All of our work thus far has been performed primarily with the goal of obtaining adequate heat transfer through a combination of modified epoxy encapsulants and metallic conductors. We have been careful to work within a framework where we feel reasonably certain of meeting the other resin requirements.

Past experience has indicated to us that thermal-shock requirements are difficult to meet without sacrificing other desirable resin characteristics. The main problem has been overdesign in heat sinking. This may be manifested in the form of large metallic surface plates and wells.

Inevitably, when large unbroken masses of metal meet encapsulants called upon to comply with a host of requirements, and units so constructed are cooled to -30°C or lower (as in several military specifications), the encapsulant will crack. This sort of situation arises when power transistors are mounted on aluminum base plates. Therefore, our first effort was to find a replacement for aluminum base plates, one that would not cause the thermal stresses that aluminum would, and yet have good heat-dissipation characteristics. We further decided that the next step would be to evaluate encapsulant-base plate combinations.

The procedure was as follows:

1. Determine ICO-junction-temperature relationship for each of a group of power transistors (Minneapolis-Honeywell H-6).
2. Prepare base plates of equal size ($3\frac{3}{4}$ by $3\frac{3}{4}$ by $\frac{1}{8}$ in.) from various materials with central mounting holes.
3. Mount transistors on base plates and determine relationship of transistor wattage to junction temperature for each base material. Measure the thermal conductivity of each base plate and correlate with plate performance.
4. Encapsulate transistors mounted on base plates and determine transistor wattage versus junction temperature for each encapsulant. Correlate with "K" factor.

In addition to the above steps, the following has been done:

5. Transistors have been encapsulated with various encapsulant-metallic conductor combinations in several configurations and tested.
6. The effects of cold plates and blowers have been evaluated for some cases.

DISCUSSION OF PROCEDURES

The ICO of several Minneapolis-Honeywell H-6 transistors was measured at several temperatures. Based on these measurements a nearly linear plot of log current versus junction temperature at 12 v was prepared for each transistor.

Resin Formulation. Several epoxy resins were compounded for high thermal conductivity utilizing high filler concentrations. The base resins were prepared at the lowest practical viscosity levels in order to obtain high filler concentrations and maintain room-temperature pourability. Two base resins were used for most of the formulating; both are mixtures of Epon 815 (Shell Chemical Co.) and epoxy flexibilizers X-2673.2 and X-2673.6, recently introduced by the Dow Chemical Co.

All resins were catalyzed with N-amino ethyl piperazine. We found that a combination of 60% Epon 815 and 40% epoxy flexibilizer gave resins with a good balance of characteristics. The X-2673.2 flexibilizer results in a more flexible system than the other. Further, the X-2673.2 has a definite retarding effect upon gelation and reduces the exotherm. The hardener used, N-aminoethylpiperazine, is a highly reactive amine, but with the retardation afforded by the flexibilizer and the filler, a pot life of approximately 2 hr was obtained for each system. All of the systems compounded were cured to a hard state at room temperature (6-10 hr) and then oven cured for 3 hr at 70°C.

The fillers investigated included alumina, black iron oxide, copper, zirconium orthosilicate, and magnesium oxide, all known for their effectiveness in increasing thermal conductivity. Through an arrangement with Brush Beryllium Co. a quantity of epoxy resin was filled with beryllium oxide for evaluation.* However, due to the relatively low filler concentration attained (53% by weight) and the high toxicity of beryllium oxide, the filled resin was not evaluated.

Several grades of aluminum oxide and magnesium oxide were evaluated; the grade giving maximum filler concentration at working viscosity was chosen for further evaluation. Particle size is extremely important; the filler concentrations (weight percent) attainable with different particle sizes of magnesia ranged from 40 to 72%. In general, higher loadings can be obtained with large particles (200 mesh) than with small ones (400 mesh). Filler settlement increases with particle size. Thermal conductivity is marginally higher with smaller particles at a given filler concentration. The optimum particle size for any particular filler is best determined experimentally, since surface interaction and adsorption effects vary. The formulations prepared for final testing are tabulated in Table I.

Preparation of Samples. The formulations described in Table I and several commercial resins were cast into $\frac{1}{8}$-in. plates. Several plates were cast incorporating aluminum heat-exchanger mats and honeycomb. The cast plates were cut into $3\frac{3}{4}$-in. squares, and holes drilled in the center with No. 19 drills for mounting one transistor on each plate. The base-plate compositions are shown in Table II.

Following the heat dissipation tests of the base plates above, several of the transistors were encapsulated in place on the base plates as depicted in Fig. 1, to form a $\frac{7}{8}$ by $\frac{7}{8}$ by 1 in. block of encapsulant on the center of the plate. In addition, several transistors were mounted and encapsulated in resin—metal combinations. The encapsulant combinations and configurations are described in Table III. The configurations are shown in Fig. 1.

Measurement of Junction Temperature Versus Wattage. Each transistor-plate combination was run at several wattage levels until thermal equilibrium was reached at each level. The junction temperature was measured at each wattage level by

*Beryllium oxide is unique among metallic oxides in that it possesses a very high thermal conductivity coefficient, 125 Btu-ft/M-ft^2-°F, higher than aluminum or beryllium metals.

TABLE I. Resin Formulations

Components	Parts by weight							
	1	2	3	4	5	6	7	8
Zirconium orthosilicate 400 mesh, M & T	632						1000	
Black iron oxide, Harshaw			908					
Lacauer aluminum, 90 KM, 400 mesh, Venus		55.5						
Alumina, CS25, 99% 325, Alcan					358			
Copper, MD105, Metals Disintegrating Co.						675		
Magnesium oxide, U-99 Int. Min. and Chem.								429
Araldite 6010						100		
Epon 815	100	100	100	100	100		100	100
Dow X-2673.2	66.7							
Dow X-2673.6		66.7	66.7	66.7	66.7		66.7	66.7
N-Aminoethylpiperazine	36	34	34	34	34		34	36
Triethylene tetramine						12		
Specific gravity	2.69	1.07	3.37	2.20	3.58	1.18	3.06	2.10
K, Btu-in./ft^2-hr-°F	6.6		5.4	6.2	6.8	1.6	9.6	
Weight percent, filler	75.7		82.0	64.2	77.1		83.3	68.1
Volume percent, filler	44.8		57.6	36.8	31.0		56.4	40.0
Viscosity, catalyzed, 25°C, cs						2000		2500

TABLE II. Base Plate Compositions and Performance

No.	Composition	Measurement conditions	Heat dissipation, watts/°C at 80°C	K, Btu-in. hr-ft^2-°F
0	Transistor in air	Suspended in air	0.027	
1	Formulation 1, Table I	Suspended in air	0.034	6.6
2	Formulation 1, aluminum heat exchange mat*	Suspended in air	0.067	12.5
		Resting on Al plate	0.093	
5	Bacon XP-33 sp.gr. = 2.36	Suspended in air	0.045	7.4
8	Aluminum	Suspended in air	0.17	1380
10	Aluminum, black coating	Suspended in air	0.22	1380
11	Formulation 7	Suspended in air	0.047	9.6
12	Formulation 7, aluminum honeycomb†, 1/4 in. thick	Suspended in air	0.107	22.0

*AiResearch Manufacturing Division, Garrett Corporation.
†Honeycomb Co. of America, Material No. 3003, thickness (sheet) 0.004 in., cell size 1/8 in.

means of a switching circuit illustrated in Fig. 2. In cases where two or more transistors were run simultaneously a common power supply and ICO-measuring circuit were used. Where necessary a variable resistance was placed in series with the wattmeter current coil in each transistor circuit to compensate for difference in current coil resistance. Encapsulants and encapsulant base-plate combinations were tested in the same manner as the base plates.

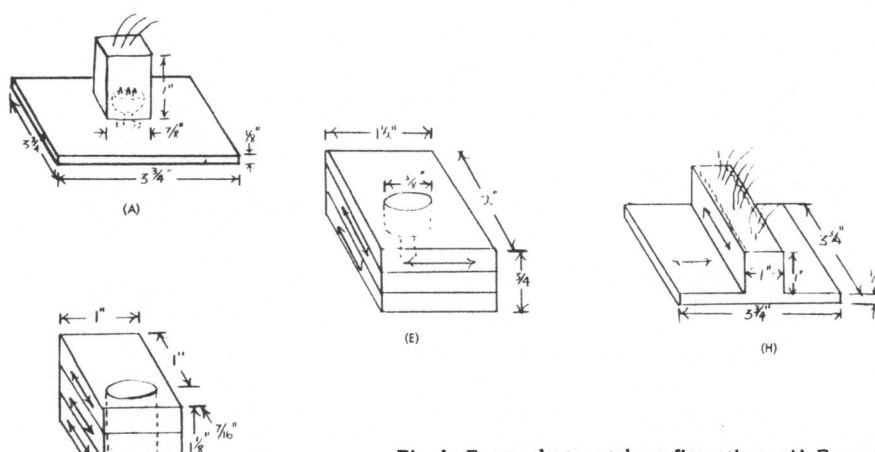

Fig. 1. Encapsulant-metal configurations. A) Base plate plus encapsulant; E) three layers of honeycomb plus formulation 8; G) three layers of aluminum heat exchanger (0.378 by 0.006 in. by 18 fpi, Type E, Trane Corp.); H) base plate (12), encapsulant including aluminum mat (AiResearch Division, Garrett Corp.).

TABLE III. Encapsulant - Aluminum Compositions, Configurations, Performance

No.	Composition	Configuration	Measurement conditions	Heat dissipation, watts/°C at 80°C
1	Base Plate No. 1, encapsulant formulation 1 containing aluminum heat exchanger*	Sample configuration A, single transistor	Suspended in air	0.072
2	Base Plate No. 2, encapsulant formulation 1	Sample configuration A, single transistor	a. Suspended in air	0.085
			b. On aluminum plate	0.120
3	Base Plate No. 12, encapsulant formulation 8 with aluminum heat exchanger*	3 transistors in 7-9-11 configuration. Sample configuration H, Fig. 1	Suspended in air	0.108
			On aluminum plate	0.148
			On aluminum plate with silicone grease coupling	0.193 each, average
		2 transistors in 7-11 configuration	Suspended in air	0.122 each, average
		1 transistor only in 11 configuration	Suspended in air	0.136
		1 transistor only in 9 configuration	Suspended in air	0.143
6	Combination of three layers of honeycomb plus formulation 8. One transistor embedded in wall. Size 2 by $1\frac{1}{2}$ by $\frac{3}{4}$ in.	Sample configuration E	Suspended in air	0.118
			In air with blower†	0.184
8	Same as above, but total size $1\frac{7}{16}$ by 1 by $1\frac{1}{8}$ in.	Sample configuration G. One transistor	On end, on aluminum plate (silicone coupler)	0.203
			Suspended in air	0.093

*AiResearch Manufacturing Division, Garrett Corp.
†Model No. 12500 Heat Gun, Veeco Vacuum Corp.

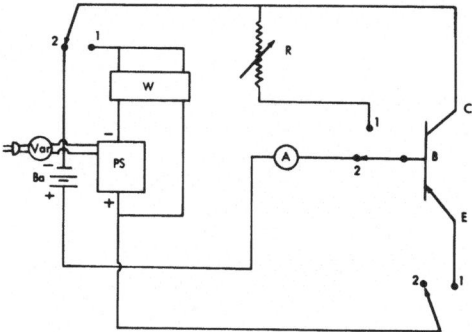

Fig. 2. Switching circuit. W – Wattmeter, sensitive research model DLW; PS – Sola constant-voltage dc power supply, 10 v dc, 4 amp; Var – General Radio Variac; Ba – four Burgess F4 BP 6-v batteries wired for 12-v output; R – General Radio Decade Resistor, Type No. 1432M; A – Esterline Angus Graphic Ammeter, Model AW, 5 ma; E – emitter; B – base; C – collector.

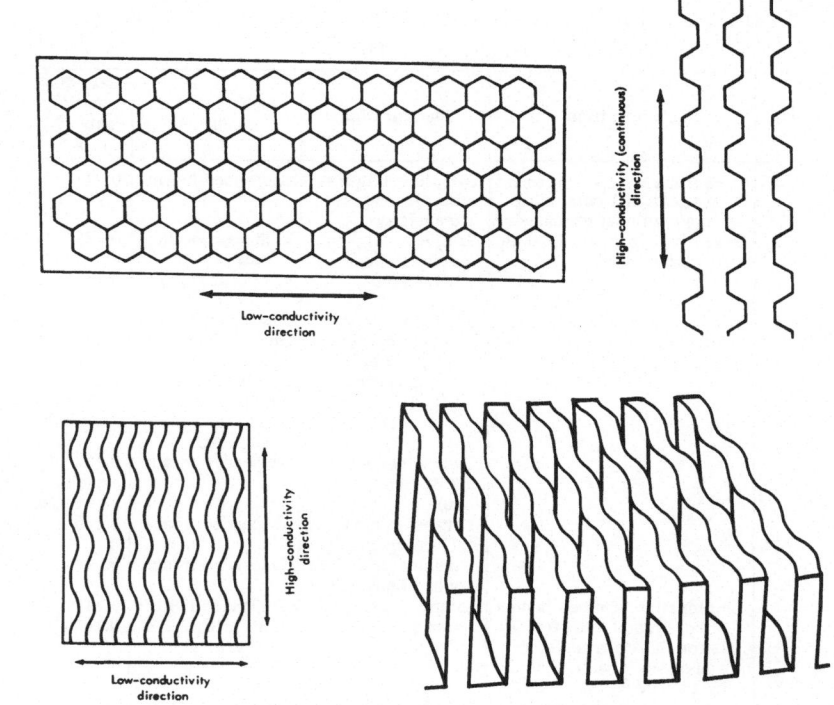

Fig. 3. (a) Aluminum honeycomb; the illustration at the right represents a delaminated honeycomb revealing continuous structure in one direction. (b) Trane heat exchanger; the illustration at the right represents the end view, showing relative path lengths.

Thermal-Conductivity Measurements. Thermal-conductivity coefficients were measured using a steady-state apparatus constructed in the laboratory. The apparatus consists of two insulated $\frac{1}{4}$ by 3 in. diameter copper plates. One plate is heated with nichrome coils, the other water-cooled with cooling coils. The temperature of each plate is monitored with 8 copper-constantan thermocouples mounted in the plate. Each plate is insulated (except for one face) with low K factor rigid urethane foam. The specimen is placed between the two plates. The thermal conductivity coefficient is calculated from the wattage dissipated in the nichrome heater at equilibrium. The calculation is based on the assumption that all of the heat energy passes through the specimen. Calculations are made as shown:

$$K = \frac{\text{(Wattage) (Thickness of specimen, in.) (12.6)}}{\text{emf}_{\text{Hot plate}} - \text{emf}_{\text{Cold plate}}}$$

This formula gives K in Btu-in. hr-ft^2. The number 12.6 groups together the conversion factors for wattage to Btu's per hour and emf difference to temperature difference, with a correction for plate diameter. The thermal-conductivity coefficients of the resins tested are shown in Fig. 1.

EXPERIMENTAL RESULTS

A representative grouping of the experimental results is tabulated in Tables II and III.

A nearly linear relationship was found for wattage versus junction-temperature rise.

Nonuniform thermal gradients were noted in base plates and encapsulants incorporating aluminum honeycomb and heat exchangers. This phenomenon is attributable to the structure of the materials, i.e., they are constructed to favor heat flow more along one axis than another. Figure 3 pictures the structural characteristics leading to this effect.

CONCLUSIONS

As a result of this study, several conclusions may be drawn. These should prove helpful in designing for volumetric packaging.

1. Substantial improvements in heat dissipation and thermal conductivity in epoxy resins are achieved by incorporating certain fillers and metallic conductors into the resins. Conductivity coefficients of 9.6 Btu-in./hr-ft^2-°F have been reached with fillers, and coefficients to 22 with filler plus honeycomb.

2. Encapsulants can be made to act as efficient heat-transfer media despite their relatively low thermal-conductivity coefficients. Resinous base plates in free air can dissipate 25% as much heat as equivalent aluminum plates using a central heat source.

3. The incorporation of aluminum honeycomb and heat-exchangers in resins results in greatly increased heat-transfer rates. A honeycomb-epoxy plate showed a heat-transfer rate, q, 50% as high as aluminum.

4. Mounting of modules on cold plates results in substantial improvements in heat transfer over still air. In some cases a 50% improvement was noted. An additional 20% can be realized by using a silicone-grease coupling agent.

5. The use of blowers effectively increases the heat dissipation rates of modules, depending upon the surface temperature. In general the use of a blower is more effective with a good conductor than a poor one, since the good conductor will have a lower temperature gradient and a higher surface temperature. By

the same token, the closer a component is to the surface the greater the effectiveness of the blower since the close proximity results in a higher surface temperature.

THEORETICAL CONSIDERATIONS IN
VOLUMETRIC PACKAGING HEAT TRANSFER

We have found that filled epoxy resins and combinations of epoxy resin and aluminum are effective heat-transfer media in volumetric packaging modules. Further, it has been shown that these materials approach metallic aluminum in over-all heat-dissipation effectiveness. The best of the epoxy-metal combinations has a conductivity coefficient of 22 Btu-in./°F-ft^2-hr while aluminum has a coefficient of 1400 Btu-in./°F^{-1}-ft^2-hr.

The reasons for the seemingly incompatible conductivities and heat-transfer rates may be found through a theoretical analysis of the problem.

In a 3-D module we have essentially a heat source surrounded by a layer of resin. This in turn is surrounded by the ambient atmosphere, whatever it may be. If we assume that the atmosphere is still air, we can then say that the heat dissipated is lost through conduction in series with a combination of radiation and convection:

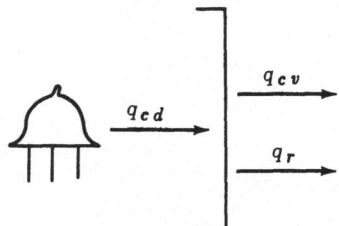

$$q_{total} = q_{conduction} = q_{convection} + q_{radiation}$$

If a drawing of temperature as a function of position were made, it would look something like this:

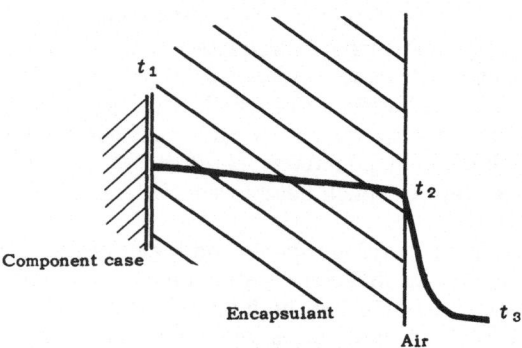

The exact calculation of the thermal conditions existing in a module is exceedingly difficult. However, by introducing some simplifications we can arrive at approximate answers which will serve to demonstrate the situation. Our model will consist of a cylindrical module with a rod shaped component at the center.

First we shall determine the convection relationship. For free convection from vertical plates and cylinders it has been shown that the film coefficient h has the following value:

$$h = 0.29\left(\frac{\Delta t}{L}\right)^{\frac{1}{4}} \frac{Btu}{hr - ft^2 - °F}$$

where

Δt is the surface temperature minus ambient, °F

L is the height of plate or cylinder, ft

This formula applies only where laminar flow conditions exist, a condition likely in volumetric packaging where surface temperatures are low.

For horizontal plates the following relationship holds:

$$h = 0.27\left(\frac{\Delta t}{L}\right)^{\frac{1}{4}} \text{ for plates facing upward}$$

$$h = 0.12\left(\frac{\Delta t}{L}\right)^{\frac{1}{4}} \text{ for plates facing downward}$$

The over-all convection equation is

$$q = hA(t_2 - t_3)$$

The heat flow rate q has been calculated as a function of temperature for $L = 1$ in. This plot is shown in Fig. 4.

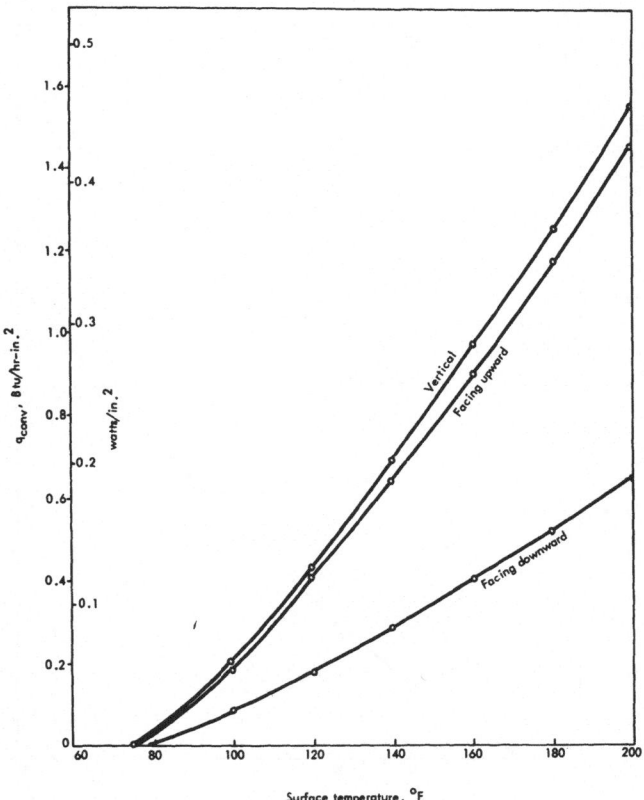

Fig. 4. Heat dissipation by convection from heated surfaces in air. Pressure = 1 atm; ambient temperature = 75°F.

Radiation from a surface in a large enclosure is essentially the same as gray-body radiation being transmitted to a black body. The relationship is defined by the following modification of the Stefan−Boltzmann law:

$$q_r = 0.174\epsilon A\left[\left(\frac{t_2}{100}\right)^4 - \left(\frac{t_3}{100}\right)^4\right]$$

The quantity ϵ is the emissivity of the surface. Measurements have shown this to be equal to approximately 0.8 for filled epoxy resins. Figure 5 is a plot of q_r versus surface temperature for an emissivity of 0.8.

The general equation for thermal conductivity is

where
$$q_{cd} = \frac{kA(t_1 - t_2)}{x}$$

k is the thermal-conductivity coefficient
A is the area of each of two parallel faces opposite each other
x is the distance between faces, ft
t_1 and t_2 are the face temperatures

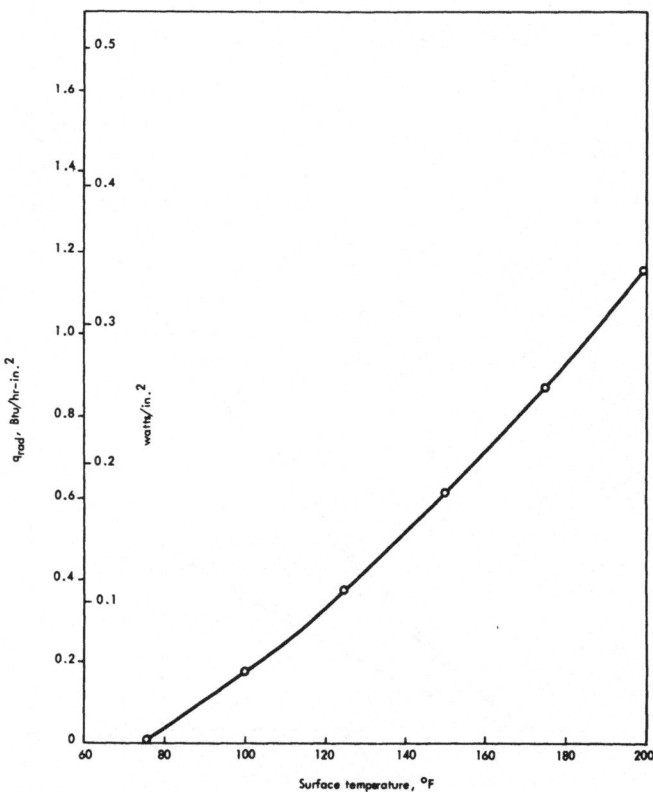

Fig. 5. Heat dissipation by radiation from a surface in a very large enclosure. ϵ = 0.8; ambient temperature = 75°F.

For a cylinder

$$t_1 - t_2 = \frac{1.152\, q_{cd} \log (r_2/r_1)}{\pi l k}$$

where

t_1 is the temperature of inner surface, °F

t_2 is the temperature of outer surface, °F

q is the heat flow rate, Btu/hr

r_1 is the inner radius, ft

r_2 is the outer radius, ft

k is the thermal-conductivity coefficient, Btu-ft/hr-°F-ft^2

l is the length of cylinder, ft

For a combination of conduction through a cylinder wall in series with convection and superimposed radiation,

$$q = q_{cv} + q_r = q_{cd}$$

$$q_{cd} = \frac{k\pi l(t_1 - t_2)}{1.152 \log (r_2/r_1)}$$

$$q_{cv} + q_r = hA(t_2 - t_3) + 0.174A\epsilon\left(\frac{t_2}{100}\right)^4 - \left(\frac{t_3}{100}\right)^4$$

This relationship is cumbersome and difficult to use. The quantity q is usually known, and t_1 and t_2 are unknown. For convenience the radiation loss may be shown as

$$q_r = h_r A(t_2 - t_3)$$

where h_r is defined as the radiation coefficient. Then the combined equation simplifies to

$$q = \frac{k\pi l(t_1 - t_2)}{1.152 \log (r_2/r_1)} = (h_{cv} + h_r)A_2(t_2 - t_3)$$

The combined coefficient h_{cv+r} has been calculated as a function of surface temperature and is shown in Fig. 6. This curve holds for vertical and top surfaces in large enclosures. Downward-facing surfaces have over-all coefficients about 30% lower.

It is possible to approximate t_2 by picking an average value for h_{cv+r} and solving for t_2. The error in this approximation would be no more than $\pm 10\%$ for $t_2 - t_3$, corresponding to ± 5°F for t_2 in most situations.

If we consider a module consisting of a cylinder of encapsulant containing at its center a rod shaped power transistor, we can approximate each of the variables t_1 and t_2 provided we know $q, k, A, r_2, r_1,$ and l. For purposes of comparison we will neglect end heat losses. Assume

$$q = 2 \text{ watts} = 6.82 \text{ Btu/hr}$$

$$k = 0.8 \text{ Btu - ft/ft}^2 - °F - hr$$

$$r_2 = 0.5 \text{ in.} = 0.043 \text{ ft}$$

$$r_1 = 0.25 \text{ in.} = 0.0213 \text{ ft}$$

$$l = 1.0 \text{ in.} = 0.085 \text{ ft}$$

$$A = 2\pi r_2 l = 0.0228 \text{ ft}^2$$

$$h_{cv+r} = 3.0 \text{ Btu/hr - ft}^2 - °F$$

$$= 0.88 \text{ watt/ft}^2 - °F$$

As shown above,

$$q = (h_{cv+r})A(t_2 - t_3)$$

First we solve for t_2:

$$t_2 - t_3 = \frac{q}{(h_{cv+r})A}$$

$$t_2 = \frac{q}{(h_{cv+r})A} + t_3$$

The next problem is to equate A to surface area of the cylinder. The area of a cylinder (sides only) is $2\pi r_2 l$; this must equal A.

$$A = 2\pi r_2 l = 0.0228 \text{ ft}^2$$

$$t_2 = \frac{2}{(0.88)(0.0228)} + 75$$

$$t_2 = 175°F$$

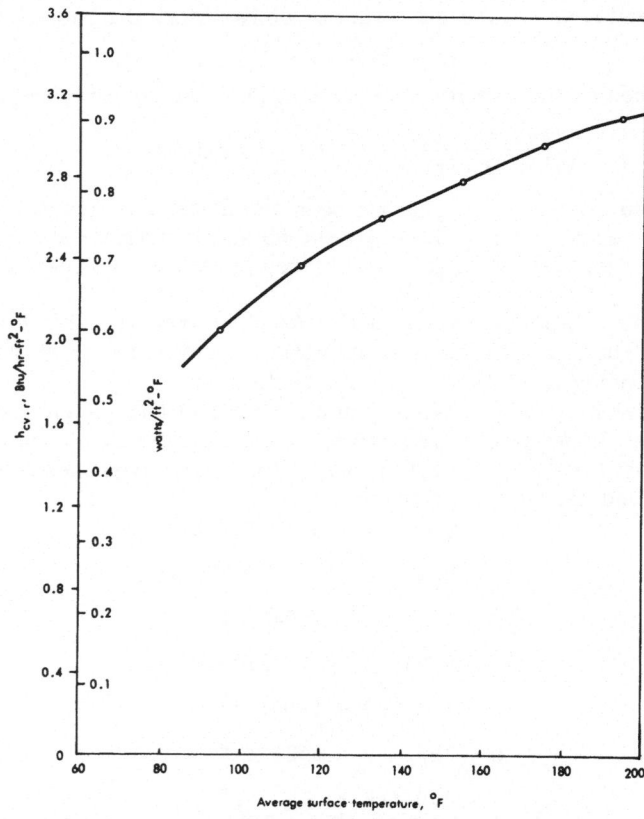

Fig. 6. Variation of h_{cv+r} with surface temperature. Pressure = 1 atm; ambient temperature = 75°F; ϵ = 0.8.

Now solve for t_1, assuming $k = 0.8$:

$$q = \frac{k\pi l (t_1 - t_2)}{1.152 \log (r_2/r_1)}$$

$$t_1 = \frac{1.152 q \log (r_2/r_1)}{k\pi l} + t_2$$

$$t_1 = 11.1 + 175 = 186.1, \quad \Delta t = 111°F$$

If $k = 115$, as in aluminum:

$$t_1 = 0.07 + 175 = 175.1, \quad \Delta t = 100°F$$

If the transistor is simply suspended in air:

$$\text{Case temperature} = t_2 = \frac{2}{0.88(0.0115)} + t_3 = 198 + t_3$$

$$t_2 = 273°F$$

Figure 7 has been prepared to show the effect of the thermal conductivity of the encapsulant upon the total Δt. The model used in the calculations was a 1-in.-diameter cylinder 1 in. long, containing a rod-shaped component 0.5 in. in diameter dissipating 2 watts.

Thus, it is shown that the most important role performed by an encapsulant in a small module is to increase the effective heat-dissipating surface of the component. The thermal conductivity coefficient is of secondary importance.

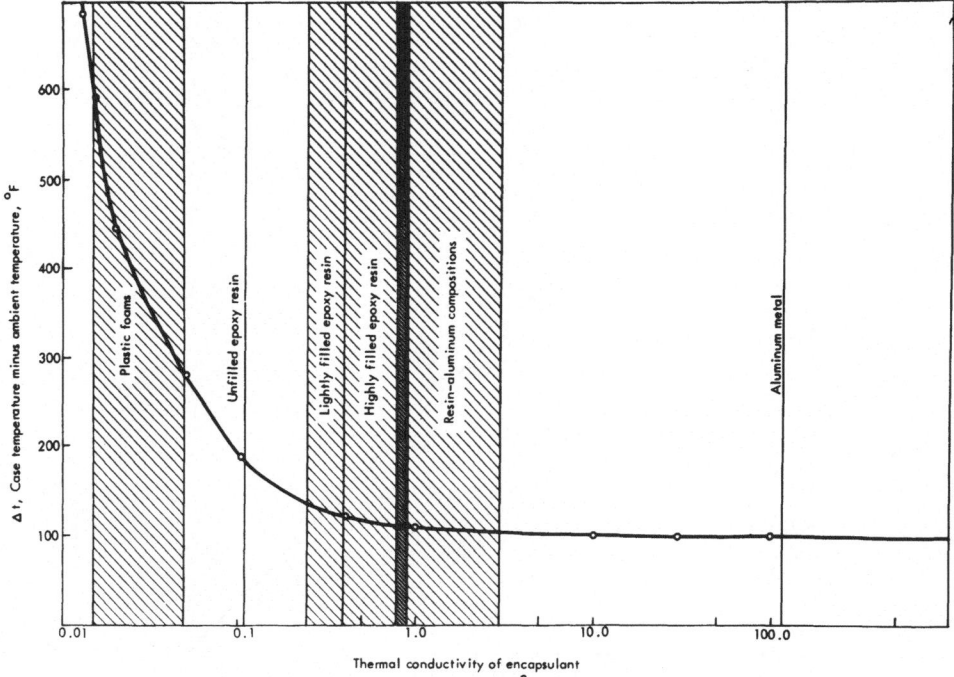

Fig. 7. Effect of thermal conductivity of encapsulant upon Δt.

This holds for encapsulant thicknesses on the order of $\frac{1}{4}$ in. As thickness increases, the thermal conductivity of the encapsulant assumes more importance. Thus, for a 2.75 in. thickness of encapsulant (6 in. diameter)

$$t_2 = \frac{2}{(0.88)(0.13)} + 75 = 92.5°F$$

$$t_1 = \frac{(1.152)(6.82)\log 12}{(0.8)(\pi)(0.085)} + t_2 = 132.2°F$$

$$t = 57°F$$

With aluminum:

$$t_1 = \frac{(1.152)(6.82)(1.079)}{(115)(\pi)(0.085)} + t_2 = 92.8°F$$

$$t = 18°F$$

Certain conclusions may be drawn as a result of these theoretical considerations:

1. The largest thermal resistance encountered in modular volumetric packaging is from the module surface to the ambient atmosphere.
2. The over-all component-to-ambient Δt is affected little by the thermal-conductivity coefficient of the encapsulant in small (~1 in.^{3}) modules using filled epoxy encapsulants. One could anticipate much larger Δt values using urethane foam encapsulation because here the k value is more than an order of magnitude smaller. The relative effect of thermal conductivity in the typical module (1 in. cylinder) described above is illustrated in Fig. 8.

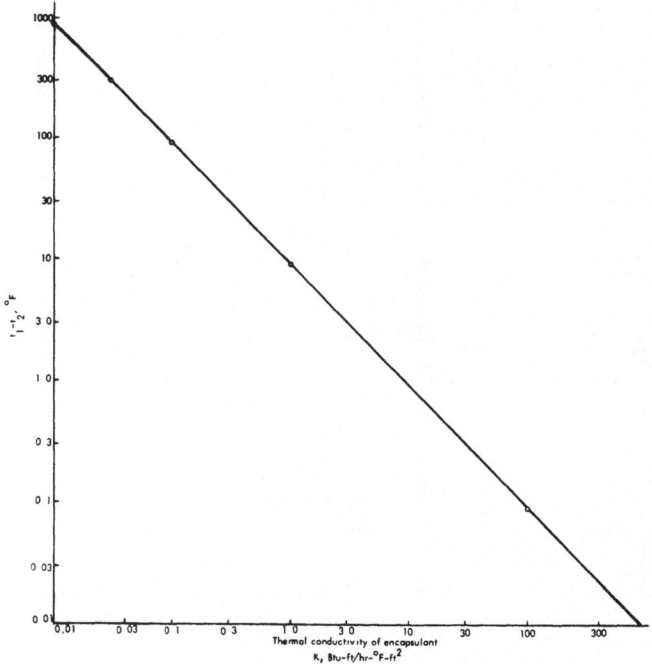

Fig. 8. Variation of $t_1 - t_2$ with thermal conductivity of encapsulant.

3. In very small modules heat dissipation is directly related to surface area; in large modules (~30 in.^{3}) thermal conductivity is more important.

4. In the thermal conductivity equation t_1 - t_2 must remain constant provided the other parameters do not change. Therefore, if t_2 is decreased by Δt, t_1 will also be decreased by Δt, provided the reduction is achieved by some external means.

5. Since t_1 - t_2 is constant for a given heat load, it is practical to reduce t_1 simply by reducing t_2. This can be accomplished by two methods:
   a. Forced air cooling
   b. Cold plate

   Basically most heat transfer systems employ one method or the other. The cold plate is simply a metallic surface upon which the module rests. The plate may be cooled by a secondary system, i.e., thermoelectric, blower, or change-of-state heat exchanger.

6. The more t_2 is reduced, the more critical the thermal conductivity of the encapsulant becomes. This is illustrated in Fig. 8.

In conclusion, we feel that it has been demonstrated that epoxy resins can be modified to perform efficiently as heat-transfer media in volumetric packaging. Further we hope that our work will contribute to the over-all knowledge and understanding of heat transfer in electronic packaging.

DISCUSSION

Question: Would you elaborate on the work you have done with beryllium oxide?

Answer: We worked out an arrangement with Brush Beryllium to have them make up samples for our evaluation because we did not wish to incorporate all the safety facilities that we would have to use to evaluate this material. We sent some resin out to them and they filled it with beryllium oxide and sent it back. Unfortunately the loading they achieved was only 53%. Now as filled epoxies go, one could say with good certainty that this compound would not have a very high thermal conductivity. So for safety reasons and because we were fairly sure it would not work out, we did not evaluate it further. Now Brush is working on new forms of beryllium oxide aimed directly at the epoxy field.

Question: Loye Pierce, Gulton Industries. On the different packages that you had there you said you had so many degrees centigrade per watt. Your surface areas seemed to differ. Heat dissipation is a function of area. Now are these areas all equal? They did not appear to be.

Answer: In the case of the test modulus they were not equal. The base plates were of equal size.

Question: How would they actually prove anything?

Answer: The base plates were all the same size, so there we did have a basis for comparison. The major point shown by the module tests was that the heat conduction can be made very directional. We think what we proved was more' relative. Epoxies can dissipate heat effectively despite their low k factors. We did not attempt to pin things down to specifics as much as to prove the major point.

Question: I have a question on your 20% improvement in heat transfer. The test we ran indicated that heat-transfer conditions are primarily dependent upon surface finish and flatness rather than silicone grease or any other contact medium. We also feel that, wherever possible, the encapsulants should be avoided in power application. You probably concluded this yourself.

Answer: I think we hit all the conclusions. Generally you are not dealing with completely flat surfaces. For instance, with castings, you might have some surface irregularities. Usually a contact medium will help you. When you run a thermal conductivity test, you usually employ a contact medium. This is the reason we tried it. As to the use of aluminum as opposed to epoxy to aid in heat transfer, the whole reason for our project was to see just how good epoxies were in order to get away from the thermal problem we introduced by using large aluminum heat sinks in the packaging.

Comment: Jake Rubin, General Electric. I would like to add a clarification in terms of the use of the encapsulant as a good means of cooling equipment as has been indicated in your paper and as many people here might feel worthwhile trying. It should be emphasized that the reason that the epoxies look so good is that you are dealing with natural convection and radiation which at the powers and temperatures we deal with in packaging are rather poor ways of getting rid of heat. If we go to a system where we use forced cooling by either blowing air or use of a heat-exchange fluid of some sort, then the coefficient at the surface becomes so much improved that the thermal conductivity again becomes very important. Or, as you had it on your slide, the approach to your plateau no longer exists once you have gotten rid of that thick boundary layer of air which is an inherent part of natural convection.

Answer: Your last statement is very true. Obviously, where you have to get heat out, no holds are barred. You may be forced to use aluminum or other metals to remove heat. They are more effective than epoxy. But in most applications you don't have the power densities that require you to go to these means and it is better from a production standpoint and epoxy formulative standpoint to eliminate the large masses of aluminum.

Comment: (Mr. Rubin). Then we might say that it isn't that epoxy is a good means of heat transfer; it's that the natural convection is such a poor means that the epoxy doesn't really hurt it much.

Answer: It has been amply stated in the paper and discussion exactly what our goals and findings were. It is the negative approach to heat transfer through encapsulants, i.e., "epoxies really don't hurt much," that we are attempting to eliminate. Epoxy encapsulants are adequate for encapsulation and removal of heat in most high-density, high-power dissipation circuits. Certain other benefits usually accrue, namely, simplified formulating and processing and lower weight. While we obviously cannot say that epoxies can be made to conduct heat as well as aluminum, we are trying to show that (1) heat conduction of the heat sink or encapsulant is never the only consideration in total heat transfer and (2) the total heat-transfer rate in most cases is not affected greatly whether filled epoxy or aluminum is used. The reason for the latter statement is simply that the thermal resistances in series with the encapsulant are usually comparatively large.

Question: I would like to know what you mean by the statement on page three, Fig. 1? I would like to know what kind of thermal shock problems you have?

Answer: What I meant there was that the aluminum heat sinking usually will give you more heat dissipation than you need. In other words you have some unknown overage, and in incorporating this aluminum you are leading to problems later when you try to encapsulate this with epoxies because epoxies do not react well to large masses of metal.

Question: What causes these thermal-shock problems?

Answer: Let us look at it more from a length of metal. It is a problem that arises because of the difference in the coefficients of expansion of epoxy and aluminum. Therefore the longer the planes that adhere to each other, the more trouble you will run into when you change temperatures. You might run into trouble with the length of aluminum of four or five inches. Then again it depends on other considerations. If you have aluminum faces separated by spacers you can get separation of the resin from one spacer in thermal shock.

Question: Bernie Litwack, Bell Labs., New York. In connection with beryllium oxide, have you explored anything regarding the uses of it in replacing aluminum for heat sinks?

Answer: We are looking at it now as a possible replacement where we don't want electrical paths. As you know, it has a higher thermal conductivity than aluminum. It has a higher conductivity than even beryllium metal, and this is unique among oxides.

Question: Ahmet, Erdogan Engineered Electronics Co., Santa Ana, Calif. I believe you said that to cool the modules you use honeycomb or cold plates. If we go into outer space how are we going to cool these? If we cool by means of forced air how are we going to get the air if we go into outer space?

Answer: Most systems are made to work in controlled ambients, even in space conditions. But you may be in a situation where you have to mount on cold plates of one sort or another, or resort to another means such as thermoelectric cooling. Actually this sort of problem is not in the domain of the module designer, but would be considered on the system and missile levels.

Question: Have you or another company worked on this problem?

Answer: We are working on it.

Question: Have you had any results?

Answer: I can cite some work that we did a few years back with computer circuitry with fairly low power densities. We embedded circuits in urethane foam. Again we utilized fairly high component density; we had no trouble with the vacuum, but power densities were fairly low. As a matter of fact, the heat dissipation was almost as high in the vacuum as at atmospheric pressure. So actually, your radiation losses amount to almost half of your total surface loss with a fairly high emissivity.

POTTING PROBLEMS RELATED TO PACKAGING DESIGN

Benjamin A. Davis

The Bendix Corporation, Kansas City Division

INTRODUCTION

Originally, epoxy resins were used as adhesives because they had unique adhesive-cohesive characteristics. Today they are widely used as potting materials for electronic components and assemblies because they possess excellent electrical and physical properties. However, their tendency to crack under severe conditions of temperature and vibration is a serious drawback to encapsulating and potting operations because a loss of effective moisture and chemical seal over the components may result and, except where repair is unobjectionable, cracked units must usually be scrapped.

The generation of fissures occurring in cured epoxy resin-hardener combinations is commonly ascribed to stresses. This does not adequately explain how the stresses are formed, nor what specific relationship the stresses hold to cracking. True, the scientific aspects of chemical and physical properties of the potting materials are well known, but technologists must try new methods and ways for applying scientific knowledge to the art of potting so that further enlightenment many ensue.

Of course, the reason for the lack of knowledge about cracking and stress development within potting materials becomes clearer when it is realized that epoxy resin mixtures undergo volume changes due to chemical heat or exotherm, to curing temperature environments in a forced convection oven, to shrinkage wrought by polymerization, and to shrinkage resulting from cooling to room temperature after curing. Clearly, a more intimate understanding of the order of reaction by the factors causing the fault would be helpful to design engineers and production personnel alike.

Several months ago, a chance observation of a peculiar recurrent cracking of a potted assembly led to the suspicion that the adhesive-cohesive nature of epoxy resins played an important role in the condition. As a result, an investigation was undertaken to determine whether a relationship existed between these forces and fissure formation; this paper presents the preliminary study made of that matter.

Fissures. A distinction among the fissures encountered is made on the basis of their location about a potting. The types defined are: (1) The crack at the surface of cured material; (2) the fine fissure which occurs in the meniscus formed at the edge of a potting; (3) the crack seen within the potting material. The surface crack is the edge or surface extremity of a crevice of indeterminate length and depth. Usually it forms over embedded components that are covered with a thin layer of potting material, a fact that has led to the commonly held idea that the difference of the product of a temperature change and the coefficient of linear thermal expansion of the potting and potted components causes cracks on potted surfaces. The crack sometimes terminates at the corner of a metal insert located in the potting surface and occasionally it spreads the length of a potted assembly. The small crack appearing in the radius of the meniscus usually forms perpendicular to the stress lines in the surface of the liquid. Crazing occurs when these hairline cracks appear in numbers. The fissure inside a mass, seen only in clear epoxy, is usually minute and appears as a localized separation. It appears to be striated, and possibly this characteristic results from light rays reflecting from the edge of laminar surfaces, causing them to appear as light and dark bands. Probably, these interior fissures are more numerous than is generally known, because they are not readily seen by the eye nor detected by means of X-ray.

Stresses. It is possible to trace the progress of stress development after gelation in a clear epoxy resin hardened with diethanolamine by photographing the stresses in the material with a camera and oriented Polaroid sheets. If the material is held in a glass-sided mold coated with mold release to permit full shrinkage from nonwetting effects, the photoelastic effect of the cured mass is clearly evident, as shown in Figs. 1, 2, and 3.

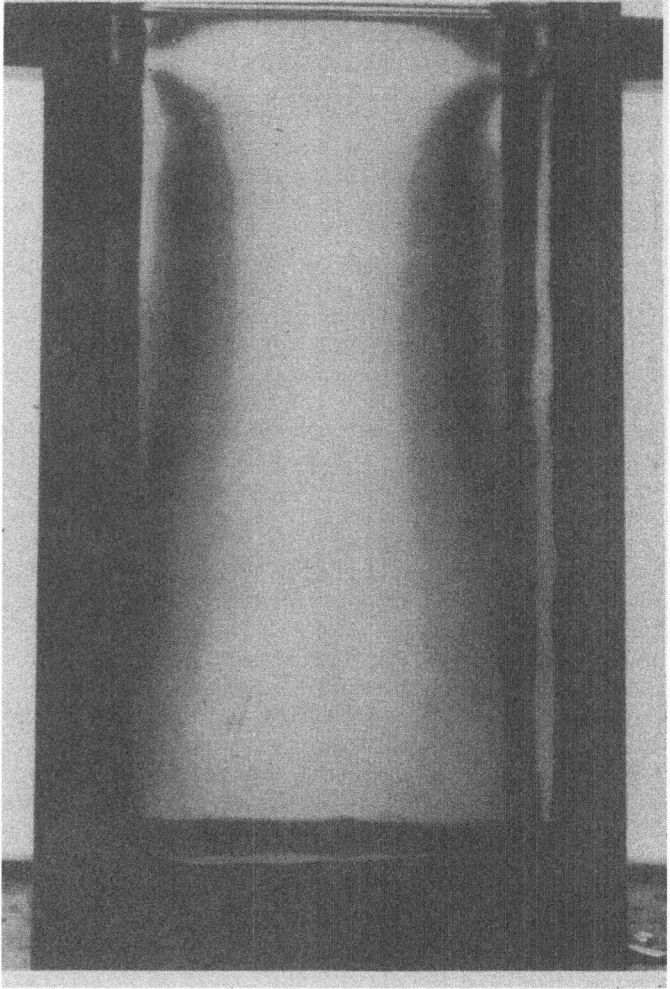

Fig. 1. Stresses caused by polymerization and shrinkage 4 hr after mixing hardener with epoxy resin. Photographed through Polaroid sheet with parallel orientation.

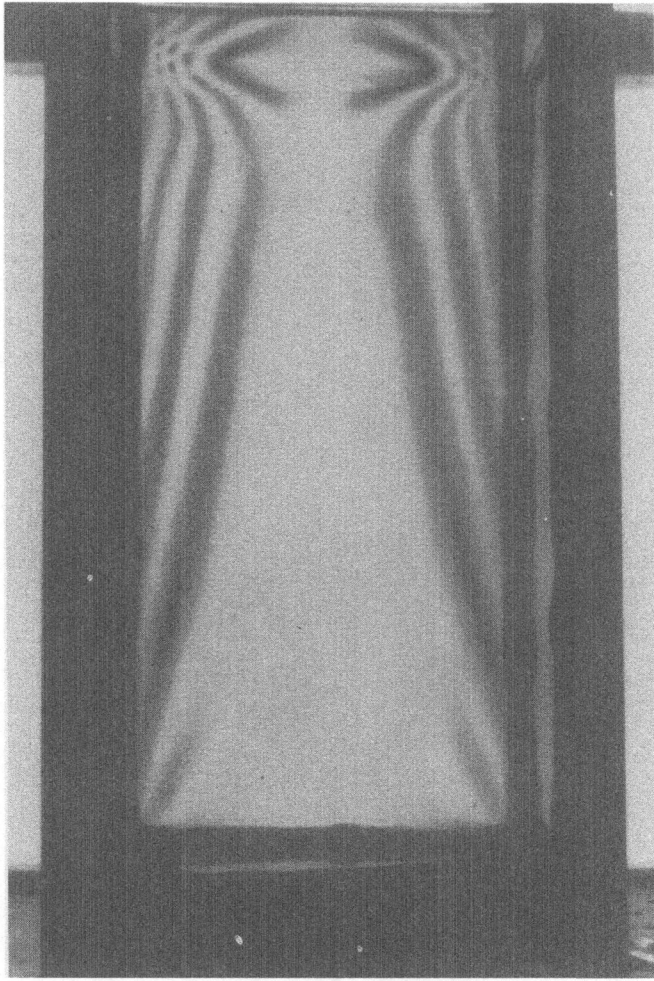

Fig. 2. Stress formation after 5 hr.

The stresses of Fig. 1 appear about 4 hr after mixing the hardener with the epoxy resin, and Figs. 2 and 3 show the development at approximately 5 hr and 7 hr later. Near the upper surface seen in Fig. 3a lateral shrinkage developed while shrinkage occurred in a vertical plane. Shown at one side in this figure is a vortex, formed by the cooling effects of the air at the point where the air currents strike the mold. When these vortices occur near mold lands, voids form [2].

The molecular chains along the darkened strata are more highly stressed than along the adjacent layers. The strata are formed by convection currents produced by heated material before gelation. While the outside surface of the material remained cooler than the interior, it flowed downward and then upward along the edge of the exothermic mass near the center. This phenomenon was traced by means of epoxy dyes placed at the top and bottom of an epoxy mixture cast in the mold shown in the figures [2].

Adhesion. The term "adhesion" takes into account the phenomena of capillarity and wetting that determine the bonding force between an adherent and an adhesive.

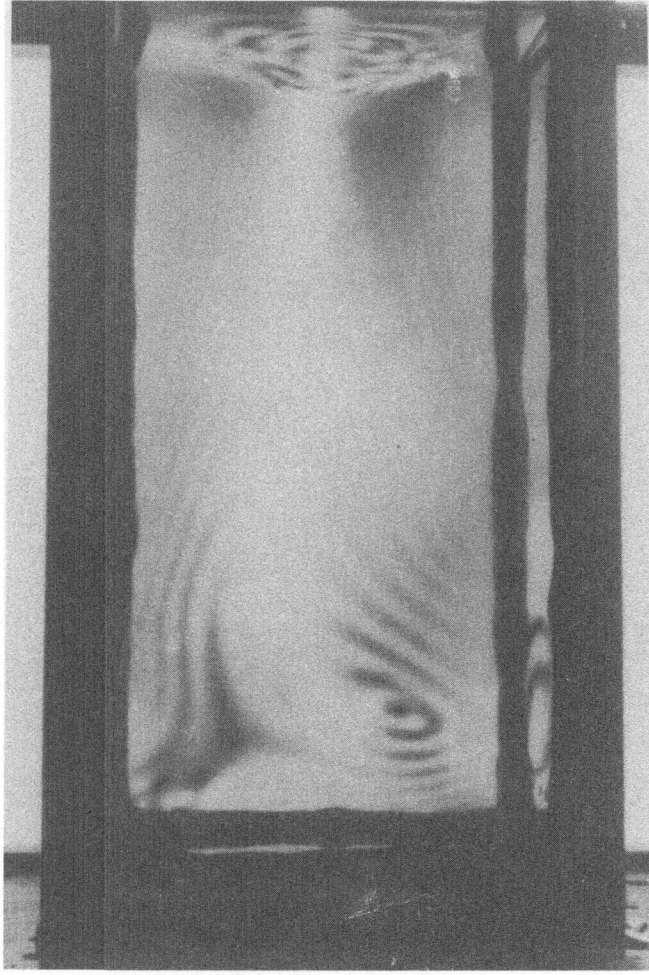

Fig. 3. Seven hours after hardener addition, shrinkage from polymerization
is complete.

Whether the basic forces are physical or chemical, such forces are produced by electronic motions within the molecules, and they influence the surface properties which determine the attraction between solids and liquid adhesives.

The surface tension of a pure liquid and the contact angle between the solid and the liquid adhesive are the qualities which determine whether an adhesive will penetrate into the capillary pores of a porous solid. Measurement of surface tension is simpler than a determination of the contact angle, being dependent upon the surface molecules only. However, surface tension is not independent of wetting, for the free energy measured at the surface of a liquid results from capillary rise of the liquid.

Wetting is determined by the measurement of the contact angle between liquid and solid, and is dependent upon the attractive forces of the solid for the liquid molecules; i.e., wetting occurs if a solid has reactive areas on its surface that can combine with the reactive molecular centers of the liquid. According to the theory of wetting, when a solid comes into contact with a liquid, the system can undergo changes if the free energy can decrease. Thus, for wetting it is necessary that the liquid be in contact with the molecules of the solid so that a change in potential energy is produced. The result of this energy change is the disappearance of the surface of the solid and the appearance of the interface.

A practical example of the importance of the interface is found in potting electrical components and assemblies enclosed by a metal housing. In this case the problem is not to make an encapsulant capable of withstanding large external loads, but to avoid the formation of residual stresses sufficiently large to produce spontaneous disruption. Consequently, to secure good adhesion, it is preferable for both the mixture and the resultant film to "wet" the surfaces. Furthermore, moderately roughened surfaces give anchorage to the adhesive agent; and indeed, the evidence indicates that in the practical use of adhesives, mechanical embedding of the plastic in the irregularities of the surface is of major importance.

REVIEW OF LITERATURE

Several means have been used to solve the problem of cracking, on the premise that stresses are neutralized by slight interior flow of the cured resin, but none of these is completely successful. For instance, overcatalyzing of a resin with a polyamine [5] curing agent is claimed to improve the resultant material so that the MIL-I-16923 hex-bar test can be passed. This test is subjection to a low-temperature environment that is severe and capable of separating the less resistant systems from the better ones. Another technique uses less than the stoichiometric amounts of catalyst [7]; and one-component resin systems have been developed in order to eliminate errors in mixing ratios of two-component resin systems [4].

A later approach uses flexible amine curing agents described as straight chain derivations, double-terminated with amine groups [9]. These materials are claimed to provide great latitude in flexibility, especially when they are blended with the common aliphatic polyamines.

In a study concerning the high reliability of transformers, Dixon states the predicament caused by cracking when he suggests that compromises must be made between the resin systems that have superior thermal-shock resistance accompanied by lowered electrical properties, and those conversely endowed [4]. Since high reliability of an electrical component is, in the last analysis, determined by its behavior under conditions of severe thermal shock, both high and low temperatures, rigid vibration, and exposure to moisture, and by the quality of electrical performance, systems susceptible to cracking in thermal shock and vibration are least reliable.

One aspect of the stresses encountered in epoxy systems was studied by means of strain gauges [1,8]. The studies were performed to determine the internal forces exerted by filled epoxy systems during free shrinkage so that selection of cast resin systems could be made in relation to mechanical damage of embedded components. The forces developed were attributed to polymerization of the epoxy resin during development of cohesive strength. Compression stress values were obtained from castings poured and cured in a Teflon mold so that adhesion to the mold was prevented.

It was successfully demonstrated that for rigid systems the compression stress varied inversely as the temperature throughout the capability range of the gauges used in the study. Flexible systems increase internal pressure as the temperature is lowered to a point typical of the system under study, and thereafter the slope of the internal stress developed is comparable to that found in rigid systems.

EXPERIMENTAL PROCEDURE

The rigid epoxy system of this study was composed of a low molecular weight bisphenol-A epichlorohydrin condensate[*] and a commercial diethanolamine. Two types of fillers were used: (1) A ground muscovite mica[†] and (2) hollow, spherical, inorganic (glass) microballoons[‡].

The mica-filled system was formulated in the ratio of 50-50-6 parts by weight epoxy, mica, and diethanolamine, respectively. In the case of microballoons, the formula was 25-100-9 parts by weight filler, epoxy, and diethanolamine, respectively. Initial cure temperatures were 130°F for 15 hr and an additional cure of 8 hr at 160°F.

Thermal-shock specimens were prepared in paper containers 3 in. high by $2\frac{3}{4}$ in. in diameter. Inserts were made of cold drawn seamless steel tubing 2 in. long by $1\frac{1}{4}$ in. od, and 0.120 in. wall. A 0.030 in. radius was machined on the ends of the tubing wall at both the inside and outside diameter positions. A $\frac{1}{16}$-in. hole was drilled through the pipe at $\frac{3}{16}$ in. from one end, for insertion of a small copper wire used for suspending the tube in the resin. Finish on inside and outside walls was 50 μin. rms.

For comparative purposes and to establish nonwetting surfaces adjacent to solvent cleaned surfaces, a mold release agent was applied to a small area of a Pyrex test tube. A steel cylinder was completely coated with mold release. These pieces, together with an acetone-washed but untreated steel cylinder, were oriented in the surface of a mica-filled epoxy mixture which was then cured. Later these specimens were cycled for 4 hr at -65°F.

Measurement of the surface tension of epoxy resin and filled epoxy resin-hardener mixtures was undertaken in order to find out whether filler additions had any effect upon the surface tension values, for it is known that generally, liquid impurities and dust change the surface effects of liquid.

For this purpose the torsion balance, sometimes referred to as the du Nuöy Tensiometer, was used. Mixtures of epoxy and filler were prepared, evacuated at 1 to 3 mm Hg absolute pressure for 2 min, then stabilized at atmospheric pressure. A thermometer was suspended in the mixture for temperature readings, and a stop watch indicated the time of measurements. The results of the evaluation of the surface tension of several epoxy resins and the filled systems are found in Table II.

An excellent indicator of state-of-cure is the coefficient of linear thermal expansion and its value is influenced by heat and additions of small amounts of chemical promoters. In order to study the effect of conditions to promote a higher cure and its influence upon the coefficient value, two slabs of material $\frac{3}{4}$ by 7 by 14 in. were prepared; to the material of one slab a small amount of a difunctional epoxy silicone[§] was added to increase the state-of-cure. Thermocouples were placed at the edge, center, and bottom sections of the curing material to ascertain the exothermic temperature during the initial cure cycle.

[*]Epon 828, Shell Chemical Company, New York, New York.
[†]4X Mineralite; Mineralite Sales Corp., Mineola, New York.
[‡]Eccospheres; Emerson & Cuming, Inc., Canton, Massachusetts.
[§]Syl-Kem 90; Dow Corning Corp., Midland, Michigan.

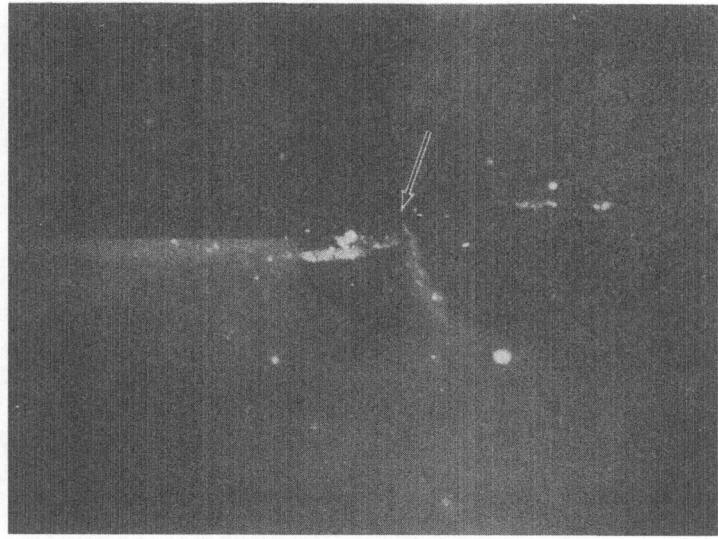

Fig. 4. Test tube in epoxy surface showing overstressed
area caused by stresses of tension and compression.

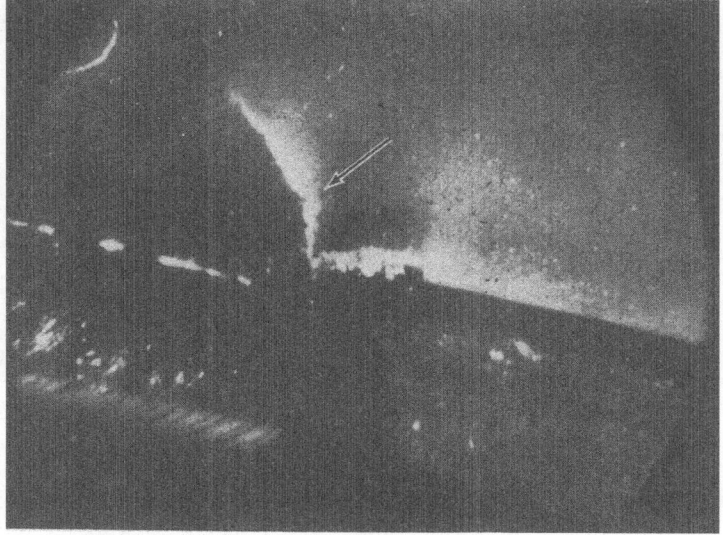

Fig. 5. After 4 hr at -65°F, the overstressed area of Fig. 4
has become an open fissure.

After curing, one-half of each block was sectioned and machined into 12 specimens
4 in. long by ½ in. diameter for a determination of the coefficient.

EXPERIMENTAL RESULTS AND DISCUSSION

That wetting is related to fissure formations is readily demonstrated by the
over-stressed area at the interface of the wetting and nonwetting regions along
the glass test tube shown in Fig. 4. After cycling the specimen for 4 hr at -65°F,

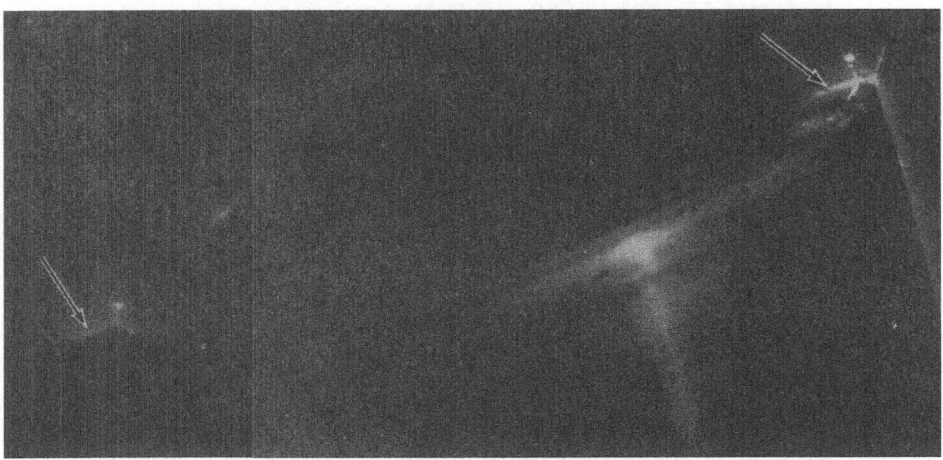

Fig. 6. Crack in filled epoxy at radius of a steel cylinder coated with mold release. Crack appeared 8 hr after start of initial cure.

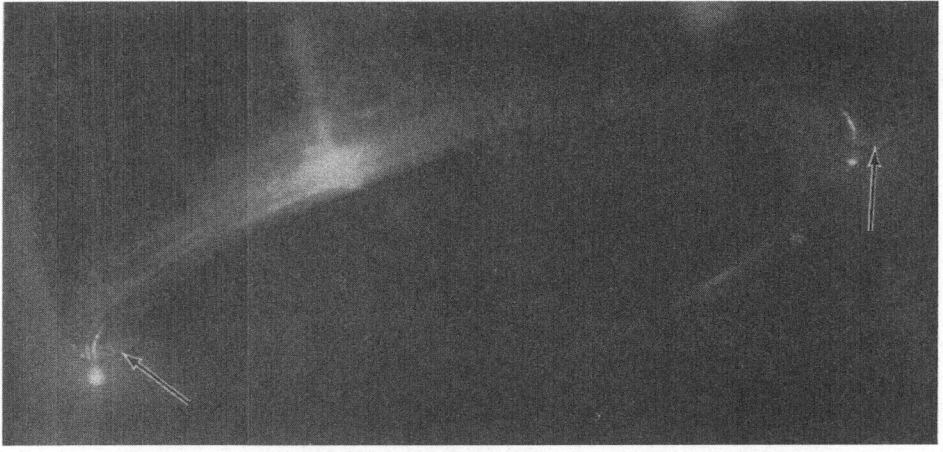

Fig. 7. Wetting of the steel cylinder by the epoxy mixture has prevented cracking at the radii.

a fissure appeared as shown in Fig. 5. The treated and untreated steel cylinders, similarly conditioned, are shown in Figs. 6 and 7, respectively.

There are two views that can be advanced for the mechanism of fissure formation. In the first case, possibly when wetting occurs, heat conduction is more rapid at the wetted surfaces; consequently, a state-of-cure takes place ahead of that which develops at the nonwetted region. The resultant unequal state-of-cure causes unequal stressing because the different values of the co-efficient of linear thermal expansion prevent an equal magnitude of expansion and contraction of adjacent layers of material. This view is supported by the coefficient values shown in Fig. 8.

In this figure, specimens 1A, 2A, 1C, and 2C occupy areas cured at lower temperatures than the remainder. The heat of the mixture reached a maximum

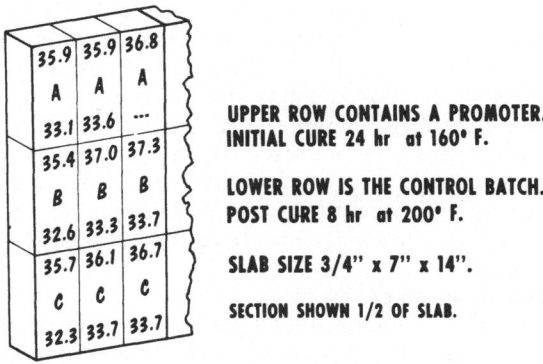

UPPER ROW CONTAINS A PROMOTER.
INITIAL CURE 24 hr at 160° F.

LOWER ROW IS THE CONTROL BATCH.
POST CURE 8 hr at 200° F.

SLAB SIZE 3/4" x 7" x 14".

SECTION SHOWN 1/2 OF SLAB.

Fig. 8. Coefficients of linear thermal expansion of a
filled epoxy resin.

at the center of the material and thence flowed toward the periphery of the mold where it was cooled by the lower oven temperature. It is readily seen that stresses are developed in a cured mass that are characteristic of the shape of the mass in relation to its environment during the cure cycle.

In the second view, wetting causes the resinous mixture to adhere firmly to the solid surface where curing occurs. Thereafter, the molecular chains are probably stretched, accompanied by an increase of elasticity of the material, while the material shrinks. This effect is tensile stress, and unlike the compression stress that results when the material does not wet, tensile stress resists deformation. This fact can be ascertained by examining Fig. 5 where the fissure opens tangent to the material under compression stress. The resultant of the dissimilar stresses acts as a notch effect which is associated with stressed rubber materials.

In theory, flexible or semiflexible epoxy systems react to environmental stresses by slippage of their molecular chains past each other [3]. Rigid systems should respond in the same manner, provided a more highly elastic state results from wetting, and this state seems likely as evidenced by the separation of the material in Fig. 5. Also, the photographed lines of induction-oriented forces of Figs. 1, 2, and 3 indicate the elastic nature of the forces within the mass and at the wetted interface of a solid surface and the liquid epoxy mixture.

In the cured matrix of an encapsulated section where tensile stress and compression stress have occurred in contiguous areas, a near-instantaneous application of energy, such as extreme cold or high impact, cannot be corrected by internal flow because the interface is not sufficiently free to move uniformly in resistance to the deforming forces; consequently, a break occurs along the highly stressed interface of the material.

Furthermore, the probability of numerous low free-energy surfaces caused by oil, plasticizers, rosin flux, nonoxide surfaces, and other low wetting surfaces occurring adjacent to high free-energy surfaces in the normal production electronic packaging unit opposes the idea that an ideal continuum would ever exist. It appears necessary, therefore, that areas to be in contact with potting mixtures must be suitably prepared to be wetted or nonwetted, but not differential-wetted, if the design engineer is to avoid cracking.

A typical example of the practice of changing high free-energy surfaces to low free-energy surfaces in order to prevent cracking is the application of a nonwetting material to a glass surface. Breakage is almost certain to occur if the glass component is an unprotected glass tube, unless a silicone or another low-wetting material is applied. Though the covering is believed to be necessary to prevent the compressive force of shrinkage from breaking the glass, probably damage does not occur wholly from compressive stress. For the high free-energy surfaces of the glass readily combine with the reactive areas of the resin to form an interface [6]. When the interface shrinks, a thin layer of glass is removed by the adhesive-cohesive forces and the weakened glass surface shatters. A thin boot of foamed silicone about the glass component, or applications of very thin layers of mold release, prevents breakage because the wetting action of the resin is made noneffective.

Fillers have been added to epoxy resin-hardener combinations because their presence lowers the coefficient of linear thermal expansion, and it is assumed their presence also reduces the tendency toward cracking of cured mixtures. If this practice is viewed in the light of the effects of wetting presented here, an entirely different understanding of the function of these materials arises. Two examples from recent studies aid to clarify this statement.

Glass microballoons have been employed as a filler in epoxy resin-diethanolamine mixes that successfully withstood a -110°F temperature through one cycle. At first the samples had been cycled five times at -65°F, plus six times at -85°F, before the lowest temperature was tried. Each cycle consisted of holding the specimens at the desired temperature for 4 hr, followed by 4 hr at room temperature.

In another study, microballoons were sieved to separate out the sizes that would pass through a 200 mesh screen only. These particles were mixed with epoxy resin and diethanolamine, and then thermal-shock specimens were prepared. The larger particles retained on the screen were similarly prepared and both lots of specimens cured. When the cured thermal-shock specimens were cycled at -65°F, cracking appeared in the specimens with the larger particle sizes, as well as several specimens prepared with mica that were included as controls. That the cracking of the specimens probably was not caused by differential expansion in this case can be determined by the coefficient values of Table I. One should note that the coefficient of linear thermal expansion of microballoons larger than 200 mesh is smaller than the coefficient determined

TABLE I. Coefficient of Linear Thermal
Expansion, in./in./°F

Filler material	-65°F to 78°F	78°F to 165°F
4X Mica	$18.4 \cdot 10^{-6}$	$24.1 \cdot 10^{-6}$
Microballoons (+200 mesh)	$17.9 \cdot 10^{-6}$	$20.9 \cdot 10^{-6}$
Microballoons (-200 mesh)	$18.3 \cdot 10^{-6}$	$24.2 \cdot 10^{-6}$

for the smaller microballoons. Also, the coefficient values of the mica-filled epoxy included in the table for comparison are similar to the values of the microballoon-filled specimens that did not crack.

The results obtained in these specimens deserve special attention, for they suggest possible answers for the cause of cracking. First, wetting of the glass microballoons serves to produce a homogeneous internal distribution of the stresses so that some part of the cured mass is less likely to be more stressed than others, permitting breakdown to start at the overstressed point. Moreover, the cracking due to contraction of an epoxy resin-hardener combination is offset by the microballoons acting mechanically to bring about an even distribution of the strains, thus avoiding crack formations. Another feature which must not be overlooked is that the filler reduces the length of the molecular chains of epoxy resin, thereby decreasing the possibility of flaws appearing along these chains,

TABLE II. Surface Tension Measurements — Unfilled Mixtures

Epoxy material vendor identity	Temperature, °F	F, dynes/cm
A	77	48.5
B	77	49.6
C	77	49.4
D	77	49.7

Formulation:	Epoxy resin Diethanolamine	100 pbw 12 pbw
Minutes after mixing	Temperature at break, °F	F, dynes/cm
10	77	49.9
20	79.3	49.3
70	95.0	50.9
130	77	62.0

Formulation:	Epoxy resin Diethylenetriamine	100 pbw 5.5 pbw
Minutes after mixing	Temperature at break, °F	F, dynes/cm
10	77	48.5
25	108	47.6
35	117	47.2

TABLE III. Surface Tension Measurements — Filled Materials

Formulation:	Epoxy resin Mica Diethanolamine	100 pbw 100 pbw 12 pbw
Minutes after mixing	Temperature at break, °F	F, dynes/cm
10	90.3	66.5
20	87.4	69.0
30	86.7	73

Formulation:	Epoxy resin Microballoons Diethanolamine	75 pbw 25 pbw 9 pbw
Minutes after mixing	Temperature at break, °F	F, dynes/cm
10	85.2	62.3
20	84.2	71.9
30	84.2	75.6
40	81.8	94.0

as well as increasing the material strength through the shortened chains or distances between wetted surfaces.

If the cracking of the specimens is viewed by the explanation given above, then those units that cracked must have contained bodies or surfaces that produced stresses either by a nonwetting effect or a nonhomogeneous internal distribution of an applied stress related to particle size or shape. An examination of the microballoons in the lot associated with cracking revealed aggregates present which were formed by cohesion of several smaller particles. In addition, there were portions of broken microballoons. The irregular shapes and sizes are believed to have been centers for the production of a notch effect during the extreme cold environment.

There is some support to the view that an increase in surface tension is brought about by adding appropriate fillers to epoxy resin mixtures. The higher tension values might be related to a lower contact angle produced by the fillers, or else the length of the epoxy molecular chains is shortened by the fillers. The latter situation would reduce the flaws per unit length and thus contribute to a higher surface strength.

Table II contains values for several commercial resins and two epoxy resin-hardener combinations. Mixing hardeners with the resins causes an exotherm that prevents surface tension measurements from being taken at comparable temperatures. However, the data obtained indicate that wetting increases after hardener addition.

Table III shows values obtained for filled systems. Fillers increase surface tension above the epoxy resin-hardener mixtures; although the data do not clearly show the relationship, microballoons seem to produce a higher surface tension value than mica.

SUMMARY AND CONCLUSION

In a preliminary investigation of the phenomenon of fissure formation in a rigid epoxy-curing agent combination it was found that nonuniform shrinkage of the cured mass is related to wetting effects of the liquid material prior to gelation.

When wetting takes place, the material is prevented from maximum shrinkage during polymerization, such as occurs in cured mixes that do not wet a surface. The resultant is an overstressed area in the material which forms a fissure upon sudden applications of low temperature.

Surface tension measurements indicate that the wetting effect of epoxy resin is increased when hardeners and fillers are added. An explanation for the results obtained is that either an increase of the reactive areas of the resin occurs, or that the fillers reduce the length of epoxy resin molecules which in turn decrease the occurrence of flaws along the resin molecules.

Wetting probably initiates a change in the rate of cure of resin-hardener mixtures. This condition possibly accounts for the differences of the coefficient values of linear thermal expansion of cured material; however, the exotherm also influences the coefficient values. Thus, unequal stresses are further developed that could cause fissures under suitable conditions.

In conclusion, studies must still be made of surface-tension effects produced by hardeners, fillers, and the materials normally associated with components and their processing. Though this inquiry is already long overdue, such studies are important today because they appear to hold the key to fissure formations.

ACKNOWLEDGMENT

The writer acknowledges the help and cooperation of his colleague, D. J. Caruthers, Materials Engineer, The Bendix Corporation, Kansas City Division, for permitting the use of certain hitherto unpublished material contained in this report.

REFERENCES

[1] A. J. Bush, "Measurement of Stresses in Cast Resins," Modern Plastics, (Feb., 1960) p. 143.
[2] B. A. Davis, "Effects of Temperature on Filled Epoxy Encapsulation Materials," J. Soc. Plastics Engineers (Dec., 1960) p. 1333.
[3] N. A. DeBruyne and R. Houwink, "Adhesion and Adhesives," (Elsevier Publishing Company, New York, 1951).
[4] L. A. Dixon, "Epoxy Resin Systems in High-Reliability Transformers," Electrical Manufacturing (July, 1960) p. 132.
[5] H. Lee and C. Hackney, "Selecting an Epoxy Resin for Encapsulating Motors," Insulation Magazine (Feb., 1961) p. 17.
[6] F. Moser, "Approximating the Attractive Forces of Adhesion for Glass and Other Surfaces," ASTM Bull., (Oct., 1960) p. 62.
[7] B. H. Mueller and C. A. Harper, "An Improved Epoxy-Resin System for Electronic Embedments," Electrical Manufacturing (Feb., 1960), p. 119.
[8] R. N. Sampson and J. P. Lesnick, "Evaluation of Casting Resins Employing Strain Gauge Techniques," Modern Plastics (Feb., 1958) p. 150.
[9] "Flexible Amine Curing Agents for Epoxy Resin," Experimental Product Technical Data Sheet (The Dow Chemical Company, Midland, Michigan).

DISCUSSION

Question: Martin Camen, Bendix Corporation. Did you vary the types of mold release agents as opposed to the use of one type?

Answer: We used one type, Garan 225. One obtains the same results from an oily thumb print on a component.

Question: We encapsulate small electronic assemblies that involve an appearance problem. We have a shell which we should like to encapsulate without putting any cover on the top. The problem is that a meniscus forms which detracts from its appearance. Do you know of any technique that we could use to minimize the effect of the meniscus?

Answer: One of the ways to improve appearance is to pour the potting materials to about one-quarter to three-eighths of an inch below the top of the mold, let gelation proceed and then add a topping to the level required. This method is better because, instead of a meniscus that is concave, you are able to get one that tends to be flat or slightly convex.

Question: Bill Noble, G. E., Syracuse. Could you tell me how clean you think the components must be? Must they be chemically clean?

Answer: Based on our study they should be chemically clean.

Question: Bill Noble, G.E., Syracuse. Does this include ultrasonic cleaning?

Answer: In general, we don't clean ultrasonically, but it is the proper thing to do in some cases. However, there is another angle about cleaning that I did not discuss and I had no intention of doing so because it was brought out this morning in another paper: what effect cleanliness has on a diode. For instance, if a glass surface is chemically clean and you pour an epoxy potting mixture on it, the glass breaks. So one must put a mold release on the glass surface or a very thin boot of another nonwetting material so that the wetting effects of the epoxy are overcome. In some papers there is the veiled idea that it is the compressive force alone of the catalyzed epoxy mixture during shrinkage that breaks the glass, but our work convinces us that it is not completely true. It is the force developed from shrinkage at the glass-epoxy interface which actually breaks the surface of the glass and causes a collapse. The simple application of mold release overcomes the wetting effects of epoxy.

Question: If you add a flexibilizer to the system, do you feel that this might improve the situation and decrease the cracking?

Answer: I would like to reserve any commitments on the question. But I say that I question that flexibilizers will completely overcome the problem of crack formation.

Question: Ken Plant, Honeywell Aero, Minneapolis. Two questions, please. On the polarized light study of stress analysis that you were showing, have you attempted any evaluation of the process of annealing plastics by heating and cooling in order to relieve this stress? The second question is an academic one regarding the surface tension studies where you showed a decrease in temperature during the tests on a normally exothermic reaction. I wondered about the matter.

Answer: Well, the answer to the first question is, No. In answer to the second one, I think the cooling is probably due to the fact that we had a small mass of material. Normally, the material sets on a small table of the torsion balance. I think some heat was lost to the table although we had a small pad underneath the material to prevent the heat loss. However, we may have lost more heat to the air, since the dish containing the material was not protected. You mentioned the exotherm. There is a higher exotherm in material where the mass is quite large and it becomes more critical than when you handle a small mass. The bigger the mass the higher the heat, that is, to a point, because you can't readily rid the mass of its heat.

Comment: Oswald, Kearfott, Clifton, New Jersey. I think I have an answer to the problem that was asked a few moments ago in regard to the meniscus that is formed during encapsulation. We, at Kearfott, have an epoxy shell that we encapsulate. We put a base plate over the open end of the shell and on one side of the case we put a rectangular cutout. We fill the module and encapsulate it and purposely leave an overflow. Then we grind the section down and put a nameplate over it.

PACKAGING OF SEMICONDUCTOR NETWORKS[*]

Ernest C. Singletary

Apparatus Division, Texas Instruments Incorporated, Dallas, Texas

The benefits promised by microminiaturized devices can be realized only when the equipment designer and manufacturer overcome problems peculiar to miniaturized systems. This paper discusses assembly, interconnection, tooling, temperature control, connectors, and handling of the Solid Circuit[†] semiconductor network.

The Solid Circuit semiconductor network offers the system designer significant advantages. It permits systems to be much smaller and lighter and allows savings through reduction in structures, enclosures, shielding, and other supporting elements.

System capability and reliability may enjoy improvement through use of redundant circuitry. Reliability is also improved because systems built with semiconductor networks require fewer interconnections and components. For example, a semiconductor network flip-flop may be regarded as a finished piece of equipment when compared with a flip-flop kit comprising an assortment of conventional components.

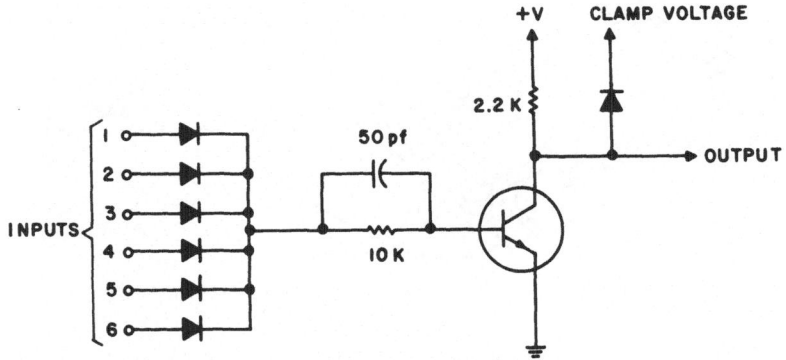

Fig. 1. NOR network circuit diagram.

The semiconductor network, or more simply, network, is essentially indivisible and nonrepairable. Although packaged in an advanced form, a network performs a conventional circuit function and may be represented with conventional circuit symbols. A typical network circuit diagram is shown in Fig. 1. Physical dimensions of the Solid Circuit, manufactured by Texas Instruments Incorporated, are shown in Fig. 2. Construction of the Solid Circuit is shown in Fig. 3.

The interconnection scheme described in this paper emphasizes high density and good manufacturability. Other concepts that have been explored and applied

*Many of the assembly techniques described herein were developed during performance of Manufacturing Methods Contract AF33(600)-42210, with Wright Air Development Division, Dayton, under the technical cognizance of the Electronic Technology Laboratory.
†Trademark of Texas Instruments.

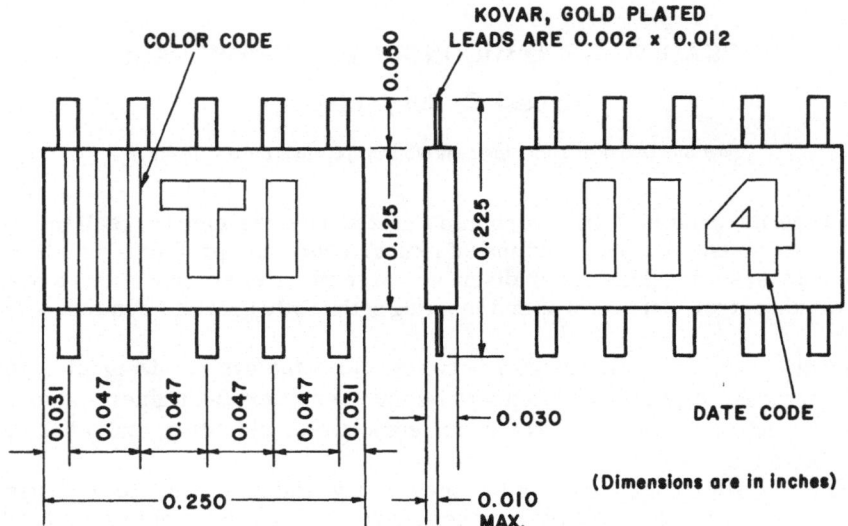

Fig. 2. Outline dimensions of solid circuit.

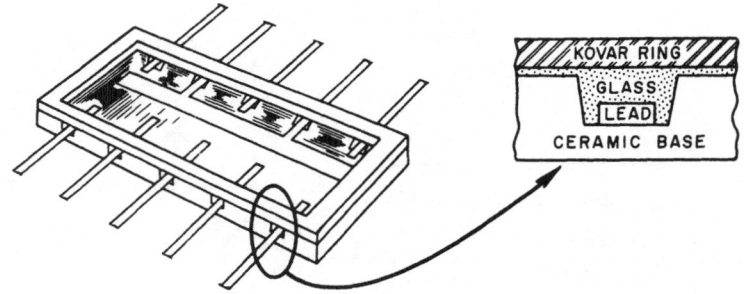

Fig. 3. Semiconductor network hermetically sealed package assembly.

emphasize other combinations of design trade-offs. For example, alternate designs individually feature automated assembly, optimum density, repairability, use on printed circuit boards, and high circuit speed.

I. NETWORK INTERCONNECTION

A. Design

The extreme miniaturization of the basic semiconductor network (approximately 12,000 equivalent components per cubic inch) presents the challenge of practical assembly.

To preserve their characteristic density, semiconductor networks are combined into dense stacks of networks, interconnections and insulators. The stack, or module, is the basic grouping of networks. The number of networks in a module might be determined by considerations of volume, external pin capacity, heat dissipation, or economics. A typical stack is a sandwich of 10 semiconductor networks, 20 interconnections, and 25 insulators. Generally, the more complex electrical groupings require more interconnections and insulators and consequent-

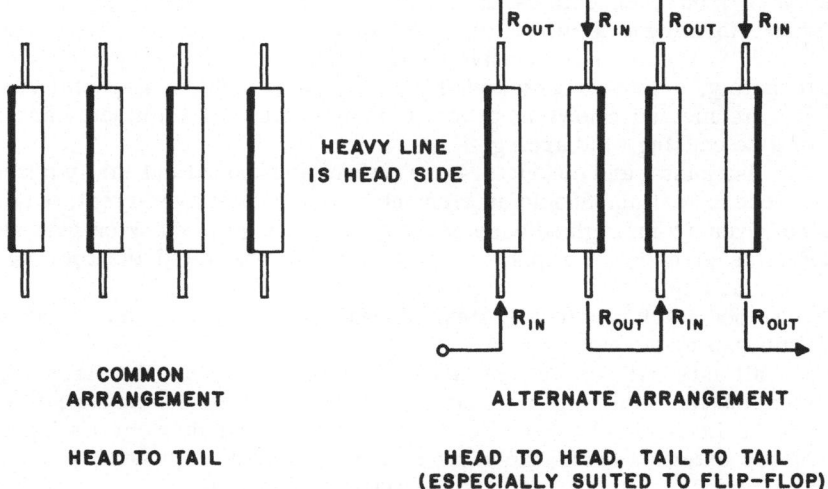

Fig. 4. Network arrangement.

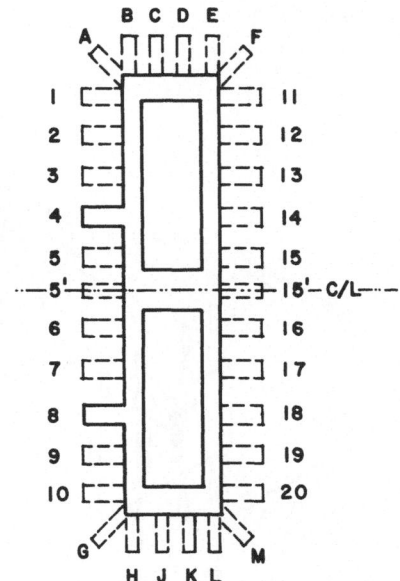

EXAMPLE: INTERCONNECTION
TRIMMED FOR LEADS 4,8

Fig. 5. Trimming of universal interconnection.

ly, permit correspondingly fewer networks per module. Simpler groupings (i.e., flip-flop counters) allow more networks per module (see Fig. 4). Interconnections are planar conductors made from a universal fingered pattern etched from thin Kovar sheet. Unwanted fingers are trimmed from the universal pattern to leave the required interconnection (see Fig. 5). Conductors are separated by rectangles of thin Mylar.

Networks, interconnections, and insulators may be glued together in a compact stack, forming a laminate with leads projecting on three or four sides (see Figs. 6 and 7). The stack is relatively easy to handle after gluing.

Alternately, a loosely assembled stack may be held in a special vise until welded. This method allows in-process straightening and eliminates any possibility of glue causing weld arcing.

All leads, joints, and conductors are externally exposed and easily accessible for in-process testing. Should an element be found defective in test, it is easily removed, even from a glued assembly, by separating the stack on both sides of the defective element. A replacement element is substituted and splicing welds restore the assembly.

The welded stack is minutely flexible, enabling it to withstand any mechanical strain imposed by the encapsulant.

After all internal connections and elements are tested, the stack is placed in a combination mold-jig for addition of the 0.018-in. diameter gold-plated Kovar leads. Leads are held in a Teflon header during welding. Following attachment of the external leads, the stack is encapsulated in a protective compound. This may be done while the stack is electrically operative.

Modules are manufactured in two configurations, as shown in Fig. 8. Configuration B is preferred for plug-in applications. A module may be used unencapsulated in either configuration.

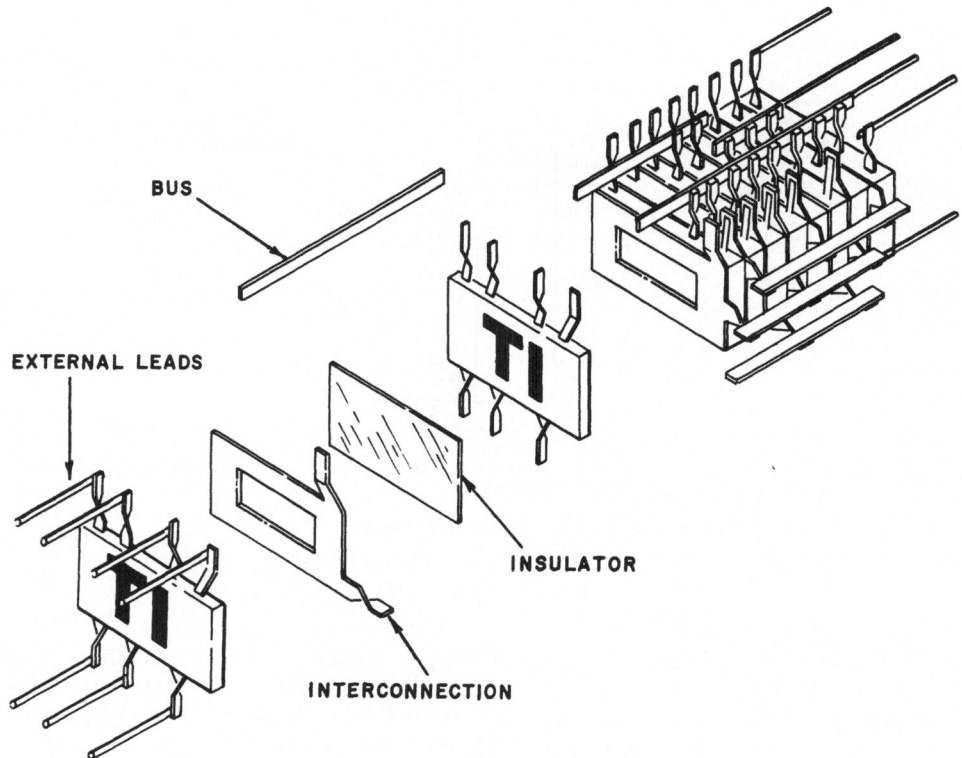

Fig. 6. Semiconductor network stack assembly.

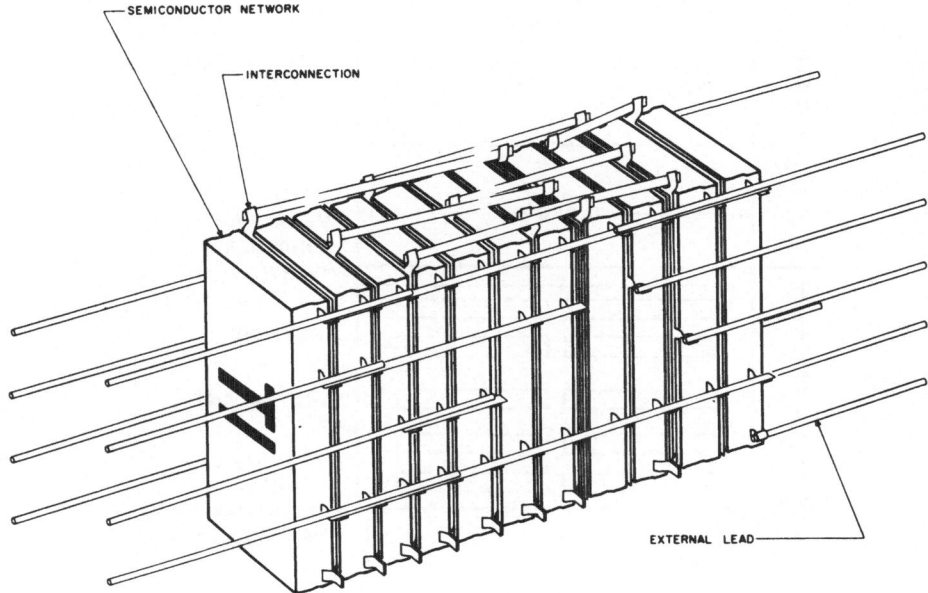

Fig. 7. Welded stack module design.

CONFIGURATION A
LEAD CAPACITY — 12 LEADS EACH END

CONFIGURATION B
LEAD CAPACITY — 26 LEADS ONE END

Fig. 8. Module configurations.

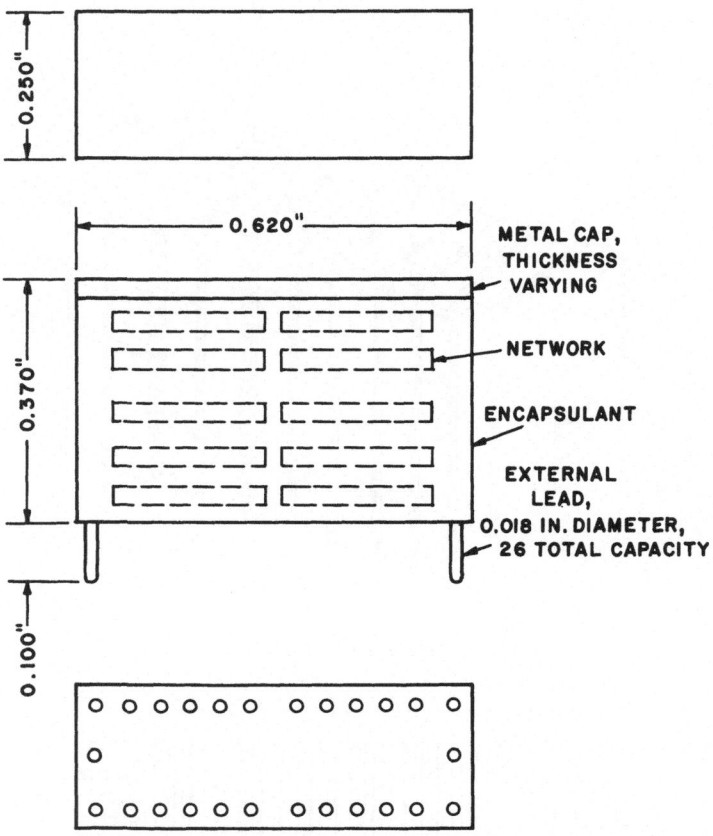

Fig. 9. Semiconductor network plug-in module.

A module occupies approximately 0.057 in.^{3}, weighs approximately 0.04 oz and may house as many as 270 equivalent components. See Fig. 9 for dimensions of a typical module.

B. Welding

Capacitor discharge resistance welding is used for joint making because of its reliability, and because it does not damage adjacent components or leads. Capacitor discharge welding provides highly localized heat input which does not raise the temperature of components. Welding also lends itself to uniform joints and high production rates.

The Kovar conductors and leads used in semiconductor network assembly are well suited to welding. The joint combinations used in the stack assembly require only one weld machine setting. Tweezer welding is currently used.

The leads of the stack are welded in several combinations. Network leads are welded to network leads and to interconnections. Longitudinal conductors of 0.020- by 0.002-in. Kovar are welded, as required, to network leads and to interconnections. Network leads and interconnection tabs are twisted 90° to make a flat connection with the longitudinal conductor.

C. Encapsulation

Characteristics essential in any encapsulation material for networks are electrical insulation, environmental protection, and nonabuse of circuits (by mechanical strains or high exothermic temperatures).

From compounds satisfying these primary requirements a selection is made to optimize the following secondary characteristics: thermal conduction; repairability advantages (transparency for locating module elements; solubility and/or ease of removal for access to parts); moisture absorption; age degradation; manufacturing advantages; curing time; safety to user; and handling ease. This selection necessarily involves compromises determined by the module's application.

Often it is desirable to somewhat alter the characteristics of an encapsulation compound. In particular, the relatively low thermal conduction of encapsulation may be improved by additives.

The epoxy resin used at Texas Instruments has a specific gravity of 1.17, slight flexibility, 60 Shore D hardness, less than 1% shrinkage, and a thermal conductivity of $3.2 \cdot 10^{-4}$ cal-cm/cm^2-°C-sec.

D. Temperature Control

Until the device circuit designer can reduce power requirements considerably below present levels, temperature control in networks will be an extremely important design consideration. The designer does have the advantage of relatively short heat paths provided by the small dimensions of networks. Equipment form factors will exploit this fact.

Temperature control of network systems will depend on the design, environment, and operating life of the equipment. Conduction will be the important mechanism for absorbing heat from the networks, although convection can be used in lower density factor designs. Radiation will probably be only a minor temperature-control mechanism. Thermally, a module has an equivalent thickness of 0.020 in. of encapsulation on each of four sides (Fig. 8, configuration B). The top incorporates a metal contact plate which aids heat transfer. The bottom, or pin side, contributes little to heat removal.

The major element in heat control is the copper egg-crate housing, which is precisely machined for close fit with the module sides. The crate is both heat sink and structure.

Heat may be rejected from equipment in one or more conventional ways, depending on system requirements. If the cooling method selected requires additional volume, the added volume is considered when the size advantage of network equipment is calculated.

A digital device package (Fig. 10) was mounted on a copper plate 12 by 12 by 0.125 in. The following data were recorded:

Total weight of computer package	9.0 oz
Total heat dissipation	16.0 w
Average network dissipation.	27.0 mw
Average module dissipation	340.0 mw
Crate heat sink weight.	6.0 oz
Maximum network design temperature	80.0°C
Network temperature rise above average plate temperature.	5.0°C

Network temperature rise above ambient air temperature	20.0°C
Network temperature rise above crate heat sink	4.0°C
Area of computer package mounted on plate heat sink.	2 ft²
Average distance of network to plate heat sink	0.170 in.

Although less important in general application, a network group might require heating or even oven control. An example of the latter might be a precision time source with all elements, crystal, oscillator, and countdown stages in a single or double oven.

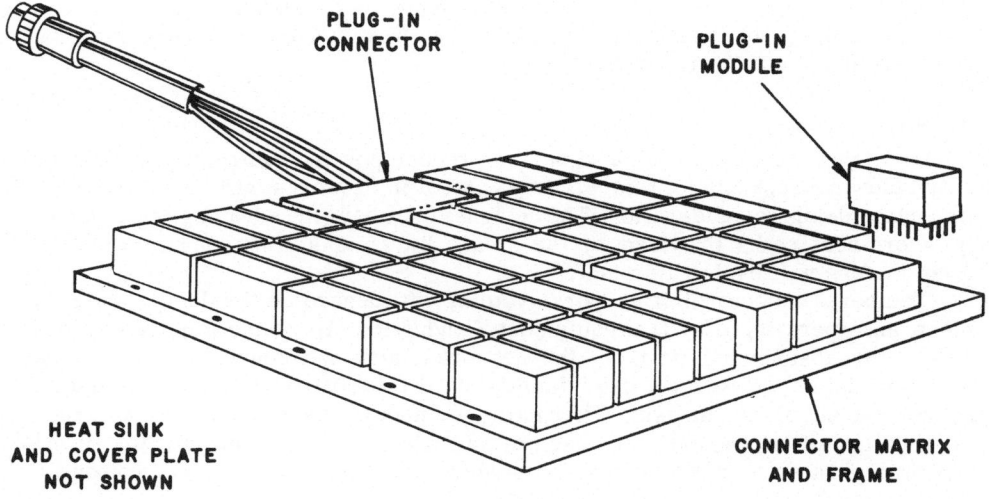

Fig. 10. A digital device.

E. Documentation

The documentation of semiconductor network systems requires new methods. Representation is network-centered rather than component-centered. Description of network interconnection is by standardized coded layouts called practical wiring diagrams, and pictorial representation is less important than it is in conventional component work.

II. MODULE INTERCONNECTION

A. Connector Design

Connectors for network systems are appropriately miniaturized. The 0.018-in. diameter male contacts extend from the encapsulated module. The female contact is a formed spring ribbon that wipes the male contact on two sides during insertion and withdrawal. The female contact is held in a thin block of plastic. Each connector block accommodates five modules. The connector contact extends

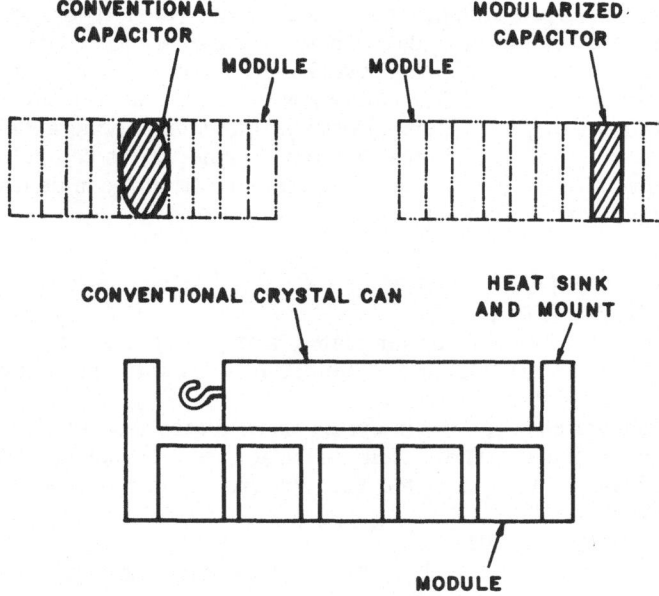

Fig. 11. Examples of hybridization.

to the reverse side of the block in a weld tab. The matrix formed by the weld tabs is back-panel wired with 0.008-in. diameter nickel wire insulated with Teflon.

B. A Digital Device

The digital device contains approximately 575 semiconductor networks of three basic circuit types: a flip-flop, a NAND circuit, and a driver circuit (see Fig. 10). The computer package houses 8600 equivalent components in 6.0 in.^{3}. Its conventional component counterpart, built as a size comparison, occupies 978 in.^{3}, or 163 times as much volume. The computer package includes 46 network modules and a clock module.

The clock module is a hybrid, including components encased in network envelopes and conventional microminiature components. Flip-flop circuits form the memory of the computer. All modules plug into a connector matrix that forms a single platform. The matrix is back-panel wired with welded connections. The computer's input/output portion is in a separate manual control unit of conventional components.

C. Equipment Design Considerations

1. Hybridization

In many applications, semiconductor network equipment will include conventional electrical components not packaged in the standard envelope of Fig. 2. Hybrids may take three forms: (a) conventional components or circuits combining with networks within the module (Fig. 11); (b) special form components, including semiconductor networks, designed for dimensional compatibility with the network (for example, a capacitor of 0.12 in. width, 0.25 in. height, and 0.070 in. thickness); (c) combinations of network groups and conventional forms.

The decrease realized in hybrid equipment depends directly on the percentage of the equipment suited to semiconductor network application. Assuming a 50-to-1 size advantage of semiconductor networks over conventional miniaturization components, a system having a 50% suitability to semiconductor networks can be reduced to 51% of the original volume. However, a system having a 95% suitability can be reduced to 7% of the original volume. Further gains can be made where the voids of conventional bulk devices (e.g., memory drums) can be used to house semiconductor network equipment.

2. Electrical Effects of Stack Packaging

a. Capacitative Effects

Signal coupling of interconnection plates through capacitive action depends on plate area, insulator thickness, dielectric constant, and signal frequency components.

At 500 kc, capacitance for two interconnection plates separated by a 0.002-in. Mylar insulator is 8.0 pf. The resultant capacitative reactance at 500 kc is an acceptable 38 kohms. It is expected that the present concept can be easily extended to operation at 5 Mc.

b. Radio-Frequency Effects

As in the case of conventional circuitry, shielding designs will be empirically determined rather than calculated.

Conducted noise is treated as it is treated in conventional equipment. Radiation shielding design benefits from the small volume of network equipment. In conducted and radiation transmission, the low power level of network equipment is to be noted.

3. Nuclear Radiation

The nuclear radiation shielding requirements of semiconductor network equipment probably will not be essentially different from those of conventional transistorized electronic equipment. The network user would have the decided advantage of less volume to protect. Nuclear radiation shielding might serve also as a heat sink.

D. Module Economics

Cost — The price of networks will become lower as the technology matures. However, as with any component, special characteristics and small quantity production increase unit cost. An economical use of networks might indicate smaller groupings for more expensive networks.

Repairability — Although networks are essentially nonrepairable, larger groupings of networks, including modules, will have some degree of repairability.

Equipment value — This is the cost of replacing a network grouping.

As network cost (and consequently, equipment value) decreases, repair will be less attractive. For example, repair might be done on the module level in early equipment, but in later equipment, an assembly might be repaired more inexpensively by replacing a defective module. The cost of systems using networks in place of conventional components involves more than the cost of circuit components. A strict comparison can be made only when values can be assigned to function, reliability, size, and weight.

III. MANUFACTURE AND USE OF SEMICONDUCTOR
NETWORK SYSTEMS

A. Tooling

Manufacture of circuits using semiconductor networks is unusual in several respects. The small size of the elements forces the assembler to use special handling devices and methods. The highly dense assembly makes repair somewhat more difficult and costly than in the case of conventional components. In-process inspection by the operator and by quality control personnel is made more difficult because of small size and high density.

The common tools of the module assembler will be microscope, tweezers and suction devices. Rarely, if ever, will the individual network, insulator, or interconnection be touched by human hands. Almost every operation will be done under a microscope (a magnification of 10 has been generally used).

B. Testing

The recommended test vehicle for a network device is the standard test card shown in Fig. 12. The card may be used for transport, incoming test, and bread-

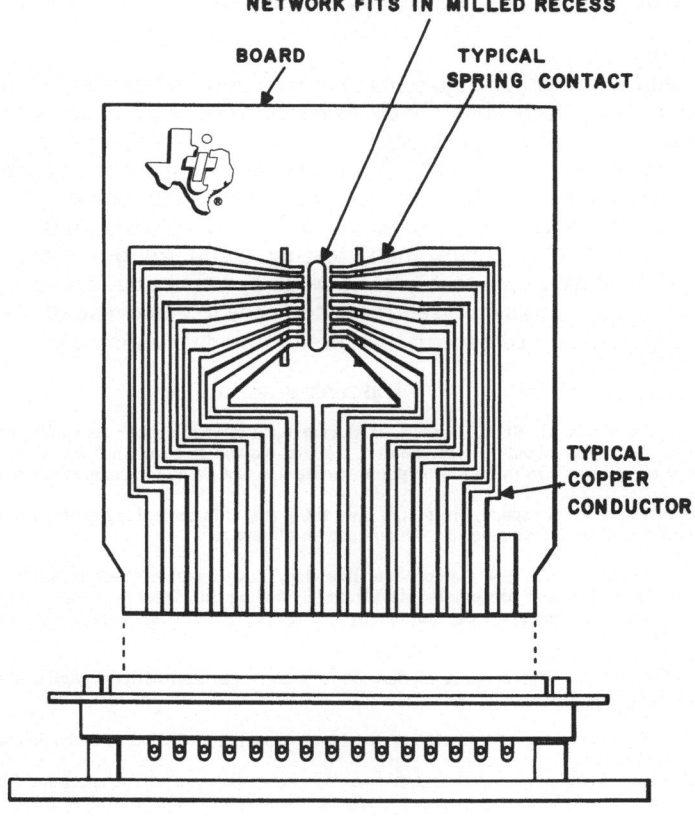

Fig. 12. Breadboard and test card.

boarding. Conventional equivalent circuits also may be mounted on similar cards for breadboard interchange.

C. Quality Control

Because of the small size of network systems and the techniques used in assembly, quality control requires new inspection techniques. Although visual inspection is not eliminated, it is not sufficient and must be supplemented by other methods.

In particular, welded joints are difficult to assess, even under magnification. A combination of qualification of weld operators, daily certification of welding equipment, sample welds, and standardized weld joints is used as a control.

D. Repair

Repair of a module, by removing and replacing damaged or imperfect networks or interconnections, will be less convenient than replacing a conventional component but is not difficult until the stack is encapsulated. Even after encapsulation, repair is feasible, and usually is achieved with a combination of mechanical removal of the system, and dissolving of the encapsulating material. Repair of welded conductors is made by splicing over the cut conductor.

E. Human Factors

The assembly of network systems requires new techniques, procedures, and documentation. The compactness of the assemblies and size of parts handled place new demands on assembly personnel. The somewhat reduced repairability and relatively high cost-to-size ratio of network systems has its effect upon the operators. Although the skill level required is not high, some workers are unsuited by skill or preference. Selecting and training appropriate personnel is a process of selection from preliminary testing, training with mechanical models, and qualification testing. Assembly of network systems attracts many applicants from personnel assembling conventional components. Because of their dexterity and objectivity, women are especially suited to module assembly.

DISCUSSION

Question: Harold Brill, RCA, Camden. With reference to your copper heat sink, that grid-type arrangement, what do you do about the differential thermal expansion between it and your modules? Either you have to have an air space in there at one point or else you are going to crush at the other point.

Answer: We have an air space, in effect, and we think that it is something of the order of one mill cut in two parts, half a mill on either side. So, we do have an air space.

Question: Lou Muenkel, IBM Components Division, Poughkeepsie, New York. Is there any danger that you are going to become extremely pin limited by adding this flexible connector on the end of your interconnected module? In other words, how many inputs and outputs can you expect from the end of this module?

Answer: We have 26 pins now. A module does become pin limited on occasion when the electrical connections are very complex.

Question: Not on the module itself but the interconnections of all these modules with this matrix?

Answer: From the matrix to the outside we have about 60 external connections. These are accommodated in the space that is occupied by three modules. We have not been limited in this respect.

Questions: Leo Fiderer, RCA, Van Nuys. In one wire matrix, do you use identical modules or do you use different modules? Do you have a large variety of different modules which constitutes a problem of maintaining spare components?

Answer: It is not a problem; it is a requirement. We do have a large variety of modules. Of the 46 modules represented in this digital device, we probably have 35 different types. Something on that order.

Question: Fred Cohen, RCA, Sommerville, New Jersey. In the first package you described, a network package, you indicate a Kovar ring as your upper edge and you mentioned sealing this to a copper plate. How do you accomplish this?

Answer: The Kovar leads are contained within a glass seal. There is a Kovar ring beyond this. The Kovar is gold plated and to that plating the copper lid is soldered.

Question: Do you solder this in vacuum?

Answer: I am not familiar with that process. This is handled in our semiconductor division. I just cannot answer that question.

Question: One further question. Can you tell me why you use a glass-ceramic combination rather than a straight ceramic box with a lead feed-through?

Answer: This again is an area that I am not comfortable in. This is done in our device department. I just cannot answer that. I am sure that there are Texas Instruments representatives here who might answer it.

Question: This is Harold Brill again. I would like to get back to that heat sink. I am fascinated by it. If you have one-mil clearance in total and you must have some manufacturing tolerances on your little modules, what kind of control do you hold on each little cutout area of that heat sink?

Answer: Let me tell you how we do that. This is a broaching operation. It's not a very cheap operation. We broach these rectangular holes after we have cut them as closely as we can. I don't know the tolerances, but this broaching is very close. The module itself is cast to approximate size and then, with crocus cloth finishing, brought down to the dimensions that give it a press fit in the rectangular hole.

Comment: Joseph Ritter, Electronic Modules Corporation, Timonium, Maryland. In the beginning of your paper, you stated that you would tell us later on how you repair these modules. I don't think you have covered this point.

Answer: I have written down here that repairability depends on what you mean by repairs and I stand by that. If you would admit that replacing a module is repair I'll say that's very easy.

Question: That would depend on the price of this module. You have approximately 15 or so networks in there. If this module is bad, are you intending this to be a throw-away module?

Answer: At some time in the future this would be so.

Comment: I want you to get to that point also. On this future curve, you assume components that are standard are going to stay static over this five or ten year period while the price of your module will come down. This does not seem reasonable to me.

Answer: Well, the history of the transistor, of course, is something that we are very close to in my company and we have drawn heavily upon that history in extrapolating the eventual cost of the networks. The other cost, the conventional component cost, we have probably predicted with some diminished accuracy. Let me finish answering the repair question though. In the assembled stacks, repair is not difficult until the stack is encapsulated. An then it is fairly impractical, although it has been done. Repair of the individual network is, of course, out of the question.

Question: Orest Meykar, Westinghouse, Air-Arm. I am a competitor of yours. The 40°C temperature differential that you have between your networks and the ambient air is remarkable. But I would like to know what potting compound you use in your modules?

Answer: It is an epoxy. If you will see me after the meeting I will be glad to tell you the number.

Question: On this test card, you show the modules on the card. How is that module attached to the network? Is that soldered or welded or what?

Answer: If you will come down to the front after the meeting I can show you the test card. These beryllium copper fingers press upon the leads of the network. So it's a finger contact arrangement. We have a better arrangement now and I will also describe that to you if you come down front.

Question: In the talk, at the beginning, you compared the time it took to weld these modules with the time it took to solder them. You consider soldering on a point-to-point basis. Most people who are making soldered components now with soldered-type modules do it by dip or flow soldering. A comparison on a point-to-point basis is meaningless.

Answer: You have a good point.

Question: H. J. Scagnelli, Bell Telephone Laboratories. What distributed parameter effects do you expect with this innerconnection technique?

Answer: We have performed tests using what we thought was the worst case condition where we took two of the universal innerconnections and sandwiched those with the insulator between and we obtained a measurement of 8 picofarads. That doesn't mean anything to me but this comforted our electrical people.

THERMAL PACKAGING FOR TRANSIENT OPERATION

Jacob G. Bartas

General Electric Company, West Lynn, Massachusetts

INTRODUCTION

This paper contains a brief study of thermal packaging of equipment for transient operation. The problem considered was this: Assuming no heat sink is available, how do we package equipment so that useful equipment operating time is a maximum?

The various components of the system may or may not generate heat. However, it is assumed that each component has mass (thermal capacity) and a defined maximum operating temperature. The components are arranged in a thermal network in which each component may be in contact, theoretically, with all other components. It is required that the network conductances be such that each component will reach its maximum operating temperature simultaneously.

It is assumed that all conductances are independent of temperature. It is assumed, also, that a definite maximum operating temperature for each component is specified. These assumptions are open to considerable argument, but they are necessary for an analytical solution of the problem without lengthy trial-and-error calculations. It is assumed, also, that all components are at the same temperature when operation begins.

The conductances, in the thermal network that has been postulated, are solved for analytically. There is an infinity of possible solutions for the more complex networks; the procedure in this paper leads to only a few solutions since numerical values of a number of constants must be assumed in order to solve the equations.

The problem is complicated by the fact that some networks have no solution for simultaneous failure of all system components. For example, a component which generates no heat and has a maximum operating temperature higher than that of any other component will not fail at the time the network generally ceases to function.

A simple procedure is used to recognize these "unsolvable" networks. A network can still be devised for a maximum time of operation within the limits of these constraints.

A number of elementary concepts are included in the preliminary sections of this paper. They are necessary to establish definitions and to provide a basis for those not acquainted with heat-transfer terminology.

THE HEAT BALANCE EQUATION*

Consider a thermal network such as shown in Fig. 1. Each nodal point in the network may be in direct thermal contact with all other nodal points. Thus, it is possible for a system of n nodes to have a total of $n(n-1)/2$ conductances.

*Definitions of the symbols used will be found near the end of the paper.

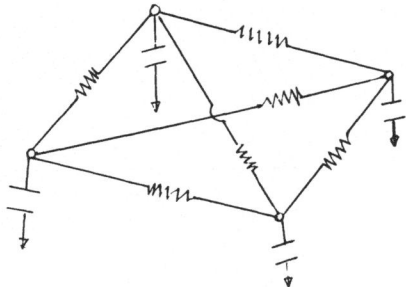

Fig. 1. Thermal network.

A heat-balance equation, a statement of the conservation of energy, can be written for each node:

$$\text{heat}_{\text{in}} + \text{heat}_{\text{generated}} = \text{heat}_{\text{out}} + \text{heat}_{\text{stored}} \tag{1}$$

The net heat flow ($\text{heat}_{\text{in}} - \text{heat}_{\text{out}}$) between node i and any node j is

$$\text{heat}_{\text{in}} - \text{heat}_{\text{out}} = C_{ji}(T_j - T_i)$$

where C is the conductance and T is the temperature.

The heat generation rate at node i is represented as G_i. The heat storage rate at node i is defined as

$$\text{Heat stored} = D_i \frac{dT_i}{dt}$$

where D_i is the thermal capacitance and t is the time.

The resultant differential equation is

$$\sum_{j=1}^{n} C_{ji}(T_j - T_i) + G_i = D_i \frac{dT_i}{dt} \tag{2}$$

DEFINITIONS

1. Existence Time. The existence time, t_e, of an isolated component is defined as the total time of satisfactory operation of the component. We integrate the heat-balance equation

$$G = D \frac{dT}{dt} \tag{3}$$

with $C = 0$ from time $t = 0$ to time $t = t$.

$$Gt = D\left(T - T_{(0)}\right) \tag{4}$$

The time required for the component to reach its maximum allowable temperature is the existence time, t_e.

$$t_e \equiv \frac{D}{G}\left(T_{\max} - T_{(0)}\right) \tag{5}$$

A plot of temperature vs. time is shown in Fig. 2.

2. Time Constant. Consider a component which is in thermal contact with a constant-temperature environment, T_a. The heat-balance equation integrates to

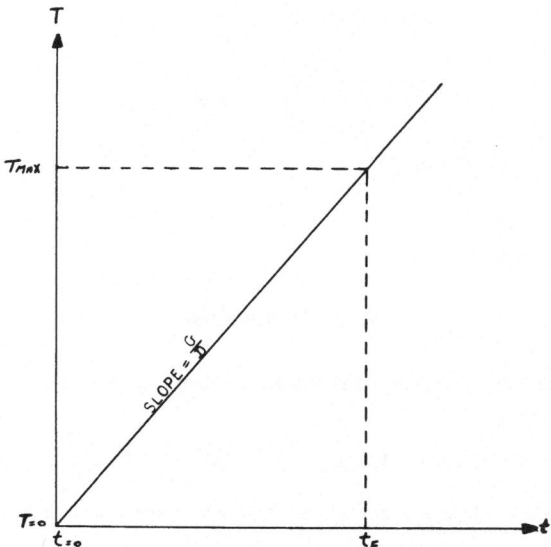

Fig. 2. Isolated component characteristics.

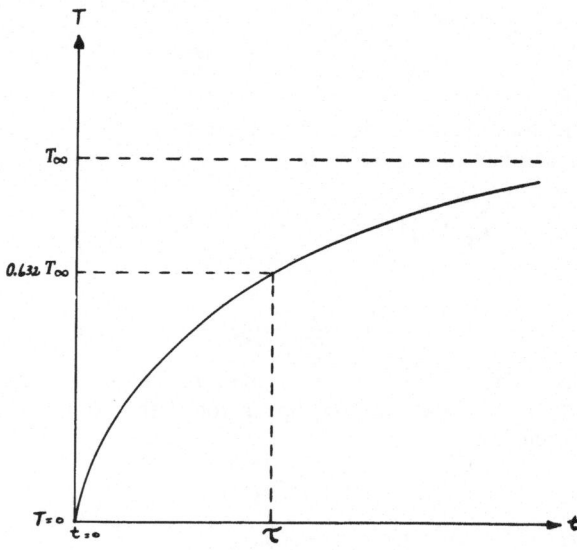

Fig. 3. Time constant.

$$T - T_{(0)} \; = \; \left(\frac{G}{C} + T_a - T_{(0)} \right) \left(1 - e^{-t/\tau} \right) \qquad (6)$$

where τ is the t i m e   c o n s t a n t. In this simple example $\tau = D/C$. A temper-ature—time plot of equation (6) is shown in Fig. 3.

  3. Useful Operating Time. The u s e f u l   o p e r a t i n g   t i m e, t_u, is defined as the time during which a component operates satisfactorily.

The useful operating time will be less than the existence time if environmental conditions are unfavorable. The useful operating time will be "infinite" if the environmental conditions are sufficiently favorable.

THE TWO-COMPONENT PROBLEM

Consider two isolated heat-generating components whose temperature histories are plotted in Fig. 4. The component with the more rapid temperature rise is labeled I, while the other component is labeled II.

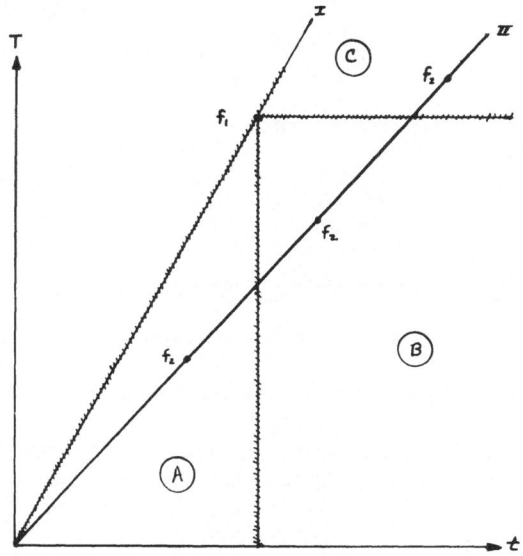

Fig. 4. Relationship of failure points.

Let us assume that the maximum temperature component I can tolerate occurs at point f_1. The failure point, temperature-wise, of component II, f_2, can be in region A, B, or C. (A failure point on either the vertical or horizontal boundary line is considered to be in region A or C, respectively.) The problem is to determine the value of a conductance C connecting the two components such that the useful operating time of the system is a maximum.

Case I: Point 2 in Region A. In this case, component II fails sooner and at a lower temperature than the concurrent temperature of component I. Any contact between components I and II will accelerate the heating of component II and, therefore, the optimum arrangement for the system is zero conductance.

Case II: Point 2 in Region C. Here, component II fails later at a higher temperature than component I. The conductance should be designed to transfer the maximum amount of energy from I to II, i.e., $C \to \infty$. Component I, however, still will fail first.

Case III: Point 2 in Region B. A proper choice of the conductance C will result in a simultaneous failure of both components (see Fig. 5). The useful operating time is computed from

$$t_u = \frac{\sum_{k=1,2} D_k \left(T_{\max} - T_{(0)}\right)_k}{\sum_{k=1,2} G_k} \tag{7}$$

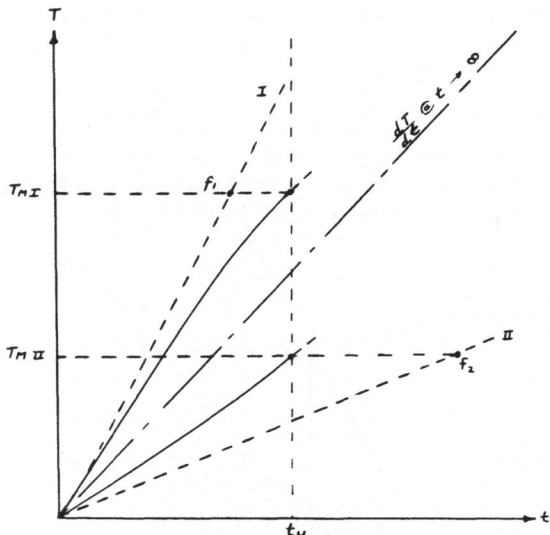

Fig. 5. Two-component system.

which represents the total system heat capacity divided by the total system heat-generation rate.

General Considerations. It is instructive to investigate the foregoing distinctions further. The useful operating time is computed using equation (7). An "average" maximum system temperature T_s is calculated, assuming $T_k(0) = 0$ using the equation

$$T_s = \frac{\sum G}{\sum D} t_u \tag{8}$$

A plot of t_u and T_s on the temperature—time graph divides the field into four quadrants (see Fig. 6). If failure points f_1 and f_2 are in quadrants II and IV, the system is that of Case III above and has a finite value of optimum conductance. If the failure points are in quadrants I and III, the optimum conductance is either 0 or ∞ as in Case I or Case II, respectively.

The foregoing considerations are helpful in the analysis of an n-component system. If any component-failure points appear in quadrants I and III, then the system may not have a solution for simultaneous component failure.

For example, suppose the component H with the maximum operating temperature has a failure point in quadrant I. In order to r e d u c e the operating time of component H, heat must be added to the component. However, it is assumed, a priori, that no other component may be continuously hotter than component H. Therefore, no continuous heat source is available to component H, and the component useful operating time is longer than that of the system. Further discussion of "unsolvable" systems is carried on below.

The Two-Component Problem Solution - Case III. The solutions of the two simultaneous differential equations for the two-body system are

$$T_k = \frac{\sum G}{\sum D} t + \frac{G_k/D_k - \sum G/\sum D}{C_{12} \sum D/D_1 D_2} \left[1 - e^{-C_{12}(\Sigma D/D_1 D_2)t} \right]_{k=1,2} \tag{9}$$

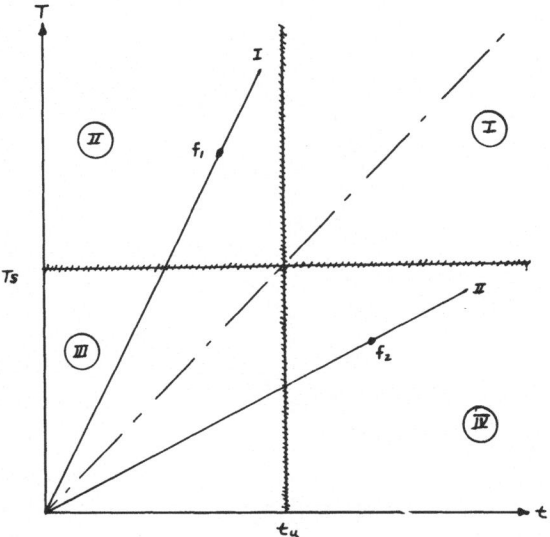

Fig. 6. Quadrant locations.

If we let $t = t_u$ then $T_k = T_{k\max}$, and we may solve equation (9) ($k = 1$ or 2) for C_{12}. Equation (8) is of the form

$$T = Bt + K\left(1 - e^{-pt}\right) + A \tag{10}$$

The temperature T will increase continually with time since there is no coupling to a steady-state ambient or environment. The time constant, $1/p$, indicates how rapidly a "steady-state" gradient is approached.

THE THREE-COMPONENT PROBLEM

Consider a three-component system whose characteristics are shown in Fig. 7. All failure points are in quadrants II and IV; the system components can fail simultaneously.

The system has a finite number of solutions for sets of C_{1-2}, C_{1-3}, and C_{2-3}, which satisfy the problem requirement. For example, near one limit, $C_{1-3} = 0$, an infinite number of solution pairs for C_{1-2} and C_{2-3} can be found which will result in a simultaneous failure of all system components. Similarly, $C_{2-3} = 0$ is another such limit; however, $C_{1-2} = 0$ is not a permissible limit for this system.

Thus, we are at liberty to select arbitrarily system values of C_{1-2}, C_{2-3}, or C_{1-3} within broad limits. Whether the chosen value of the conductance is permissible for the given system is determined by the end result, i.e., the physical reality of all conductances calculated for the system.

General Considerations. The number of degrees of freedom F in an n-component system is given by the relation

$$F_n = \frac{(n-1)(n-2)}{2} \tag{11}$$

A four-node system has three arbitrary selections; a five-node system has six selections, and so forth. Note that all conductances must have real positive values!

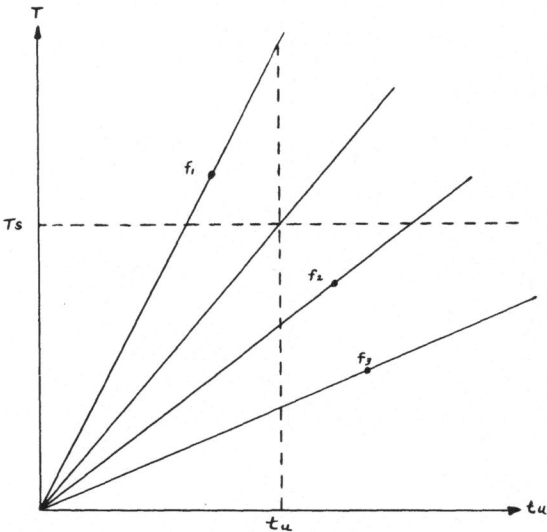

Fig. 7. Three-component problem.

Conventional Solution. The three differential equations [equation (2)] describing the three-component system, solved simultaneously, lead to three equations (one equation of the three is redundant) for the three node temperatures. However, this method has several drawbacks. 1. The roots of a cubic equation must be found. 2. The conductance must be computed using trial-and-error methods since the resulting equations are not linear in C_{12}, C_{13}, or C_{23}.

Hence, the conventional approach has obvious difficulties. These difficulties prove insurmountable when five or more components comprise the system since a five-component system requires the solution of a quintic for which no general solution is known. This limitation compels us to try alternate methods of solution.

Proposed Solution. Consider the three-component system in Fig. 7. The equation for the temperature at any node is of the form, of equation (10).

$$T_k - T_k(0) = B_k t + K_{k_1}\left(1 - e^{-p_1 t}\right) + K_{k_2}\left(1 - e^{-p_2 t}\right) \tag{12}$$

We define a function Γ^* as

$$\Gamma = \frac{1 - e^{-pt}}{p t_u} \tag{13}$$

and replace $T - T(0)$ by θ

$$\theta \equiv T - T(0) \tag{14}$$

Substituting the foregoing definitions into equation (12), we arrive at

$$\theta_k = B_k t + E_{k_1}\Gamma_1 + E_{k_2}\Gamma_2 \tag{15}$$

The equations for a three-component system are

$$\theta_1 = B_1 t + E_{11}\Gamma_1 + E_{12}\Gamma_2$$
$$\theta_2 = B_2 t + E_{21}\Gamma_1 + E_{22}\Gamma_2 \tag{16a, b, c}$$
$$\theta_3 = B_3 t + E_{31}\Gamma_1 + E_{32}\Gamma_2$$

*See Appendix I for a tabulation of $\dfrac{1-e^{-x}}{x}$ vs. x.

Only two of these equations are independent. The physics of the problem requires that the temperature gradients of all components be equal as $t \to \infty$. Therefore,

$$B_1 = B_2 = B_3 = B = \frac{\sum G}{\sum D} \tag{17}$$

Now, consider two independent equations for θ.

$$\theta_1 = Bt + E_{11}\Gamma_1 + E_{12}\Gamma_2$$
$$\theta_2 = Bt + E_{21}\Gamma_1 + E_{22}\Gamma_2 \tag{18a, b}$$

At time $t = t_u$, $\theta = \theta_{max}$, $\Gamma = \Gamma_u$, and we have

$$\theta_{1m} = Bt_u + E_{11}\Gamma_{1u} + E_{12}\Gamma_{2u}$$
$$\theta_{2m} = Bt_u + E_{21}\Gamma_{1u} + E_{22}\Gamma_{2u} \tag{19a, b}$$

As shown above there is one degree of freedom in a three-component system. Solving for the E's and Γ's in terms of the conductances and then attempting to evaluate equations (19), after assuming a value for one conductance, was shown to be unfeasible. Instead, we attack the problem by arbitrarily setting one of the E's equal to zero.

For example, assume $E_{12} = 0$. Differentiate equation (18a)

$$\frac{d\theta_1}{dt} = B + E_{11}\frac{d\Gamma_1}{dt} \tag{20}$$

where

$$\frac{d\Gamma_1}{dt} = \frac{e^{-p_1 t}}{t_u} \tag{21}$$

Setting $t = 0$, we arrive at

$$\left(\frac{d\theta_1}{dt}\right)_{t=0} = B + \frac{E_{11}}{t_u} \tag{22}$$

But from equation (2) we have

$$\left(\frac{d\theta}{dt}\right)_{t=0} = \frac{G_1}{D_1} \tag{23}$$

We combine equations (22), (23), and (17) and solve for E_{11}

$$E_{11} = t_u\left(\frac{G_1}{D_1} - \frac{\sum G}{\sum D}\right) \tag{24}$$

The value of Γ_{1u}, i.e., the pseudo time constant $1/p_1$, is calculated from equation (19), plus Appendix I

$$\Gamma_{1u} = \frac{\theta_{1m} - Bt_u}{E_{11}}$$

or

$$\frac{1 - e^{-p_1 t_u}}{p_1 t_u} = \frac{\theta_{1m} - (\sum G/\sum D)\, t_u}{(G_1/D_1 - \sum G/\sum D)t_u} \tag{25}$$

Next, combine the three differential heat-balance equations

$$\sum_{k=1}^{3} D_k \frac{d\theta_k}{dt} = \sum_{k=1}^{3} G_k \tag{26}$$

Substitution of equation (15) into equation (26) leads to

$$\left(D_1 E_{11} + D_2 E_{21} + D_3 E_{31}\right)\Gamma_1 + \left(D_1 E_{12} + D_2 E_{22} + D_3 E_{32}\right)\Gamma_2 = 0 \tag{27}$$

Since equation (27) must be valid for all t, it follows that

$$D_1 E_{11} + D_2 E_{21} + D_3 E_{31} = 0$$

$$D_1 E_{12} + D_2 E_{22} + D_3 E_{32} = 0 \tag{28a, b}$$

In general,

$$\sum_{k=1}^{h} D_k E_{kj} = 0 \tag{29}$$

From equation (2)

$$D_k \frac{d\theta_k}{dt} = B + \sum C_{kj}\left(\theta_j - \theta_k\right) \tag{30}$$

Upon combination of equations (15) and (30), we have for Node 1:

$$D_1 E_{11} p_1 \Gamma_1 + D_1 E_{12} p_2 \Gamma_2 = \left\{ C_{12}\left(E_{11} - E_{21}\right) + C_{13}\left(E_{11} - E_{31}\right) \right\} \Gamma_1$$

$$+ \left\{ C_{12}\left(E_{12} - E_{22}\right) + C_{13}\left(E_{12} - E_{32}\right) \right\} \Gamma_2 \tag{31}$$

A similar expression is obtained for Node 2. Since equation (31) must be valid for all t, we have

$$D_1 E_{11} p_1 = \left(C_{12} + C_{13}\right)E_{11} \qquad\quad - C_{12} E_{21} - C_{13} E_{31}$$

$$D_1 E_{12} p_2 = \left(C_{12} + C_{13}\right)E_{12} \qquad\quad - C_{12} E_{22} - C_{13} E_{32}$$

$$D_2 E_{21} p_1 = \qquad\quad - C_{12} E_{11} + \left(C_{21} + C_{23}\right)E_{21} - C_{23} E_{31}$$

$$D_2 E_{22} p_2 = \qquad\quad - C_{12} E_{12} + \left(C_{21} + C_{23}\right)E_{22} - C_{23} E_{32} \tag{32a, b, c, d}$$

Equations (28a, b) and (32a, b, c, d) are manipulated as follows: (Note that $E_{12} = 0$.) Eliminate E_{31} and E_{32} using (28a, b).

$$D_1 E_{11} p_1 = \left\{ C_{12} + C_{13}\left(1 + \frac{D_1}{D_3}\right) \right\} E_{11} - \left\{ C_{12} - \frac{D_2}{D_3} C_{13} \right\} E_{21}$$

$$0 = \qquad\qquad 0 \qquad\qquad - \left\{ C_{12} - \frac{D_2}{D_3} C_{13} \right\} E_{22}$$

$$D_2 E_{21} p_1 = -\left\{ C_{12} - \frac{D_1}{D_3} C_{23} \right\} E_{11} + \left\{ C_{21} + C_{23}\left(1 + \frac{D_2}{D_3}\right) \right\} E_{21}$$

$$D_2 E_{22} p_2 = \qquad\qquad 0 \qquad\qquad + \left\{ C_{21} + C_{23}\left(1 + \frac{D_2}{D_3}\right) \right\} E_{22} \tag{33a, b, c, d}$$

Equations (33a, b) are used to derive the results

$$C_{12} = \frac{D_1 D_2 p_1}{\sum D} \tag{34}$$

$$C_{13} = \frac{D_1 D_3 p_1}{\sum D} \tag{35}$$

Equations (33c) and (34) are combined by eliminating

$$\frac{\sum D}{D_1} C_{12} E_{21} = -\left\{C_{12} - \frac{D_1}{D_3} C_{23}\right\} E_{11} + \left\{C_{21} + C_{23}\left(1 + \frac{D_2}{D_3}\right)\right\} E_{21}$$

Thus,

$$E_{21} = -\frac{E_{11} D_1}{D_2 + D_3} \tag{36}$$

Differentiating equation (18b) and combining with equation (2), we have

$$\left(\frac{d\theta_2}{dt}\right)_{t=0} = \frac{G_2}{D_2} = B + \frac{E_{21} + E_{22}}{t_u} \tag{37}$$

Coefficient E_{22} is calculated from equation (37). The parameter Γ_{2u}, i.e., p_2 is found from equation (19b), and the remaining conductance, C_{23}, is computed from equation (33d).

$$C_{23} = \frac{D_3 \left(D_2 p_2 - C_{21}\right)}{D_2 + D_3} \tag{38}$$

Thus, by assuming $E_{12} = 0$, we have determined a set of network conductances C_{12}, C_{13}, and C_{23}, which satisfies the requirement of simultaneous component failure. Two additional sets of conductances may be computed by assuming $E_{22} = 0$ or $E_{32} = 0$. The existence of any set depends on whether all conductances in the set have real positive values.

THE n-COMPONENT PROBLEM

The heat-balance equation for the n-component network node is

$$D_j \theta_j = G_j + \sum_{k=1}^{n} C_{jk}\left(\theta_k - \theta_j\right) \tag{39}$$

The solution for θ has the form

$$\theta_j = Bt + \sum_{i=1}^{n-1} E_{ji} \Gamma_i \tag{40}$$

Select for study any set of $n-1$ equations. In matrix notation

$$\{\theta\} = Bt + [E]\{\Gamma\} \tag{41}$$

The number of degrees of freedom from equation (11) is

$$F_n = \frac{(n-1)(n-2)}{2} \tag{11}$$

This is exactly the number of E's above the diagonal in the E matrix. The solution of the n-component problem begins by setting the E's above the diagonal equal to zero.

$$[E] \equiv \begin{bmatrix} E_{11} & 0 & 0 & \cdots & 0 & 0 \\ E_{21} & E_{22} & 0 & \cdots & 0 & 0 \\ E_{31} & E_{32} & E_{33} & \cdots & 0 & 0 \\ \vdots & \vdots & \vdots & & \vdots & \vdots \\ E_{m-1,1} & E_{m-1,2} & E_{m-1,3} & \cdots & E_{m-1,m-1} & 0 \\ E_{m1} & E_{m2} & E_{m3} & \cdots & E_{m,m-1} & E_{mm} \end{bmatrix}_{m=n-1} \tag{42}$$

The algebraic steps proceed in the order indicated in the foregoing section for the three-component problem. Although the algebra may appear complex, use of symbolic notation renders the set of equations amenable to manipulation; a pattern emerges, and the end results are as follows:

For diagonal E's

$$E_{jj} = \left(\frac{G_j}{D_j} - B \right) t_u - \sum_{i=1}^{j-1} E_{ji} \tag{43}$$

For nondiagonal E's

$$E_{ji} = - \frac{E_{ii} D_i}{\sum\limits_{k=i+1}^{n} D_k} \qquad j > i \tag{44}$$

For the p's (and Γ's)

$$\Gamma_{ju} = \frac{\theta_{j\,max} - B t_u - \sum\limits_{i=1}^{j-1} E_{ji} \Gamma_{iu}}{E_{jj}} \tag{45}$$

For the conductances

$$C_{jk} = \frac{D_k}{\sum\limits_{m=j}^{n} D_m} \left[D_j p_j - \sum_{i=1}^{j} C_{ji} \right] \qquad k > j \tag{46}$$

It may be shown that equations (43) through (46) are indeed solutions of the problem by substituting these equations into equations (39) and (40).

EXAMPLE

A network contains four components whose thermal characteristics are as follows:

Component	G	D	θ_m	T_e
a	4	2	3.4	1.7
b	3	2	3.1	2.06
c	2	3	2.1	3.15
d	1	3	1.9	5.7

The values are in any set of consistent units. It is required that several sets of conductances be determined such that system operating time is maximized.

Solution:

    I. *Calculate System Parameters*

$$t_u = \frac{\sum D\theta_m}{\sum G} = 2.5$$

$$T_s = \frac{\sum D\theta_m}{\sum D} = 2.5$$

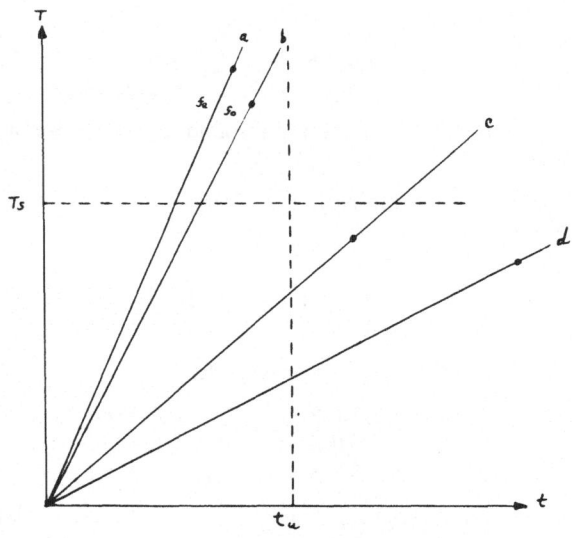

Fig. 8. Example.

    II. *Plot System Temperature Time Graph - Fig. 8.* The system is well behaved; all failure points are in quadrants II and IV and the system has one or more solutions.

    III. *Solution No. 1.* a) Arrange θ-equations in the following order.

Node No.	Component	G	D	θ_m
1	a	4	2	3.4
2	b	3	2	3.1
3	c	2	3	2.1
4	d	1	3	1.9

b) Compute E's from equations (43) and (44).

$$E_{11} = 2.5$$
$$E_{21} = E_{31} = E_{41} = -0.625$$
$$E_{22} = 1.875$$
$$E_{32} = E_{42} = 0.625$$
$$E_{33} = 0.416$$
$$E_{43} = 0.416$$
$$E_{44} = 0 \quad \text{(check)}$$

c) Compute p's using equation (45) and Appendix I.

$$\Gamma_{1u} = 0.36 \qquad p_1 t_u = 2.57 \qquad p_1 = 1.028$$
$$\Gamma_{2u} = 0.44 \qquad p_2 t_u = 1.95 \qquad p_2 = 0.78$$
$$\Gamma_{3u} = 0.24 \qquad p_3 t_u = 4.10 \qquad p_3 = 1.64$$
$$\Gamma_{4u} = \frac{0}{0} \quad \text{(check)}$$

d) Compute C's using equation (46)

$$C_{12} = 0.4112 \qquad C_{23} = 0.4308$$
$$C_{13} = 0.6168 \qquad C_{24} = 0.4308$$
$$C_{14} = 0.6168 \qquad C_{34} = 1.9362$$

IV. *Solution No. 2.* Another solution is found by considering another column arrangement of the nodes. For example:

Node No.	Component
1	d
2	c
3	b
4	a

The conductances for this arrangement are

$$C_{12} = 0.9552 \qquad C_{23} = 0.4385$$
$$C_{13} = 0.6168 \qquad C_{24} = 0.4385$$
$$C_{14} = 0.6168 \qquad C_{34} = 0.9581$$

V. *Other Solutions.* The foregoing method can be used to determine a total of 12 independent solutions for a four-component system. The physical existence of each set is dependent on the existence of real positive values for all C's in the set.

In general, the number of possible independent solutions, S_n, is given by

$$S_n = \frac{n!}{2}$$

where n is the number of components.

DISCUSSION

The foregoing procedure leads to realistic solutions for a well-behaved system. (The realistic solutions may be impractical for mechanical reasons, electrical reasons, and so forth, but nevertheless, they are thermally feasible.) However, many systems are not well behaved. These systems require considerable study before a suitable network is found. The following occur often:

1. More than one component with the same maximum operating temperature, θ_m — the components are combined ($C \rightarrow \infty$) and considered as one.
2. Component with maximum operating temperature, H, has a failure point in quadrant I — The ideal maximum system operating time

$$t_{max} = \frac{\sum D\theta_m}{\sum G}$$

cannot be realized (see discussion above for the two-component system). Component H is eliminated from the system, and the remaining subsystem is solved. Then it is necessary to study the possibility of allowing thermal contact between component H and any components with steeper slopes on the T-t diagram. This study will determine if the useful operating time of the subsystem can be extended.
3. Components, other than component H, with failure points in quadrant I — Various subsystems, consisting of these components, and components with quadrant II failure points, must be analyzed. If subsystems containing all quadrant I components have useful operating times greater than the t_u for the entire system, then the system solution proceeds as if the system were well behaved. Otherwise, the subsystems must be studied to determine the best combination of components which produce the maximum operating time.
4. Component L, the component with the minimum operating temperature, has failure points in quadrant III — the ideal maximum system operating time cannot be realized. Subsystems containing component L and components with shallower slopes on the T-t diagram must be analyzed to determine the optimum subsystem arrangement.
5. Components other than component L, with failure points in quadrant III — subsystems containing these components and components with failure points in quadrant IV must be evaluated. If all quadrant III components may be matched in subsystems whose operating times are greater than t_u for the entire system, then the entire system is well behaved. Otherwise, the subsystems must be studied to find the "best" combinations of components.

In conclusion it is apparent that the method for determining network conductances presented herein has many limitations. However, in many instances the method does lead to acceptable solutions for sets of network conductances which allow maximum transient operation of a heat generating system.

NOTATION

Symbol	Definition
B	$\Sigma G/\Sigma D$ (for system)
C	Conductance
D	Thermal capacitance
E	Coefficient
G	Heat-generation rate
p	Root of equation
T	Temperature
t	Time
Γ	Function [see equation (13)]
θ	$T - T(0)(t = 0)$

APPENDIX I

TABULATION OF $\dfrac{1-e^{-x}}{x}$ vs. x

0.00	1.00000	0.46	0.80156	0.92	0.65378
0.01	0.99500	0.47	0.79787	0.93	0.65102
0.02	0.99000	0.48	0.79421	0.94	0.64226
0.03	0.98500	0.49	0.79055	0.95	0.64554
0.04	0.98025	0.50	0.78694	0.96	0.64282
0.05	0.97540	0.51	0.78333	0.97	0.64012
0.06	0.97067	0.52	0.77977	0.98	0.63744
0.07	0.96586	0.53	0.77623	0.99	0.63477
0.08	0.96100	0.54	0.77268	1.00	0.63212
0.09	0.95633	0.55	0.76918	1.01	0.62948
0.10	0.95160	0.56	0.76570	1.02	0.62686
0.11	0.94700	0.57	0.76223	1.03	0.62426
0.12	0.94233	0.58	0.75879	1.04	0.62168
0.13	0.93777	0.59	0.75537	1.05	0.61910
0.14	0.93314	0.60	0.75198	1.06	0.61655
0.15	0.92860	0.61	0.74861	1.07	0.61401
0.16	0.92412	0.62	0.74526	1.08	0.61148
0.17	0.91965	0.63	0.74192	1.09	0.60897
0.18	0.91517	0.64	0.73861	1.10	0.60648
0.19	0.91074	0.65	0.73531	1.11	0.60400
0.20	0.90635	0.66	0.73204	1.12	0.60154
0.21	0.90200	0.67	0.72879	1.13	0.59909
0.22	0.89764	0.68	0.72556	1.14	0.59665
0.23	0.89335	0.69	0.72235	1.15	0.59423
0.24	0.88904	0.70	0.71916	1.16	0.51982
0.25	0.88480	0.71	0.71600	1.17	0.58943
0.26	0.88058	0.72	0.71285	1.18	0.58705
0.27	0.87637	0.73	0.70971	1.19	0.58469
0.28	0.87221	0.74	0.70661	1.20	0.58234
0.29	0.86807	0.75	0.70351	1.21	0.58000
0.30	0.86393	0.76	0.70043	1.22	0.57768
0.31	0.85984	0.77	0.69739	1.23	0.57537
0.32	0.85578	0.78	0.69435	1.24	0.57308
0.33	0.85176	0.79	0.69134	1.25	0.57080
0.34	0.84774	0.80	0.68834	1.26	0.56853
0.35	0.84374	0.81	0.68536	1.27	0.56628
0.36	0.83978	0.82	0.68240	1.28	0.56403
0.37	0.83586	0.83	0.67946	1.29	0.56181
0.38	0.83195	0.84	0.67654	1.30	0.55959
0.39	0.82805	0.85	0.67364	1.31	0.55739
0.40	0.82420	0.86	0.67074	1.32	0.55520
0.41	0.82036	0.87	0.66787	1.33	0.55302
0.42	0.81655	0.88	0.66502	1.34	0.55086
0.43	0.81277	0.89	0.66218	1.35	0.54871
0.44	0.80900	0.90	0.65937	1.36	0.54657
0.45	0.80527	0.91	0.65657	1.37	0.54444

1.38	0.54233	1.89	0.44917	2.40	0.37887
1.39	0.54023	1.90	0.44759	2.41	0.37767
1.40	0.53814	1.91	0.44603	2.42	0.37648
1.41	0.53607	1.92	0.44447	2.43	0.37529
1.42	0.53401	1.93	0.44293	2.44	0.37411
1.43	0.53195	1.94	0.44139	2.45	0.37294
1.44	0.52991	1.95	0.43986	2.46	0.37178
1.45	0.52788	1.96	0.43834	2.47	0.37062
1.46	0.52586	1.97	0.43682	2.48	0.36946
1.47	0.52386	1.98	0.43532	2.49	0.36831
1.48	0.52186	1.99	0.43382	2.50	0.36717
1.49	0.51988	2.00	0.43233	2.51	0.36603
1.50	0.51791	2.01	0.43085	2.88	0.32773
1.51	0.51595	2.02	0.42938	2.89	0.32679
1.52	0.51401	2.03	0.42791	2.90	0.32586
1.53	0.51206	2.04	0.42646	2.91	0.32492
1.54	0.51014	2.05	0.42501	2.92	0.32400
1.55	0.50822	2.06	0.42357	2.93	0.32307
1.56	0.50632	2.07	0.42213	2.94	0.32215
1.57	0.50443	2.08	0.42071	2.95	0.32124
1.58	0.50254	2.09	0.41929	2.96	0.32033
1.59	0.50067	2.10	0.41788	2.97	0.31943
1.60	0.49881	2.11	0.41647	2.98	0.31853
1.61	0.49696	2.12	0.41508	2.99	0.31763
1.62	0.49512	2.13	0.41369	3.00	0.31674
1.63	0.49329	2.14	0.41231	3.05	0.31234
1.64	0.49148	2.15	0.41094	3.10	0.30805
1.65	0.48967	2.16	0.40957	3.15	0.30386
1.66	0.48787	2.17	0.40821	3.20	0.29976
1.67	0.48608	2.18	0.40686	3.25	0.29576
1.68	0.48430	2.19	0.40552	3.30	0.29185
1.69	0.48253	2.20	0.40418	3.35	0.28804
1.70	0.48078	2.21	0.40285	3.40	0.28430
1.71	0.47902	2.22	0.40153	3.45	0.28065
1.72	0.47728	2.23	0.40021	3.50	0.27708
1.73	0.47556	2.24	0.39890	3.55	0.27360
1.74	0.47384	2.25	0.39760	3.60	0.27019
1.75	0.47213	2.26	0.39630	3.65	0.26685
1.76	0.47043	2.27	0.39502	3.70	0.26359
1.77	0.46874	2.28	0.39374	3.75	0.26039
1.78	0.46706	2.29	0.39246	3.80	0.25727
1.79	0.46538	2.30	0.39119	3.85	0.25421
1.80	0.46372	2.31	0.38993	3.90	0.25122
1.81	0.46207	2.32	0.38868	3.95	0.24829
1.82	0.46042	2.33	0.38742	4.00	0.24542
1.83	0.45879	2.34	0.38618	4.10	0.23986
1.84	0.45716	2.35	0.38495	4.20	0.23452
1.85	0.45554	2.36	0.38372	4.30	0.22940
1.86	0.45394	2.37	0.38250	4.40	0.22448
1.87	0.45234	2.38	0.38128	4.50	0.21975
1.88	0.45075	2.39	0.38007	4.60	0.21521

4.70	0.21083	5.50	0.18107	6.75	0.14797
4.80	0.20662	5.60	0.17791	7.00	0.14273
4.90	0.20256	5.70	0.17485	7.50	0.13326
5.00	0.19865	5.80	0.17189	8.00	0.12496
5.10	0.19488	5.90	0.16903	8.50	0.11762
5.20	0.19125	6.00	0.16625	9.00	0.11110
5.30	0.18774	6.25	0.15969	9.50	0.10526
5.40	0.18435	6.50	0.15362	10.00	0.10000

DISCUSSION

Question: Harold Ferris, Bell Helicopter. I would like to know exactly how you physically transfer heat from one component to the other?

Answer: Physically we are limited to the three basic modes: conduction, convection, and radiation. Where possible, of course, we prefer to use conduction because it is the most reliable and predictable. We use convection in pressurized applications, and radiation in space applications where weight is at a premium and conductors have tremendous weight penalties.

Question: Frank Kottwitz, Collins Radio, Cedar Rapids. It seems to me that the nature of the mathematics of this problem would suggest the possibility of the solution on something like, say, a pace general purpose analog computer. Have you people ever done anything along the line?

Answer: We have used analogs very little. The reason is that within the company we have at least eight heat-transfer programs set up for digital computers. The calculation procedure requires only that we fill out a standard form sheet specifying system constants. The form sheets go to the computer facility and, typically, in two to three hours reams of paper return containing scores of possible solutions to the problem. We never even have to leave our desks. We have not used the analog. It's too convenient to use the digital with the ready-made programs and procedures.

Comment: I understand the situation. I think this is a common failing of engineers these days.

Answer: I agree.

Question: Bernie Litwack, Bell Labs., New York. In connection with the transfer of heat do you employ encapsulants as well as wiring methods between components for your conduction?

Answer: Yes, we employ encapsulants. One of our big problems is getting data from our manufacturers as to what the thermal conductivity is.

Question: Do you use wire between the component bodies?

Answer: We haven't yet. No.

FORMULATION AND SOLUTION OF CIRCUIT CARD
DESIGN PROBLEMS THROUGH USE OF GRAPH METHODS

Uno R. Kodres
Data Systems Division, International Business Machines

INTRODUCTION

The motivation for the work presented in this paper stems from two major sources:

1. The desire to reduce the manufacturing of the computer to mass production techniques.
2. The desire to miniaturize the size of the computer so that lead lengths between the logic elements do not become the major delays in the system.

The back panel of a present-day computer cannot be called a mass-produced item. Even with the help of present-day machines which wrap individual wires around pins on the back panel, the manufacturing process is a time-consuming, bulky, and costly task. To replace the back panel by several layers of printed circuitry which could be manufactured in minutes rather than in hours at a cost of dollars rather than thousands of dollars would be a tremendous advance from the point of view of cost as well as of time. Not only would the cost of the computer be reduced this way but—what better way do we have of producing connections which are ten mils wide and can assume any desired shape ? Miniaturization certainly requires printed circuit techniques.

The design of a circuit package as large as the present-day back panel is a frightening task for any layout technician, especially since he is asked to make no errors. Because of the likelihood of (costly) human errors, it seems imperative to use a computer (more accurate than a human being) to design a circuit package of this size.

Two approaches have been taken in the effort to solve this circuit card design problem. One approach assigns locations to the circuit elements first and then generates the connections. This approach works quite satisfactorily in case one can make connections between the various planes of wiring at any desirable point. However, this approach is space-consuming, since each connection between wiring planes requires a considerable amount of space.

The second approach is more complicated. We assume that interplanar connections can be made only at a certain fixed number of locations corresponding to the input/output leads of the basic logic elements. This assumption implies that leads connecting two points must lie entirely within a single plane. To facilitate the solution to this problem we first neglect the fact that eventually each circuit element must be located in a specific location. We generate an embedding in which no crossings exist between connections falling into the same plane, then transform it into an embedding in which each basic logic circuit is located in one of the allowed locations and in which no new intersections between conductors lying in the same plane are introduced.

MATHEMATICAL PRELIMINARIES

Planar Graphs. A collection of points in a plane together with a set of curves that connect pairs of points on this plane is called a (mathematical) graph. The points are called nodes and the curves are called edges of the graph. If each edge has a direction assigned to it, then the corresponding graph is called a directed graph.

If a graph can be drawn on the plane so that no edges intersect, then the graph is called a planar graph. Figure 1 illustrates the distinction between planar and nonplanar graphs. If a color is associated with each node of the graph, then such a graph is called a chromatic graph.

a   b

Fig. 1. a) Planar graph; b) Nonplanar graph.

There are some theorems which should be mentioned at this point.

Euler's Formula: If a graph is planar, then the number of elementary areas F, the number of edges E, the number of nodes (vertices) V, and the number of connected components t of the graph obey the following relationship:

$$F = E - V + t + 1$$

The elementary areas or faces are the regions of the plane either enclosed by a set of edges, or exterior to a set of edges, i.e., a connected region containing the point at infinity.

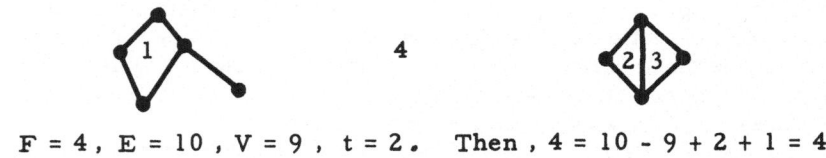

$F = 4$, $E = 10$, $V = 9$, $t = 2$.   Then, $4 = 10 - 9 + 2 + 1 = 4$

Fig. 2. Illustration of Euler's formula.

Chromatic Graphs. If we associate with each node of the graph one of n colors so that any two nodes connected by an edge have distinct colors assigned to them, then we say that the graph has been colored by n colors. The minimum number of distinct colors necessary to color the graph is called the chromatic number of the graph. The following theorem characterizes the graphs which can be colored with two colors.

König's Theorem [1]: A graph is bichromatic if and only if the graph does not contain odd cycles.

By an odd cycle, above, we mean a closed path containing an odd number of edges. Unfortunately, there exists no analogous characterization of graphs which have a chromatic number higher than 2.

SYSTEMS GRAPH

The first computer systems were described in great detail by drawings depicting a maze of interconnected resistors, capacitors, tubes, and neon lights. A great deal of time and patience was required to understand the behavior of such a collection of circuit elements. Engineers soon realized that certain parts of the circuits performed definite functions describable by Boolean algebraic connectives. It then became customary to omit from the schematic these leads which supplied power to such circuits. Only the vital logic entities were described (by means of symbols such as triangles, half disks, squares, rectangles, and so on). Each of these geometric objects represented a different logic connective between the inputs and outputs of such a block. In addition to the above geometric description, there are Boolean algebraic and functional notations in current use to describe the circuits which make up a computer [2].

No matter what method of description is used, the basic component circuits are assembled to form a number of pluggable units which then are connected together to form the computer. Many computer manufacturers use a printed circuit card on which the circuit elements are assembled. These circuit cards are inserted into a panel consisting of sockets on one side and pins on the other. The pins are then interconnected by a maze of wires.

We shall describe the circuits in a different way, concentrating on the physical structure of the circuits and ignoring their logic properties. The systems graph (which we now define) consists of a collection of points p_{i0} ($i = 1, \ldots, n$) corresponding to the circuits which will be considered as fundamental circuits in the system. We may interpret these circuits as the basic, logic, building blocks, such as AND or NOR circuits. This interpretation is the one primarily used in this paper, since we are concerned with the design of circuit cards from these fundamental building blocks. If, on the other hand, our concern is to design a back panel, then the points p_{i0} will represent circuit cards. Each fundamental circuit has input/output pins subsequently called I/O pins. These I/O pins are also represented by points p_{ij} ($j = 1, 2, \ldots, i_n$) where i_n is the total number of I/O pins on the ith fundamental circuit. We need not distinguish between the inputs and outputs of these circuits, since, physically, no difference exists. The interconnections between the various I/O pins now define the structure of our system. We represent each connection between the I/O pins by a curve connecting the corresponding points. The points p_{ij} are always regarded as being connected to p_{i0}. In case one I/O pin is connected to several other I/O pins, we shall regard all the corresponding nodes as being pairwise connected. Such a collection of points and the interconnecting edges is called a net. The inputs and outputs of our system shall also correspond to nodes, denoted by p_{0j}. With each edge of the above graph, we shall assign a weight $w(p_{ij}, p_{kl})$, where

$$w(p_{ij}, p_{kl}) = \begin{cases} n, & \text{if } p_{ij} \text{ is connected to } p_{kl} \text{ by } n \text{ edges.} \\ 0, & \text{if no edge connects the nodes.} \\ \dfrac{2}{m}, & \text{if the edge connecting } p_{ij} \text{ to } p_{kl} \text{ belongs} \\ & \quad \text{to a net of } m \text{ nodes.} \end{cases}$$

The graph with the above assignment of weights is called a "systems graph."

With each systems graph we associate a matrix called the "associated matrix." This matrix is obtained by associating with each node of the graph a row and a column of the matrix. The element a_{ij} is equal to $w(p_{kl}, p_{mn})$, where the

ith row is associated with the node p_{kl} and the jth column is associated with p_{mn}. The system's three logic blocks and the corresponding graph, and the partially filled associated matrix are shown below (Fig. 3).

Since most of the matrix elements are zero, we store only the nonzero elements in the computer in the collapsed form, row by row.

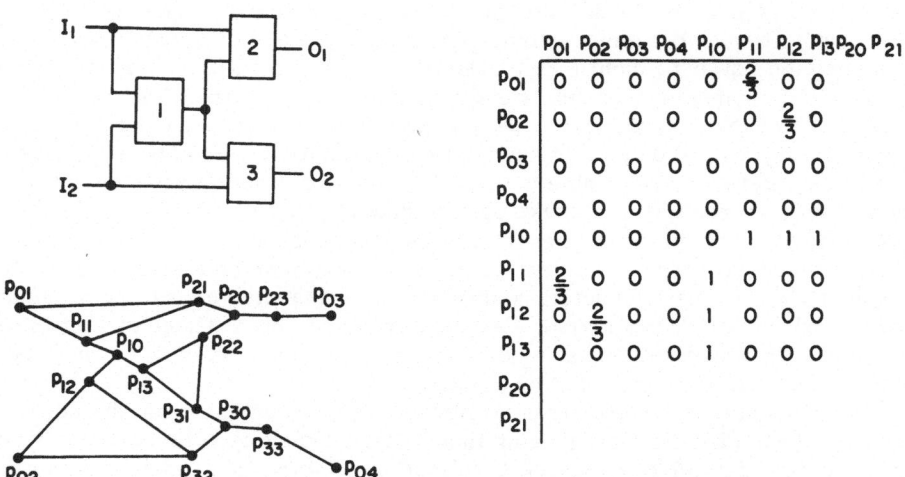

	P_{01}	P_{02}	P_{03}	P_{04}	P_{10}	P_{11}	P_{12}	P_{13}	P_{20}	P_{21}
P_{01}	0	0	0	0	0	$\frac{2}{3}$	0	0		
P_{02}	0	0	0	0	0	0	$\frac{2}{3}$	0		
P_{03}	0	0	0	0	0	0	0	0		
P_{04}	0	0	0	0	0	0	0	0		
P_{10}	0	0	0	0	0	1	1	1		
P_{11}	$\frac{2}{3}$	0	0	0	1	0	0	0		
P_{12}	0	$\frac{2}{3}$	0	0	1	0	0	0		
P_{13}	0	0	0	0	1	0	0	0		
P_{20}										
P_{21}										

Fig. 3. Logic blocks with corresponding graph and matrix.

EMBEDDING SURFACES

Ever since printed and etched circuits became economical, circuit card manufacturing technique has made use of planar surfaces of conductors. One of the most recent approaches makes use of a series of laminations of planar surfaces which are etched independently first, then put together to form a sandwich of planar surfaces, where the conducting layers are interspersed with layers of dielectric. One method for interconnecting the conducting layers is to bore holes through all the laminations and then electroplate a layer of conducting material on the surface formed by the hole. The hole technique requires that the holes themselves be fairly large, and that a conductor area surround the hole. This process wastes a considerable amount of space and, therefore, the space for the number of holes necessary to form the circuits becomes critical. Electrical characteristics also depend at least partially on compactness of packaging. A high price is also paid for the number of frames if the package is designed inefficiently (spacewise). Techniques which are applied to package circuits in the most simple-minded fashion, i.e., make holes wherever they "appear" necessary, simplify the design processes considerably. On the other hand, these methods are very inefficient (spacewise) and questionable in their electrical characteristics.

Figure 4 can be viewed as a many-layered sandwich through which holes are made for the purposes of interconnecting the planar layers of conductor patterns. The circuit elements which perform the basic logic functions are packaged into a miniature package from which leads extend which can be inserted into the holes in the sandwich to make the connection. A design should be capable of handling any circuit configuration equivalent to this figure. Present-day back panels may be viewed in this light. Instead of wiring crisscrossing the panel,

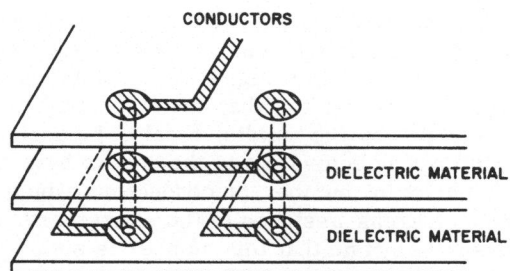

Fig. 4. Embedding surface.

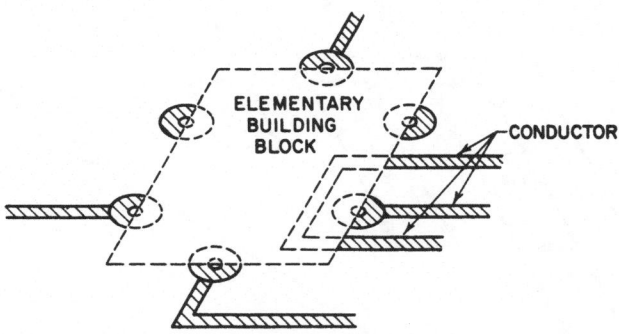

Fig. 5. Elementary building block.

conductors are etched into one of several planar layers of printed circuits which form the sandwich described above. Instead of the elementary logic elements forming the basic packages, we are dealing now with very complex circuit cards which have many conductors extending from the terminals. The significant similarity is that the leads communicating with the panel emerge from a small area of the panel.

Since we are going to be concerned with designing a circuit card instead of the panel, we shall confine our thinking to the circuit card only. The only significant difference is the fact that on a panel the circuit cards have a much larger number of conductors leaving the socket area in comparison with the corresponding socket position of the basic building block.

Suppose, in particular, that our basic building block has a maximum of three inputs and one output which then may branch into three other inputs. Suppose also that this basic building block is manufactured so that it can be assembled on a small area of the circuit card. We may regard this building block as a small piece of surface on the perimeter of which there are a certain number of holes through all the surfaces corresponding to the I/O pins of the elementary building blocks. If we visualize this piece of surface as rectangular, then we may think of the elementary building block as illustrated in Fig. 5.

In Fig. 5, we have passed a conductor through the building block to illustrate that, depending on the design method, we may have possibilities of passing a certain number of conductors through the elementary area corresponding to the building block. This will be of significance later on.

The circuit cards to be designed will then consist of several layers of planar surfaces through which holes are made at certain locations on the perimeters of areas corresponding to the circuits. There is a fixed number of locations where the elementary building blocks can be positioned. The circuit card thus may have an appearance as shown in Fig. 6. It is not necessary that these elementary areas be rectangular or, for that matter, that all circuit elements be within the area we ascribed to the building block. The design should, however, be such that the circuit card is equivalent to the one we have described.

Also, there is a maximum number of conductors which may pass between the rows and columns of areas assigned to the elementary circuits. In the following discussion we shall assume that this number is six.

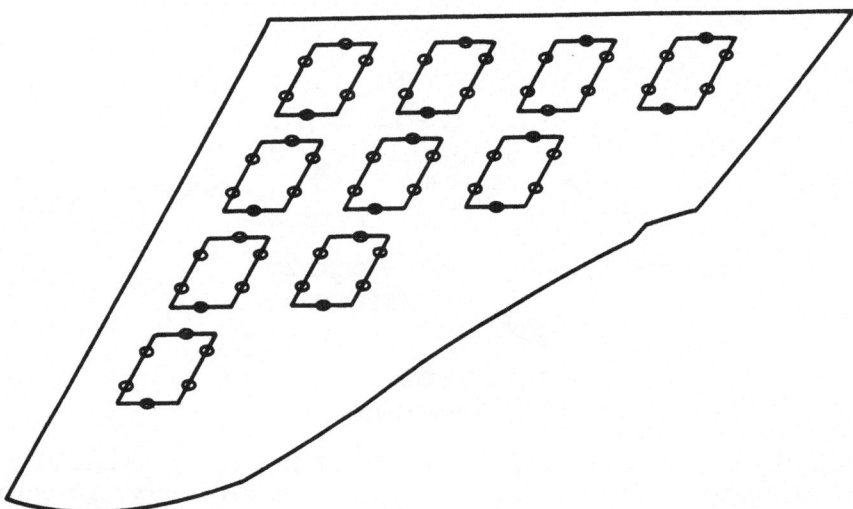

Fig. 6. A section of the circuit card.

DESIGN PROCEDURE

The circuit card design problem can now be stated more exactly as follows:

Given:
1. A collection of elementary building blocks which are to be connected in some prespecified manner.
2. The type of embedding surface, i.e., a surface in the form of a sandwich of planar conducting layers at least two of which are reserved for signal connections.
3. The rectangular array of locations into which the building blocks may be assigned.
4. A grid of channels into which the conductors may be assigned on each planar surface.

The Problem:
a. To assign locations to all the given building blocks so that no two are assigned to the same location.
b. To assign channels to all the prescribed connections between the building blocks so that in one conducting plane no two conductors produce a short circuit.

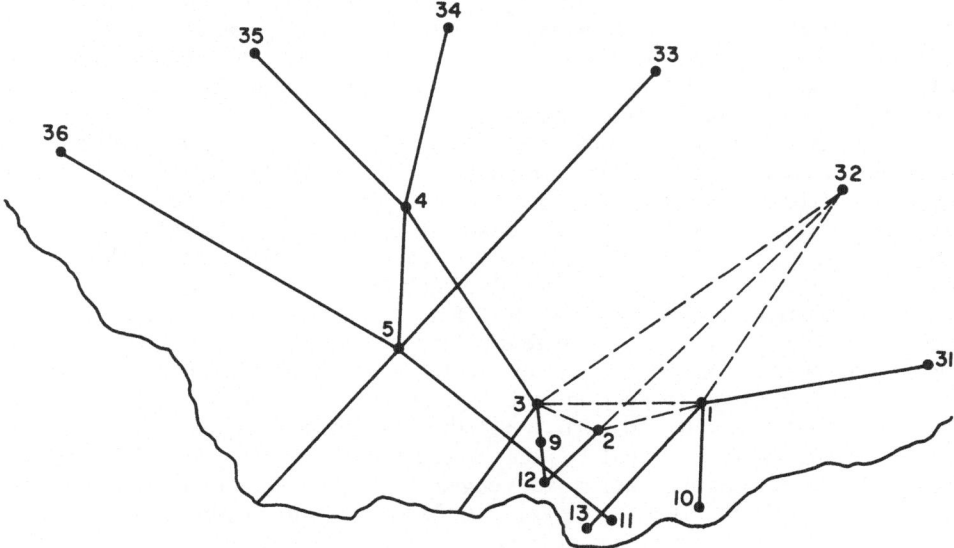

Fig. 7. Standard embedding.

We shall describe the proposed design procedure by applying it to a particular sample problem representative of the class of problems for which the design procedure is usable. The design procedure will be described in three phases:
1. Embedding Phase
2. Factorization Phase
3. Transformation Phase
The general approach to the solution of the problem is a familiar one in mathematics.

We first take the given problem and transform it into an easier one which we either know how to solve or are more hopeful of being able to solve. We solve this easier problem and transform it back into the original problem by a transformation which preserves certain indispensable properties.

For the problem at hand, we first find an embedding for the systems graph where the number of intersections between the graph's edges is sufficiently small. This constitutes the Embedding Phase. We then eliminate these intersections by factoring the given graph into two planar factors. This is called the Factorization Phase. We take this solution and transform it into terms of the original problem, preserving the indispensable property that under this transformation the systems graph should remain planar in both factor planes. We call this the Transformation Phase.

Embedding Phase

Step 1. The systems-graph embedding technique was chosen because (1) a computer can be used to find the embedding for the graph, (2) this technique enables us to construct a biplanar factorization for the systems graph, which is quite efficient. (See Appendix 1 for the justification of the term "efficient.")

With each p_{i0} node (corresponding to an elementary logic circuit) of the systems graph, we associate a pair of position variables (x_i, y_i) which describe the location of this node in a rectangular coordinate system. With each node p_{ij} $(j \neq 0)$ we also associate the same pair (x_i, y_i); i.e., we regard the I/O pins of the ι th circuit as being in the same physical location as the circuit itself. With each input/output node of systems graph, p_{0k}, we associate a fixed location (a_k, b_k). Thus, we associate a variable point with coordinates (x_i, y_i) with the nodes $p_{ij}(j = 0, 1, 2, \ldots, i_n; \iota \neq 0)$; and a fixed point (a_j, b_j) with the j th I/O node of the systems graph.

We shall now define "standard embedding." "Standard embedding" has the input/output nodes of the systems graph fixed in a counterclockwise order around the circumference of the circle, starting with the left edge of the circuit card and ending with the right. The nodes associated with (x_i, y_i) are located in the interior of this circle in such a fashion that such nodes are located at the center of mass with respect to the nodes with which they are connected; each node is regarded as having w units of mass on it if it is connected to the given node with a connection of strength w. Note at this point that nodes may have more than one edge between them and that whenever a net connects more than a pair of nodes, the weight of the edge connecting each node of this net has a weight $2/n$ (n being the number of nodes in the net). A standard embedding for (part of) the graph corresponding to our original system is seen in Fig. 7. The edges of weight 1 are drawn as full lines, whereas the edges of weight less than 1 are drawn for only one net by broken line segments. This "standard embedding" is achieved by simply solving the following set of linear equations [3]:

$$x_i \left[\sum_{k=1}^{n} \overline{w}(i,k) + \sum_{k=1}^{m} w^*(i,k) \right] = \sum_{k=1}^{n} x_k \overline{w}(k,i) + \sum_{k=1}^{m} a_k w^*(k,i)$$

$$(i = 1, 2, \ldots, n)$$

$$y_i \left[\sum_{k=1}^{n} \overline{w}(i,k) + \sum_{k=1}^{m} w^*(\iota,k) \right] = \sum_{k=1}^{n} y_k \overline{w}(k,i) + \sum_{k=1}^{m} b_k w^*(k,i)$$

where

$$\overline{w}(i,k) = \sum_{l=1}^{k(m)} \sum_{j=1}^{i(m)} w(p_{ij}, p_{kl}) \qquad (\iota \neq 0, \ k \neq 0)$$

and

$$w^*(i,k) = \sum_{j=1}^{i(m)} w(p_{ij}, p_{0k}) \qquad (\iota \neq 0)$$

$w(p_{ij}, p_{kl})$ was the weight associated with the edge connecting node p_{ij} and p_{kl}, and $i(m)$ refers to the total number of I/O pins on circuit i.

The method of solution ideally suited to solve the above system is the Gauss—Seidel iteration technique. We can solve comparatively large systems (200×200) in a matter of seconds (40 sec) on the IBM 7090 with 5 decimal place precision. The result is plotted in Fig. 7. As seen there, the fourth node is in the center of mass with respect to the set of nodes connected to it (5, 3, 35, 34). Each node carries a mass $\sum_{k,l} w(p_{4k}, p_{jl})$, $(j = 5, 3, 35, 34)$. Whenever the edges connecting a pair of nodes are of weight greater or equal to 1, we draw these edges as straight-line segments between the corresponding nodes. After drawing all such edges into our picture, we look at all these edges of weight less than 1. Each such edge belongs to one, and only one, net. These nets are given to us in some order, and we shall consider these nets in the given order.

Step 2. Each net is a c o m p l e t e g r a p h connecting all the nodes in the net pairwise. We must choose from all possible edges a set which spans the net. That is, we are interested in finding a spanning subtree of the complete graph connecting the nodes, such that the total length of the subtree is minimized. Here the length of an edge is given by the number of crossings (k) it has with edges of weight 1, plus the normalized length of the given edge:

$$(i,j) = k + \frac{d(i,j)}{1+d(i,j)} \qquad d(i,j) = \sqrt{(x_i - x_j)^2 + (y_i - y_j)^2}$$

This allows us to pick the set of edges out of the graph which gives rise to the smallest number of crossings; and if two edges have the same number of crossings, we pick the edge of smaller length. The method to compute the minimal spanning subtree has been known for some time (Kruskal's algorithm) [4]. However, we must modify it somewhat to suit our purposes, since our distances should not be computed as simply as mentioned above. Since each node corresponds to an element (as shown in Fig. 8), there are crossings between lines which do not appear under the category of ordinary crossings. These crossings will be called "induced crossings," and we shall modify the length of an edge by an additional factor. Each time we make a choice of an edge by the Kruskal algorithm, we compute the induced crossings and add the corresponding numbers to the length $l(i,j)$, already computed. Using this algorithm, we shall be able to find an optimal spanning tree (with respect to the length we defined earlier) for the first net. The edges in the spanning tree have weight 1 and we continue with the next net in the list. After we have exhausted all nets, we will have a network of nodes and straight line edges embedded in a circular region of a plane.

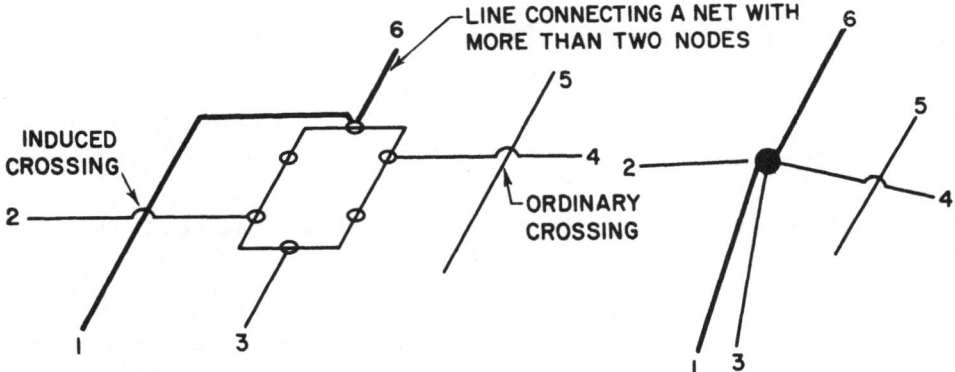

Fig. 8. Induced crossings.

Factorization Phase

Step 1. Taking the given standard embedding, we next are interested in factoring this graph into two planar factors. We do this by first defining an "intersection graph," which depends on the standard embedding. The intersection graph consists of nodes corresponding to the edges of the embedded graph and edges corresponding to the crossings (both actual and induced) between the edges in the embedded graph. That is, if two edges cross in the embedded graph, then the corresponding nodes of the intersection graph are connected by an edge. The collection of disconnected pieces of the intersection graph are shown in Fig. 9.

A factorization of the original graph into k planar graphs can be obtained by coloring the nodes of the intersection graph with k colors in such a way that if two nodes of the intersection graph are connected, then they are colored with

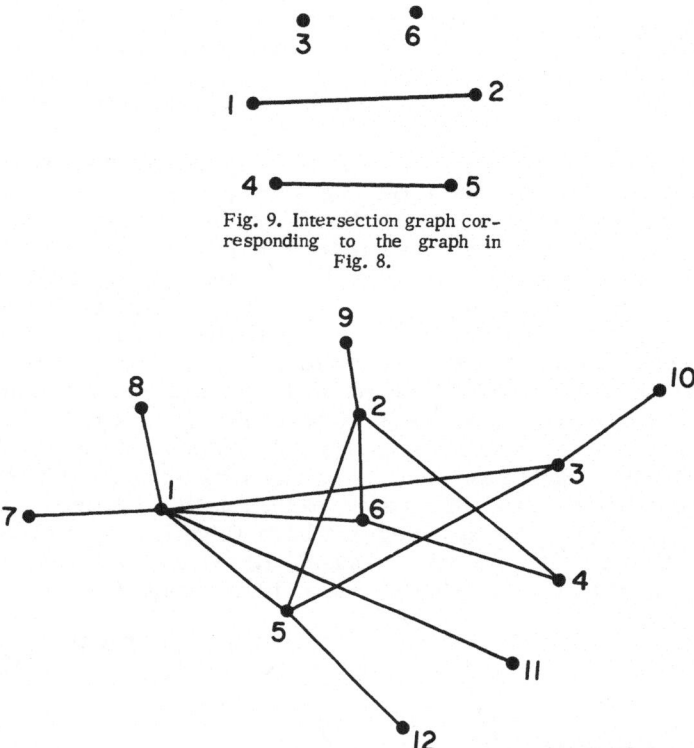

Fig. 9. Intersection graph corresponding to the graph in Fig. 8.

Fig. 10. Graph in plane 1.

distinct colors. That is to say, whenever a crossing exists between two edges of the original graph, then these edges cannot be in the s a m e plane. It is clear that if we make the number of laminations sufficiently high, we can take any graph (no matter how complicated) and factor it into planar factors. It seems that for the case at hand we need only two planes for the signal wiring if we take advantage of the fact that there exists a certain small number of paths which can pass through the area allotted to the logic building block. We are thus able to eliminate certain crossings, as seen in Fig. 11, if we recall that a node in the graph is truly a configuration of nodes close together so that a small number of paths can penetrate it.

Assuming that we have only two planes available for the embedding, our problem is to color the nodes of the intersection graph with only two colors. At this point, we are helped by König's Theorem, which states: A g r a p h c a n b e c o l o r e d b y t w o c o l o r s i f a n d o n l y i f t h e g r a p h c o n t a i n s n o o d d c y c l e s.

Because of the fact that we can eliminate crossings in the original graph, or equivalently, edges in the intersection graph, it is enough to consider the problem of removing a minimal number of edges from the intersection graph so that the remaining subgraph has no odd cycles. The technique for doing this is outlined in Appendix 2. After the procedure in Appendix 2, we color the nodes of the intersection graph with two colors. We thus have graphs in planes 1 and 2, each of which is planar except for a small number of intersections which we now remove.

Step 2. We have decided, at this time, which of the particular crossings between which pair of edges we must remove. The procedure for the intersection removal can be best described by illustration. Suppose that Fig. 10 depicts the graph in plane 1. All crossings appearing in this graph must be removed by using the allowed paths through the nodes. We first look at those edges having the most crossings, i.e., nodes in the intersection graph having the most edges removed. They are (1, 3) (2, 5) (3, 5). We choose (1, 3) arbitrarily. Is there a node from which more than one line emanates which crosses (1, 3)? Node 2 has three lines emanating, each of which crosses (1, 3). We lead the edge (1, 3) through node 2, eliminating all three crossings. Next, we look at the edge with the next highest number of crossings, which is (3, 5). We lead (3, 5) under node 4, eliminating two crossings. We have two edges, (1, 11) (2, 5), each of which has two crossings. Choosing (1, 11) arbitrarily, we feed (1, 11) through 5. Finally, we have one crossing, (2, 5) (1, 6), to eliminate. We compare the areas of the four triangles that are generated: (1, 6) and node 2; (1, 6) and node 5; (2, 5) and node 1; (2, 5) and node 6. Choosing the vertex generating the smallest area causes us to bring (2, 5) through 6. We end up with an embedding in which all intersections are eliminated. For clarity, Fig. 11 has expanded nodes.

We note that if our graph in plane 1 were much more complicated, we would not be able to eliminate the intersection by using only the paths under the nodes. If this were to happen frequently, we would go into another plane and factor our original graph into three rather than two planes. This is a function of the type of elementary building blocks used.

At this stage in the procedure we have an embedding of the graph in two planes so that no crossings appear between edges in the same plane. That is, we have two planar graphs whose superpositioning gives the original graph (corresponding to the system for which we are constructing the circuit card). We have thus far completely neglected the fact that nodes of our graph must eventually be positioned into a rectangular array on a circuit card, and that the edges must also be assigned to certain allowed channels between the node locations.

Transformation Phase

Step 1. In this phase we are interested in a transformation which transforms our stacked planar graphs in the standard embedding into stacked planar graphs on a rectangular card so that each node of our graph occupies a location in the rectangular array of allowed locations and each edge falls into some allowed channel. A transformation which is sufficiently general and which is reasonably simple for the computer to generate is a piecewise linear transformation which we shall now describe.

Let us assume that our circuit card has nine columns and four rows of locations for the logic building blocks. We shall think of the circuit card as consisting of rectangular pieces of surface as shown in Fig. 12.

We are interested in a transformation which takes the rectangle and transforms it into the region of the standard embedding in such a way that the base of the circuit card goes into the circumference of the standard embedding, and the edges of the circuit card go into a polygonal path which crosses no edges of the already embedded graph. The vertices of the rectangular pieces go into the nodes of the embedded graph so that each node of the embedded graph will coincide with exactly one vertex of a rectangle. Also, a suitable triangulation exists in each rectangular piece so that in any such triangle the transformation is linear. We require, further, that the transformation be a one-to-one mapping, which ensures the existence of an inverse transform. If we impose no further requirements, there are many distinct transformations satisfying all the above restrictions. Because of the fact that it is more economical to construct circuit cards in a uniform way, the number of channels between consecutive locations is limited to a fixed maximum number. Therefore, we would like to impose a further restriction on our transformation, namely, each polygonal path which is the image of an edge of a basic rectangle must not cross any more than a certain fixed maximum number of edges of any one planar factor of the embedded graph. If a reasonable choice is originally made in this maximum number of channels between node locations, then a transformation of the above form is obtainable, and a method of doing this is outlined in Appendix 3.

It is the inverse of the transformation which really interests us. The inverse transformation is also piecewise linear and, hence, each edge of the graph in the standard embedding will go into a polygonal path on the circuit card. A planar graph consisting of straight-line edges will go into a planar graph consisting of polygonal edges. We need not explicitly calculate the inverse transformation to determine how to assign connectors to channels. We can do this much more simply. This brings us to Step 2.

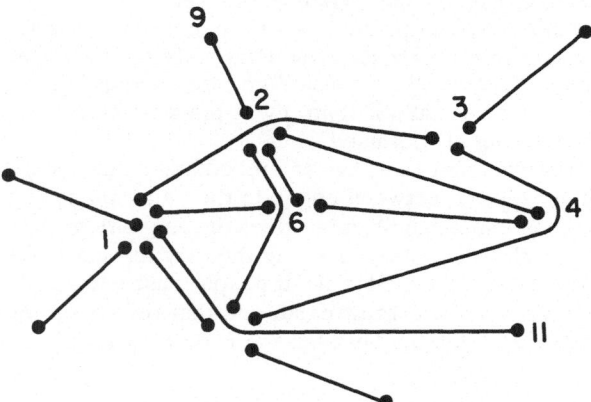

Fig. 11. Embedding with eliminated intersections.

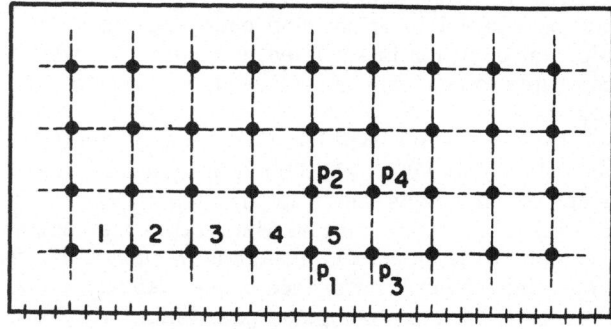

Fig. 12. Circuit card.

Step 2. We consider the transformation described above which took each basic rectangle of the circuit card and transformed it into some polygonal region inside the area of the standard embedding. Each edge of the basic rectangle transforms into some polygonal path.

In Fig. 13, the boundary $(1, 2, 3, 4)$ transforms into the polygonal path $(1', 2', 3', 4')$. We denote all the intersections between edges of the graph in the standard embedding and the boundary $(1', 2', 3', 4')$ by points $(q^1, q^2, q^3, q^4, q^5)$. Our transformation was constructed so that each path connecting two vertices could not intersect any more than six (the maximum number of channels between adjacent nodal locations) edges in one plane. The intersections q^2, q^3, q^4 could be assigned to any three of the six available channels. We assign them arbitrarily to channels labeled by p_2, p_3, p_4. Intersections q^1 and q^5 are similarly assigned. Because the inverse transform is piecewise linear, we know that connections falling into allowed channels between $p_1, p_2; p_4, p_5$; and p_3, 4 do exist. We generate these connections and at the same time locally minimize the number of bends in the connectors.

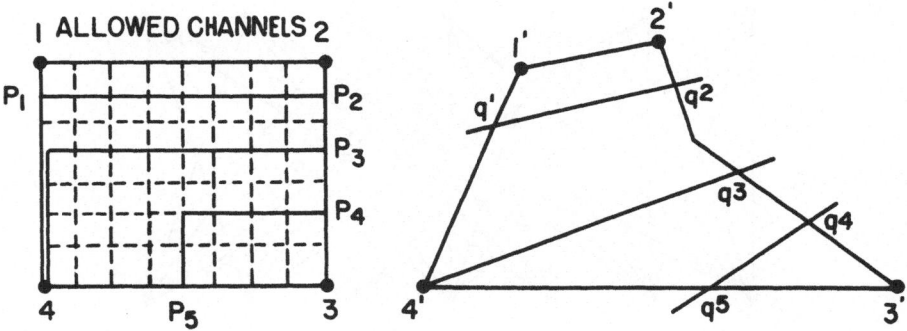

Fig. 13. Inverse transformation.

We follow the above procedure for each elementary rectangle of the type $(1, 2, 3, 4)$ and simultaneously compile information for the drafting machine, which produces the master drawings for each plane.

CONCLUSIONS

The proposed design procedure is intended to have maximum efficiency with regard to the number of planar layers used for the signal connections. The lead length of the interconnections was very much ignored in this process. It is conceivable that the length will be a very important consideration, especially in the future.

Our design procedure could be altered to be more conscious of the length of the interconnections by considering a rectangular rather than a circular standard embedding. The rectangular embedding would resemble the final embedding more closely, and, in the transformation, the lengths of interconnections would change little. The number of planes may have to be increased if one is to use this type of standard embedding, since it appears that the number of intersections between the edges of the embedded graph roughly doubles. If one is willing to make this trade, our design procedure can still be used if an extension is made in the factorization part of the design procedure.

We are hopeful that this method can be extended to the design of back panels as well as circuit cards. The central problem in the extension is an efficient graph-coloring design procedure for more than two colors. Future effort will be devoted to the solution of this problem.

APPENDIX 1—EVALUATION OF THE EMBEDDING METHOD

It is unquestionably very difficult to evaluate the embedding method in any analytic form. However, we can gain some insight into the efficiency of some parts of the given design procedure.

The central and perhaps the most significant part of the design procedure is the initial standard embedding of the graph. An interesting question to pose is: how efficiently does the embedding procedure use the two planes?

To answer this question, we test the efficacy of our design procedure in solving a relatively simple problem. As an example, consider a graph consisting of 14 nodes and 49 edges as shown in Fig. 14. We can show that this graph is not

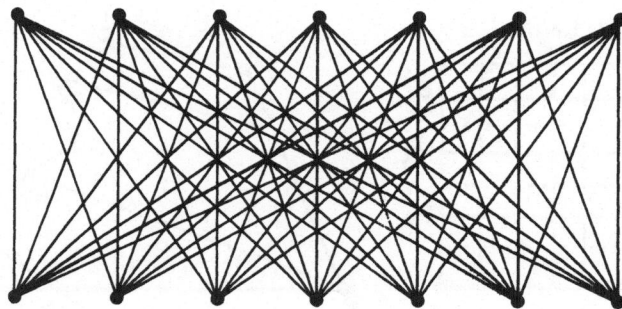

Fig. 14. Sample graph.

biplanar, i.e., it cannot be factored into two graphs each of which is planar. For if we assume that this graph is biplanar with factors G_1 and G_2, then

$$E_1 + E_2 = 49, \quad V_1 \leq 14, \quad V_2 \leq 14, \quad t_1 \geq 1, \quad t_2 \geq 1,$$

where E_i, V_i, t_i $(i=1,2)$ denote edges, vertices, and connected components of each graph, respectively. If we count edges bounding each face of the graph G_1, we get

$$2F_1^2 + 3F_1^3 + 4F_1^4 + 5F_1^5 + 6F_1^6 + \cdots + nF_1^n = 2E_1^*$$

where F_1^i denotes the number of elementary areas bounded by i edges in graph G_1, and E_1^* is the number of edges in the boundaries of the faces. $F_1^2 = 0$ and $F_1^3 = 0$, since there exist no cycles of length less than four in our graph Now,

$$4F_1^4 + 5F_1^5 + \cdots + nF_1^n \geq 4F_1^4 + 4F_1^5 + \cdots + 4F_1^n = 4F_1$$

Hence, $4F_1 \leq 2E_1^* \leq 2E_1$, where E_1 is the total number of edges in G_1. The same inequality holds for G_2, and we can write

$$F_1 \leq \frac{E_1}{2}, \quad F_2 \leq \frac{E_2}{2}$$

Since Euler's relation must hold for any planar graph,

$$\frac{E_1}{2} \geq F_1 = E_1 - V_1 + t_1 + 1 \geq E_1 - V_1 + 2$$

$$\frac{E_2}{2} \geq F_2 = E_2 - V_2 + t_2 + 1 \geq E_2 - V_2 + 2$$

Adding the two inequalities, we get

or,

$$\frac{1}{2}(E_1 + E_2) \geq (E_1 + E_2) - (V_1 + V_2) + 4$$

$$2(V_1 + V_2) \geq (E_1 + E_2) + 8 = 57$$

But $2(V_1 + V_2) \leq 56$, giving $56 \geq 57$, which is absurd. The assumption that our graph is biplanar is, therefore, wrong.

Even though the graph is not biplanar, we can make it biplanar if we are allowed to eliminate intersections by passing through the nodes of the graph. It appears that for the above graph, the minimal number of intersections which we must eliminate is four. (This conclusion is not based on proof; it is just a conjecture.)

If we employ our embedding procedure upon this graph, we are able to obtain a factorization with ten intersections which must be eliminated. Thus, we are not able to produce an embedding which gives rise to the minimal number of intersections; but we are able to obtain a biplanar embedding for this rather complex graph, which has approximately doubled the conjectured minimal number of intersections.

APPENDIX 2—DESIGN PROCEDURE TO DISCARD THE MINIMAL NUMBER OF EDGES OF A GRAPH TO MAKE IT BICHROMATIC

Problem:

Given any undirected graph, discard the minimal number of edges of this graph so that the resulting subgraph is bichromatic, i.e., colorable by two colors.

Solution:

According to König's Theorem (see page 122), the above problem is equivalent to the problem of discarding the minimal number of edges so that all odd cycles of the graph are eliminated. Let us call such a set of edges the "minimal breaking set" and denote it by B_m. To explain how we find the minimal breaking set

$$B_m = (i_1, i_2, \ldots, i_m)$$

where $i_1, i_2, \ldots, i_m$ are m edges of the graph which contains edges $(1, 2, \ldots, E)$, we first define the concept "cycle basis of a graph." Let each edge of a connected graph be labeled by u_i ($i = 1, 2, \ldots, n$). Any cycle in this graph can be expressed by the following vector:

$$c = (c^1, c^2, \ldots, c^n)$$

where

$$c^k = \begin{cases} 0, & \text{if the } k\text{th edge is not in the cycle.} \\ 1, & \text{if the } k\text{th edge is in the cycle.} \end{cases}$$

The modulo 2 sum of two cycles c_1 and c_2 can be defined as

$$c_1 \dotplus c_2 = (c_1^1 \dotplus c_2^1, \ c_1^2 \dotplus c_2^2, \ldots, \ c_1^n \dotplus c_2^n)$$

where

$$c_1^i \dotplus c_2^i = \begin{cases} 0, & \text{if } c_1^i \text{ and } c_2^i \text{ are both 0 or 1.} \\ 1, & \text{if exactly one } (c_1^i \text{ or } c_2^i) \text{ is 1.} \end{cases}$$

Intuition tells us that the modulo 2 sum of two cycles is either another cycle, the zero cycle, or two disconnected cycles, depending on whether the two cycles have some, all, or no edges in common.

A collection of cycles $c_1, c_2, \ldots, c_k$ is called "linearly independent" if, and only if, no cycle in this collection can be expressed as a modulo 2 sum of the remaining cycles. The maximal linearly independent set of cycles for a given graph is called the "cycle basis" (for this graph).

It is not difficult to determine the maximal number N of linearly independent cycles in an undirected graph. If this graph is connected, then the number

$$N = E - V + 1$$

gives us the correct number of cycles. To obtain the cycle basis for a given graph, we proceed as follows:

We first obtain any spanning subtree for this graph, i.e., a subgraph which connects all the vertices of the graph and which contains no cycles. The spanning subtree contains $V - 1$ edges. With each of the remaining $E - (V - 1)$ edges, we form N cycles $c_1, c_2, \ldots, c_n$, where

$$N = E - (V - 1) = E - V + 1$$

These cycles are formed by taking any one of the edges not belonging to the spanning subtree and forming a cycle with edges belonging to the subtree. Clearly, each cycle so formed is linearly independent from all cycles formed similarly, since each contains one edge that is in none of the other cycles. It follows that such a collection of cycles forms a cycle basis for the graph. The following example illustrates the above procedure.

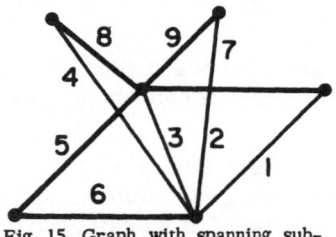

Fig. 15. Graph with spanning subtree.

Suppose our spanning subtree (see Fig. 15) consists of edges $5, 6, 7, 8, 9$. We form c_1 from edge 1 and edges $5, 6, 7$. Continuing in this way, we may express the cycle basis by the following matrix, called the "basis matrix" and denoted by C.

$$C = \begin{array}{c|ccccccccc} & 1 & 2 & 3 & 4 & 5 & 6 & 7 & 8 & 9 \\ \hline c_1 & 1 & 0 & 0 & 0 & 1 & 1 & 1 & 0 & 0 \\ c_2 & 0 & 1 & 0 & 0 & 1 & 1 & 0 & 0 & 1 \\ c_3 & 0 & 0 & 1 & 0 & 1 & 1 & 0 & 0 & 0 \\ c_4 & 0 & 0 & 0 & 1 & 1 & 1 & 0 & 1 & 0 \end{array} \qquad \text{"Basis Matrix"}$$

We now state the theorem on which our proposed design procedure is based.

Theorem:

The necessary and sufficient condition that B_m be the minimal breaking set is that the removal of corresponding columns in the basis matrix reduces each modulo 2 row-sum to zero, i.e.,

$$\sum_{i \in \overline{B}_m} c_j^i = 0, \quad j = 1, 2, \ldots, N$$

and $\overline{B}_m$ is the set of edges remaining after the removal of B_m.

Proof:

We prove first the sufficiency by showing that any cycle in the set of edges $\overline{B}_m$ that remains after removal of minimal breaking set B_m is an even cycle. Consider any cycle c^* whose edges lie in $\overline{B}_m$. Since the set of cycles $(c_1, c_2, \ldots, c_n)$ constitutes a basis, c^* can be expressed as a modulo 2 linear combination of the basis cycles, i.e.,

$$c^* = \sum_{j=1}^{n} a_j c_j$$

where $a_j = 1$ or 0, depending on whether or not cycle c_j was used in the linear combination.

To see if the number of edges in cycle c^* is even or odd, we look at the modulo 2 sum of its components:

$$c^{*1} \dotplus c^{*2} \dotplus \cdots \dotplus c^{*E} = \sum_{i \in B_m} c^{*i} \dotplus \sum_{i \in \overline{B}_m} c^{*i}$$

Since cycle c^* has all its edges in $\overline{B}_m$, each $c^{*i} = 0$ for $i \in B_m$. Therefore,

$$\sum_{i \in \overline{B}_m} c^{*i} = 0 \quad \text{and} \quad \sum_{i \in \overline{B}_m} c^{*i} = \sum_{i \in \overline{B}_m} \sum_{j=1}^{N} a_j c_j^i = \sum_{j=1}^{N} a_j \sum_{i \in \overline{B}_m} c_j^i$$

From the hypothesis, $\sum_{i \in \overline{B}_m} c_j^i = 0$ for each j. Hence

$$\sum_{i \in \overline{B}_m} c^{*i} = 0$$

and the cycle is even.

To prove the necessity part of the theorem, we assume the contradiction, namely, that the removal of the columns of our basis matrix corresponding to the edges of B_m (the minimal breaking set) does not reduce all modulo 2 row-sums to zero, i.e., that there exists at least one row j for which

$$\sum_{i \in \overline{B}_m} c_j^i = 1$$

We shall obtain a contradiction by showing the existence of an odd cycle consisting entirely of the edges of $\overline{B}_m$, thus contradicting the assumption that B_m is the minimal breaking set.

Consider the basis cycle c_j, corresponding to the row for which the row sum is different from zero. This cycle, c_j, has (possibly) some edges in common with the minimal breaking set B_m. Let the set of such edges be denoted by $K = (k_1, k_2, \ldots, k_n)$. For any such edge (k_i) we can construct an odd cycle consisting entirely of the edges of $\overline{B}_m$ and the edge k_i. If this were not so, B_m

could not be the minimal breaking set—because $B_m - (k_i)$ would be a breaking set with one less element than B_m. Let us denote such an odd cycle by b_i. We continue constructing such odd cycles for each k_i belonging to the set K.

Let us look at the linear combination

$$c = c_j \dotplus b_1 \dotplus b_2 \dotplus \cdots \dotplus b_n$$

and express the modulo 2 sum of components of such a vector in the following way:

$$\sum_{i \in K}{}'(c_j^i \dotplus b_j^i \dotplus b_2^i \dotplus \cdots \dotplus b_n^i) + \sum_{i \in \overline{B}_m}{}'(c_j^i \dotplus b_1^i \dotplus \cdots \dotplus b_n^i)$$

Because of the way we constructed cycles b_i, there is only one edge, namely, k_i, in common with the set K and, hence,

$$b_j^i = 0, \text{ if } i \neq j, \text{ and } b_j^i = 1, \text{ if } i = j$$

We can rewrite the above expression as

$$\sum_{i \in K}{}'(c_j^i \dotplus b_i^i) \dotplus \sum_{i \in \overline{B}_m}{}'c_j^i \dotplus \sum_{i \in \overline{B}_m}{}'b_i^i \dotplus \cdots \dotplus \sum_{i \in \overline{B}_m}{}'b_n^i$$

Since $c_j^i = 1$, for $i \in K$, each term $c_j^i + b_i^i$ equals zero.

This shows that the linear combination c has no edges in B_m. Since c was formed of cycles which had at least one element in common with c_j, c is also a cycle or (at worst) a combination of cycles.

We have $\sum_{i \in \overline{B}_m}{}' c_j^i = 1$ (by assumption).

Each of the sums

$$\sum_{i \in \overline{B}_m}{}' b_k^i \qquad (k = 1, \ldots, n)$$

must be zero, since b_k was constructed as an odd cycle having exactly one edge in B_m. Hence, c is either an odd cycle or a collection of cycles at least one of which is odd. We remarked that c consists of edges in $\overline{B}_m$ only; hence, B_m cannot be a breaking set. This contradicts the hypothesis and thus completes the proof of the theorem.

To find the minimal number of columns to remove in order that the resulting modulo 2 row sums are zero, we use an integer linear programing* algorithm [5]. According to this formulation, we associate with each column of the basis matrix a variable x_i ($i = 1, 2, \ldots, E$) which assumes the value 1 if the corresponding column has been discarded, and the value 0 if the column is left in. To force the modulo 2 row sums to be zero, the variable x_i's are subjected to the following restrictions:

$$\sum_{j=1}^{E} c_k^j x_j + \sum_{j=1}^{E} c_k^j = 2 y_k \qquad (k = 1, 2, \ldots, n)$$

where the y_k's are integers, $y_k \geq 1$, and c_k^j's are elements of the basis matrix C. The integer linear programing problem which gives the solution to the above problem consists of minimizing the objective function

---

*P. C. Gilmore of IBM pointed out to me that the integer linear programing technique was applicable to the solution of this problem.

$$w = \sum_{i=1}^{E} x_i$$

subject to the above restrictions, where x_i and y_i are non-negative integers.

APPENDIX 3—THE PIECEWISE LINEAR TRANSFORMATION

The piecewise linear transformation which enables us to transform the standard imbedding into the rectangular configuration of the circuit card is based on a mapping which is familiar to algebraic topologists. This transformation in our application can be quite simply described. Let p_1, p_2, p_3 be any three vectors in the plane whose endpoints are not colinear. The endpoints determine a triangle in the plane. Any point in the triangle [6] can be expressed as the endpoint of the vector written in the form:

$$z = \lambda_1 p_1 + \lambda_2 p_2 + \lambda_3 p_3$$

where λ_1, λ_2, λ_3 are the non-negative real numbers and $\lambda_1 + \lambda_2 + \lambda_3 = 1$.

Each point in the triangle corresponds to a unique triplet $(\lambda_1, \lambda_2, \lambda_3)$. These numbers are called the barycentric coordinates of the point. The linear transformation which we are going to use maps a triangle determined by three vectors p_1, p_2, p_3 onto a triangle determined by q_1, q_2, q_3 in the following way: each point in the first triangle corresponds to a unique triplet, namely $\lambda_1, \lambda_2, \lambda_3$. We assign to this point the point in the second triangle which has the same barycentric coordinates. It is not hard to show that straight-line segments in the triangle are preserved under this mapping. What is even more important, lines which do not intersect in one triangle are transformed into lines that do not intersect in the other triangle. If both triangles are nondegenerate, then the inverse transformation shares all of the above properties.

We construct the piecewise linear mapping by starting with the rectangular circuit card which is broken into the elementary squares, as seen in Fig. 12. Let us assume that we have completed the construction of the piecewise linear transformation for squares $1, 2, 3$, and 4. We next extend this transformation to square 5, whose vertices are p_1, p_2, p_3, p_4. We have already chosen the image points q_1, q_2, q_3 seen in Fig. 16. We first check whether the points $q_1 q_2 q_3$ form a nondegenerate triangle which contains no nodes of the standard embedding. If a node is found in the interior of such a triangle, we shall consider this node, q_4', as a candidate for the image q_4 of the point p_4. Whether or not we shall chose q_4' to be the image point depends on the following:

1. Whether the triangle $q_1 q_2 q_4'$ contains any other nodes. If a node is located in the triangle, that node is chosen to be the next candidate, q_4'.
2. How many edges of the graph in either plane intersect line segment $q_2 q_4'$. If the allowed number of intersections is exceeded in any one plane, we must rechoose our candidate.
3. Whether $q_1 q_4' q_3$ forms a triangle which contains no other nodes. If the answer is yes (as it is in Fig. 16), we look at the number of intersections of $q_4' q_3$ with edges in either of the two planes.
4. The number of intersections. If the number of interesections is not well balanced between $q_2 q_4'$ and $q_4' q_3$, i.e., the number of intersections is small with line segment $q_2 q_4'$ and large with $q_4' q_3$, we will consider q_4' as an intermediate point rather than the vertex. To remove singularities in the eventual mapping, we shall pick a point $\bar{q}_1$ on the line of $q_1 q_4'$ to be the point onto which we will map a point $\bar{p}_1$ on the line segment of $p_2 p_4$. Thus,

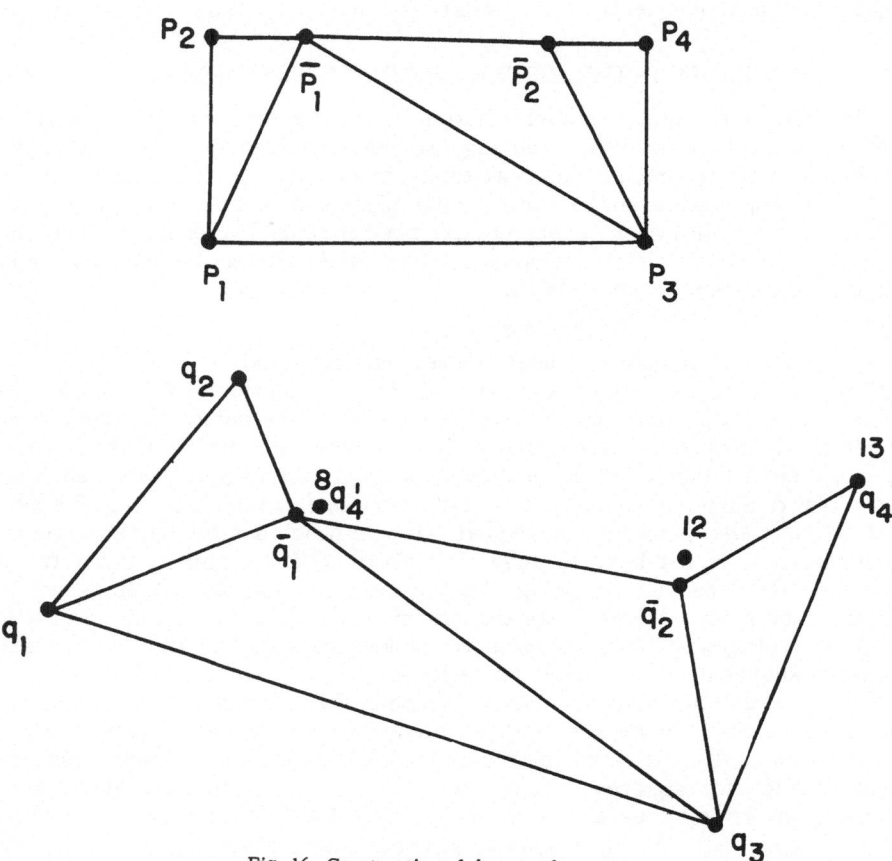

Fig. 16. Construction of the transformation.

the triangle $p_1 p_2 \bar{p}_1$ is mapped onto $q_1 q_2 \bar{q}_1$ by the linear mapping described earlier.

In our example, $p_1 \bar{p}_1 p_3$ can be mapped onto $q_1 \bar{q}_1 q_3$. We choose our next candidate point, q_4', considering how many lines that emanate from it cross $\bar{q}_1 q_3$. Again, we check whether this candidate is suitable. In our example, we continue by bypassing it with node $\bar{q}_2$. Node 13 is finally chosen as q_4. The mapping which is used to map the rectangle $p_1 p_2 p_4 p_3$ onto $q_1 \bar{q}_2 \bar{q}_1 \bar{q}_2 q_4' q_3$ is the piecewise linear mapping which takes triangles onto triangles as indicated in Fig. 16.

The question whether this algorithm can be continued to find the transformation which takes the rectangular region into the circular one in the appropriate manner must be answered somewhat unsatisfactorily. If the number of edges between adjacent nodes is not too stringently limited, then a mapping can be found which obeys the limitations (1) that it be piecewise linear, (2) that it be one-to-one, so that the inverse mapping also exists. The reasoning which supports the above statement follows from two considerations:

1. According to our construction method, any node which has not been used as an image vertex for some elementary rectangle is found outside that region for which we already have constructed the mapping.
2. The boundary curve for the region for which we have constructed the mapping in a closed polygonal path which does not intersect itself.

According to Jordan's "Curve Theorem for Polygons" [7], any two points which fall outside the above region are connectable by a polygonal path. Thus, in particular, any point outside the region can be connected to a point on the boundary of the region by a polygonal path. We can extend this path to a polygonal region by the method outlined above, making certain that no other points are included in such a region. Hence, we can at any stage extend our mapping to eventually include all nodes of our standard embedding.

REFERENCES

[1] C. Berge, "Theorie des Graphes et Ses Applications," (Dunod, Paris, 1958) pp. 31-32.
[2] R. J. Preiss, "Design Automation Survey," IBM Technical Note, TN 00.480 (1960) pp. 3-4.
[3] U. R. Kodres, "Geometrical Positioning of Circuit Elements in a Computer," AIEE CP 59-1172 (Oct., 1959).
[4] J. B. Kruskal, Jr., "On The Shortest Spanning Subtree of a Graph and the Traveling Salesman Problem," Proc. Am. Math. Soc., Vol. 7 (1956) pp. 48-50.
[5] R. E. Gomory, "An Algorithm for Integer Solutions to Linear Programs," Princeton — IBM Mathematics Research Project Technical Report, No. 1 (Nov. 17, 1958).
[6] A. H. Wallace, "Introduction to Algebraic Topology," (Pergamon Press, New York, 1957) pp. 100-102.
[7] R. Courant and H. Robbins, "What Is Mathematics?" (Oxford University Press, New York, 1941) pp. 267-269.

DISCUSSION

Question: Dave Caldwell, General Electric, Phoenix, Arizona. Is there a practical limitation to the number of layers in your circuit board?

Answer: Manufacturing difficulties seem to impose a limit of about 8-10 layers. The more layers one can produce, the easier the design problem is to solve. We do not have a very efficient technique for more than two layers. However, even inefficient techniques will still solve the problem.

Question: Dick Fetterolf, General Electric, Phoenix. Could your method be used with ordinary circuit components as well as with the standard circuit wafers you described?

Answer: Yes, it could. If we build our logical circuits out of conventional components in a way which closely resembles the building of wafers, then we can apply the present design procedure to conventional componentry as well.

Question: (Mr. Fetterolf). After you have designed the cards and the system by this method, there remain the other inherent problems with packaging and with electronic design: your heat transfer, your cross talk, and, in many cases, alterations to wiring and updating of your circuit card. In your experience has this updating been a great problem?

Answer: You hit on a very sore point. Once you have designed the circuit card, obviously, it is not very easy to change it. You would have to either produce a completely new card, or alter the old card by putting on additional wires and scratching out certain others. This is a very hard thing to do and this is one of the sore points of the entire packaging problem. However, since we will save a considerable amount of money packaging circuits in the way that I described, perhaps the amount of money saved this way can be used to produce another set of cards. We have no really good method of changing the card once it has been produced. Therefore, we had better be quite certain that what we produce is right in the first place.

Question: If I heard you correctly, you said that printed wire back panels would be cheaper than the present method of point to point wired interconnections. Upon what do you base this opinion?

Answer: The reason I say that the printed circuits are considerably cheaper is this: You can produce the printed circuits chemically by an etching process. So you can produce all connections at once, as soon as you have designed the photographic master. Each card produced from this master does not cost you very much at all. On the other hand, when you produce a back panel you have to produce each connection on the back panel separately and construction of a copy of the back panel would cost just as much as the original did. Thus, there is a considerable saving of money if we produce many copies on one machine.

Question: Jake Rubin, General Electric. This comment relates to applying this design method to ordinary electrical components: resistors, capacitors, etc., of different sizes and shapes. Would it be possible to adapt this method to the innerconnection of the individual circuit elements on a particular circuit board?

Answer: Our method can be adapted to design with conventional components only if one is willing to use space somewhat inefficiently. If we insist on using space on the circuit card with utmost efficiency, then we are forced to keep in mind the different physical sizes of the components. This makes any mathematical treatment of the problem so complicated that any general procedure of the type I described is bound to fail. However, as I pointed out before, we can think of certain collections of conventional components as equivalent to a wafer. If we are willing to use space somewhat inefficiently we can package the components belonging to one wafer close together on a piece of surface. We can think of these components as a wafer and thus adapt our procedure for wafers to conventional components.

MAINTAINABLE ELECTRONIC COMPONENT ASSEMBLIES

Harry Wasiele, Jr.

AMP Incorporated, Harrisburg, Pa.

MECA (MAINTAINABLE ELECTRONIC COMPONENT ASSEMBLIES)

Due to the new demands imposed upon industry to manufacture electronic equipments requiring but a small fraction of the space their forerunners once occupied, major problems have arisen in packaging and interconnecting. For manufacturing reasons, the industry found it to be most practical to construct the equipments in modular form. For maintenance reasons, it was desirable that the modules be plug-ins. This is exemplified by the plug-in card systems in general use today. Another factor is that of heat dissipation, which further complicates the over-all problem.

In order to achieve these smaller volumes, the industry is integrating the components into three-dimensional blocks (Fig. 1). In doing this, manufacturers are using the techniques that best lend themselves to the equipment design and manufacturing capabilities of their particular companies. This has progressed to a point where some companies are concentrating on the end system design and are encouraging vendors to supply their needs for small functional blocks.

In making a study of this new packaging problem, our company discovered that manufacturers were very cooperative in pointing out the shortcomings of their

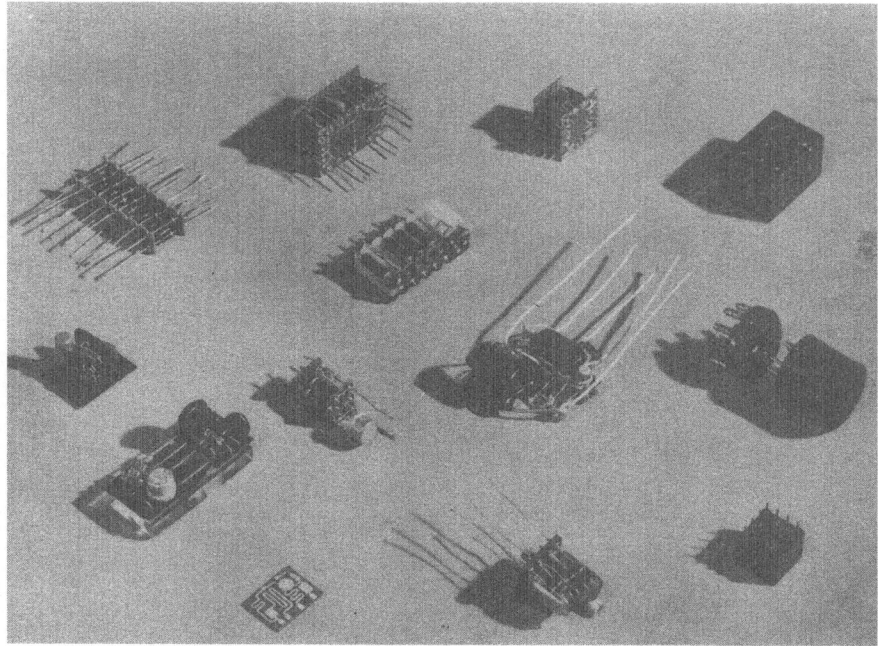

Fig. 1

existing techniques as well as their new design objectives. For the most part, it became apparent that a substantial number of designers and packaging men were remarkably consistent in their basic requirements.

PLUG-IN SYSTEM WITH INCREASED INTERCONNECTION CAPABILITY

The new MECA (Maintainable Electronic Component Assemblies) concept [1] was a direct result of this investigation (Fig. 2). It is most simply described as a modular system which will accommodate most of the basic functional circuits now being constructed by industry and will satisfy the new design objectives which require increased interconnection capability.

Fig. 2

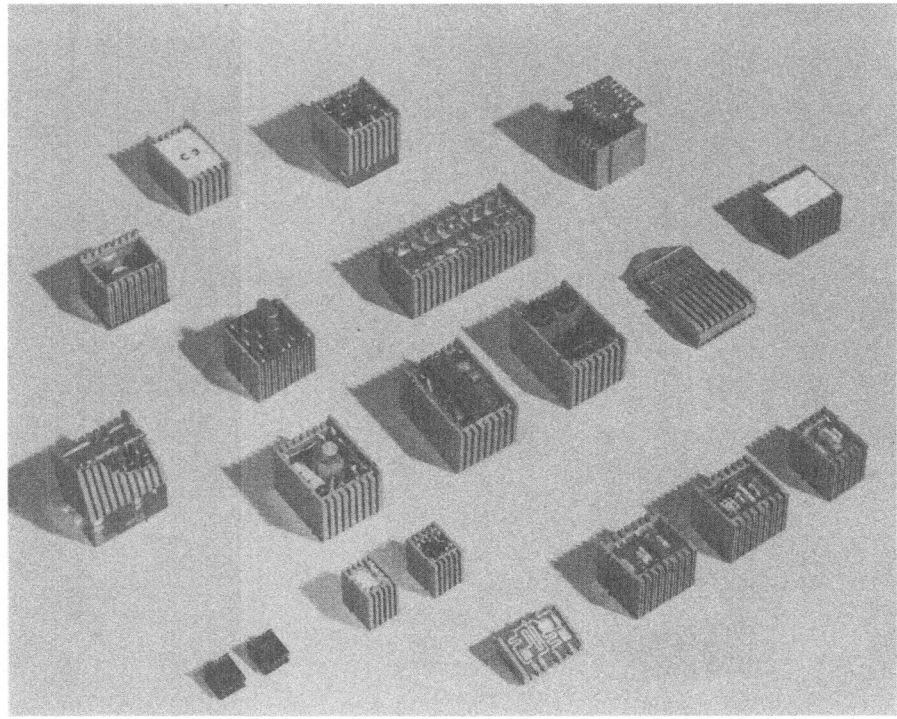

Fig. 3

ADAPTABILITY OF MECA TO VARIATIONS IN THE CONSTRUCTION OF CIRCUIT MODULES

A MECA cell connector is used to contain the circuit module (Fig. 3). Cellular height, length, and width are expandable, and the final size will depend upon the size of the circuit module and the number of input-output connections required. For applications where it is necessary to replace individual components within a circuit module, the circuit can be assembled in a way that permits the replacement of any specific component without disturbing any other connections (Fig. 4). Although probably not potted in this instance, the cell connector does provide a

Fig. 4

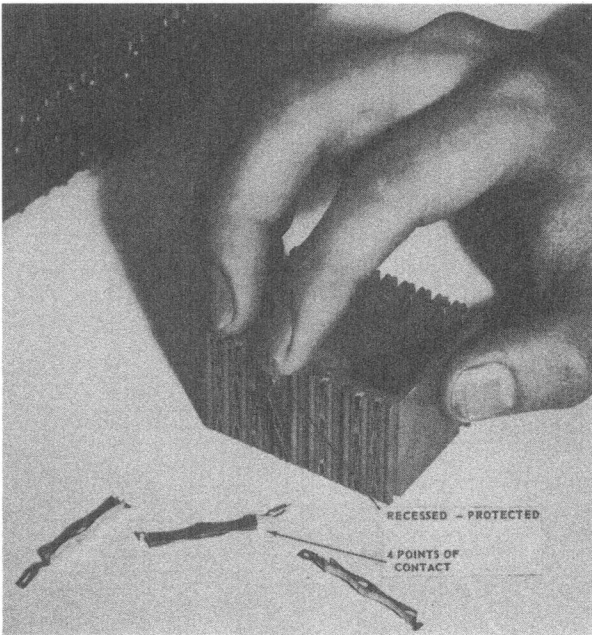

Fig. 5

protective enclosure for the components. When the anticipated down time of an equipment is of extreme importance, although it may still be desired to keep maintenance costs to a minimum, many systems designers are finding it necessary to use expendable modules, i.e., nonrepairable circuit functions as also shown in Fig. 4. To be economic in manufacture and as a throw away, they should be as small as practical without jeopardizing the system's reliability. Because the MECA cell can be as large as desired, it is very well adapted to implementation of this concept. In this instance, the cell serves as a potting fixture. Also, the utmost in packaging density can be achieved when the circuit modules and connectors can vary in size.

In the course of the original investigation, numerous equipment designers revealed their desire for a plug-in modular system as opposed to modules soldered onto circuit boards, modules that incorporated too many individual circuit functions, or modules too permanently attached to the interconnecting wiring. However, they indicated that, generally, the currently available module connectors did not offer the reliability or density required for use in today's equipments. To ensure the greater degree of reliability required, the MECA cell contact has been designed to provide four parallel current paths (Fig. 5). This four point contact is recessed within the cell wall to prevent abuse in handling.

Generally, the contact lengths are as long as the height of the modules. This permits plugging the three-dimensional module completely within it receptacle connector. This assists in increasing the ultimate reliability of the system by providing good mechanical support for the module.

INTERCONNECTING CIRCUIT MODULES

It became apparent early in the investigation that the mere miniaturization of receptacles into which modules could be inserted added nothing to the ability to

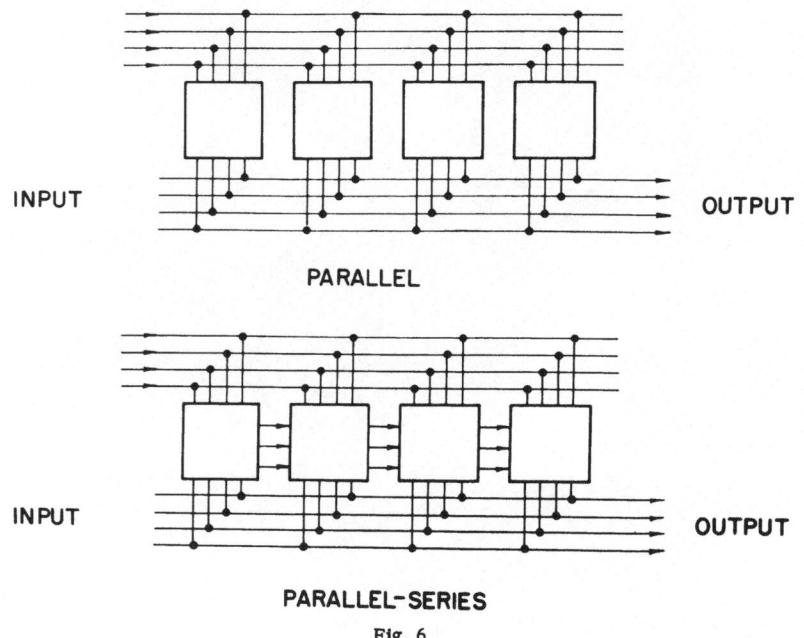

PARALLEL

PARALLEL-SERIES

Fig. 6

Fig. 7

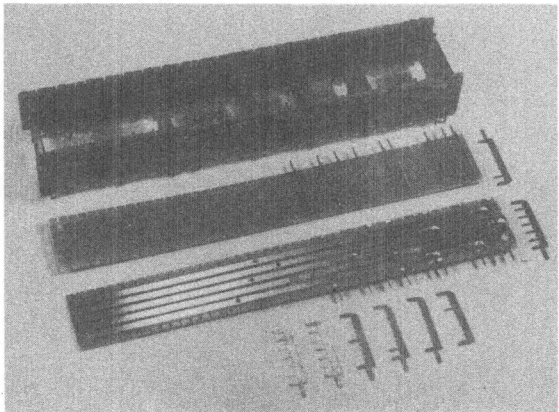

Fig. 8

interconnect these same modules. Regardless of the method utilized to make these module to module interconnections, the problem was complicated by increasing the number of input-output connections per unit volume. The very first objective of the new MECA system was to fulfill this need for increased interconnection capability.

The basic system is designed to accommodate complex-parallel or series-parallel interconnections (Fig. 6). Connections from one cell to another are uniquely accomplished through the use of rib contacts and two one-sided circuit conductor rails. The receptacle connector is made to vary in length, depending upon the number of modules desired per string (Fig. 7). Each individual connector section in the string can also vary in length, depending upon the length of the modules to be accommodated. The maximum number of conductor buses for each conductor rail is dependent upon the height of the module. For each incremental change in height, an additional conductor bus can be employed on each side.

Rib contacts are used to make connections between the base board, the module cell contacts, and the conductor circuit rails (Fig. 8). Top and bottom tines of the ribs are secured within slots on the top and bottom edge of the conductor circuit rail. Additional rib tines are located on the same spacing as the conductor circuit buses. Predrilling or punching of holes through rail and conductor make it possible for the rib contact tines to project through the rail and be bent down onto the conductor bus and soldered. The unrequired tines are cut off in an auto-

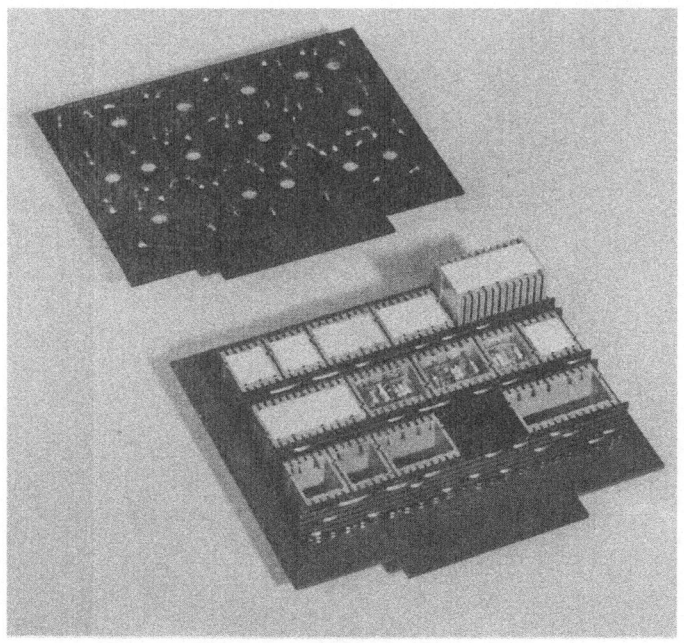

Fig. 9

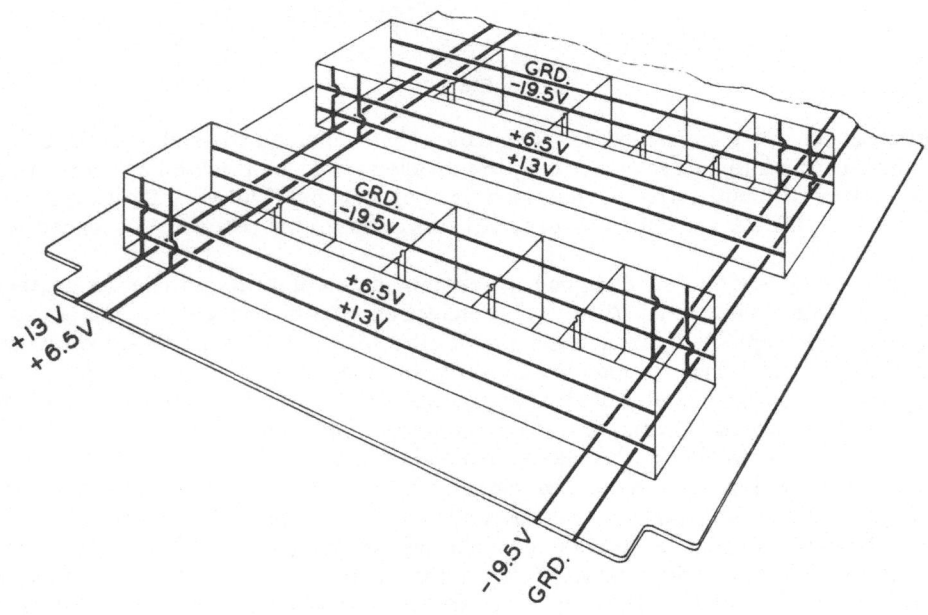

Fig. 10

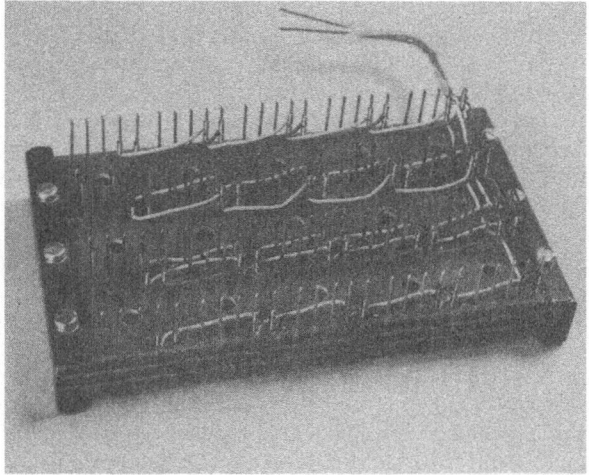

Fig. 11

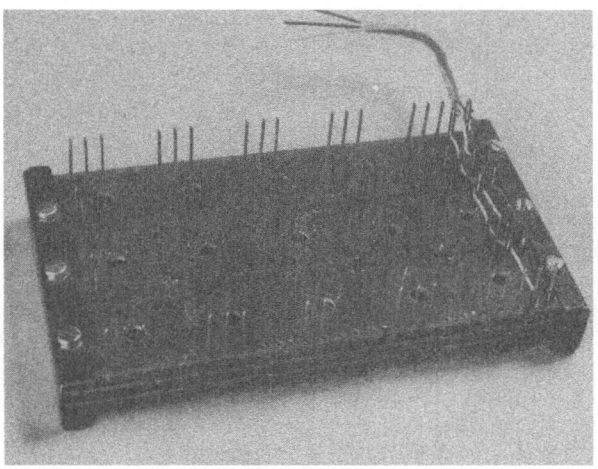

Fig. 12

matic programing machine. The two circuits are assembled together, using spacers located between each cell location. Note that the conductor buses are not necessarily continuous. Where the circuitry requires a break in continuity, the bus is spot-faced. This provides opportunities almost without limit to arrange the interconnections to suit the designer's needs. This figure also shows how a circuit path can proceed along a particular bus to a rib tine, enter a given cell at that point, and then emerge from the cell through another tine to a different conductor bus.

Connections between strings of cells are accomplished through the use of rib-base tines projecting through holes in a printed circuit board and are soldered to the underside of the board (Fig. 9). Figure 10 shows the ease with which all power and ground connections can be made to all modules in one subassembly. For instance, an input of +13 v could be made to all modules on the board by using one short conductor run on the base mother board. Module contact position No. 1

could be used for each module that requires a + 13 v input. The input voltage would be run from the conductor pad at the edge of the base board to contact position No. 1 of the first module in each string located on the board. The base tine of the rib contact in position No. 1 of the receptacle connector would relay the input through the rib contact to the cell contact of module No. 1. The same rib contact would also contain a conductor circuit rail tine, energizing one circuit bus which would serve as a common voltage bus to each cell position in the string. The location of a rib contact with a tine to pick up the voltage bus would permit an input to each cell that contains a cell contact at that specific position.

The possibility of adapting the new system to base-panel wiring other than the usual one- or two-sided printed circuit board exists. Special rib tines for wire-wrap techniques are currently being developed. They can be used in conjunction with conductor circuit rails.

Normally, when using wire-wrap to make logic and power connections to a back-plane frame which contains modules with wire-wrap tabs, it is necessary to jumper power and ground wires to each module (Fig. 11). By using the same principle of power busing through the receptacle connector as was described for use with a printed board (Fig. 12), a considerable reduction of labor and material can be realized. As an example, consider a frame containing ten circuit modules per string assembly with ten strings per frame. Only 40 (rather than 400) connections are needed with the MECA system—a 90% reduction of wire-wrap connections.

CIRCUIT LAYOUT

One of the primary advantages which has been recognized in the application of MECA, is the ease of circuit layout. Generally, using conventional techniques, it would require several weeks to draft the interconnections between numerous three-dimensional circuit modules. The use of a graph (Fig. 13), makes it possible to complete the layout in a matter of hours.

Figure 14 is a section of a representative sample of a completed interconnection diagram. It is a simple way of depicting in two dimensions conductor paths that are in reality in three dimensions. The close-spaced horizontal solid bar grids represent the actual circuit rail conductors and the vertical dash lines the base-plane connecting paths.

As in conventional layout systems, the normal numbering system for making module connections is utilized. The difference, however, is that with the graph approach one can readily detect where and how the connections can be made without the necessity of trial-and-error techniques. It will be noted that because a considerable number of connections are made with the side-rail conductors, the base-panel wiring is considerably simplified. The base-panel conductor paths are practically one dimensional and afford direct access to the card module connector.

More complete information regarding the procedure described above can be made available.

PROCESSING OF CONDUCTOR CIRCUIT RAILS AND RIB CONTACTS

Programing of the conductor circuit rails and rib contacts is done in one of two ways:

1. For maximum production, automated equipment for programing circuit rails and rib contacts is used.
2. For prototype or applications which require relatively nominal quantities of rails, a pantograph and simple fixtures are used.

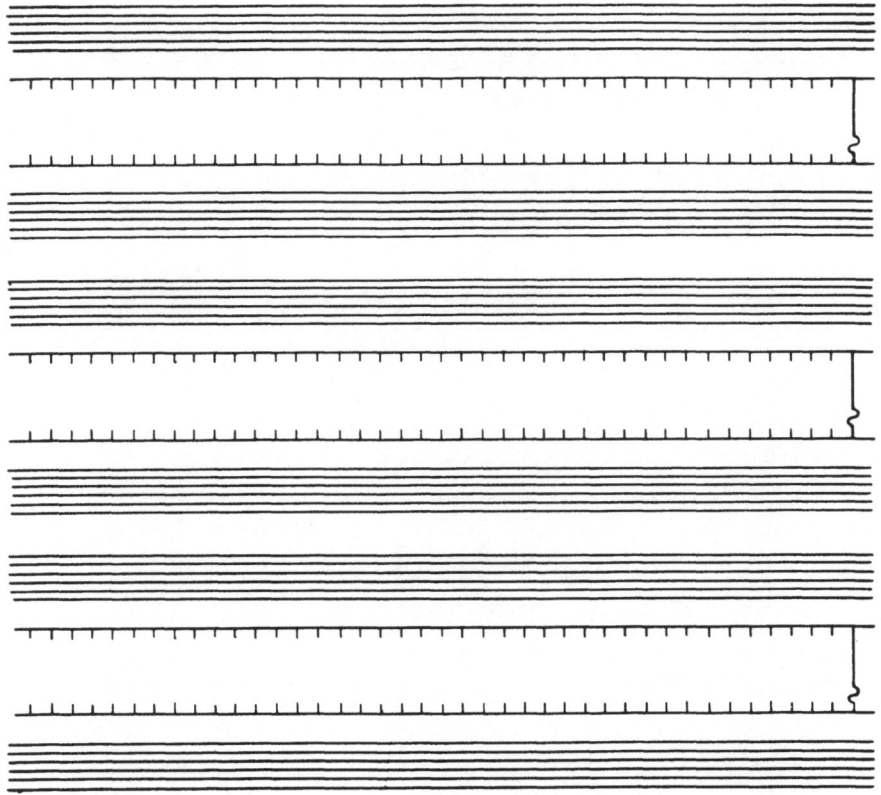

Fig. 13

The following illustrations describe the tooling which can be used for small quantity production:

Figure 15 shows a pantograph machine with an appropriate platen and drill head fixture set up on a 2 to 1 ratio. A prepunched card is laid on the platen and a conductor circuit rail is located under the drill head. By inserting the stylus into the prepunched holes of the master card, the drill head will either drill or interrupt the conductor bus, depending upon the tool used.

Figure 16 illustrates a block which has been slotted to accommodate preprogramed rib contacts. A tape at the top of the block identifies the specific programing of the rib contact tines. The preprogramed conductor rail is laid over the top of the projecting rib tines. The mechanism located at the right of the block is passed over the conductor rails and bends the tines down onto the conductor buses. This assembly is then ready for soldering.

APPLICATION ADAPTABILITY

The assembly is adaptable to specialized military or nonmilitary, simple and complex systems, as well as to large-volume systems. The following examples reveal several of the various types of applications in which the MECA principle has already been applied.

On May 22, "Aviation Week" announced that Grumman Aircraft had selected the Federal Systems Division of the IBM Corporation to develop and build a data

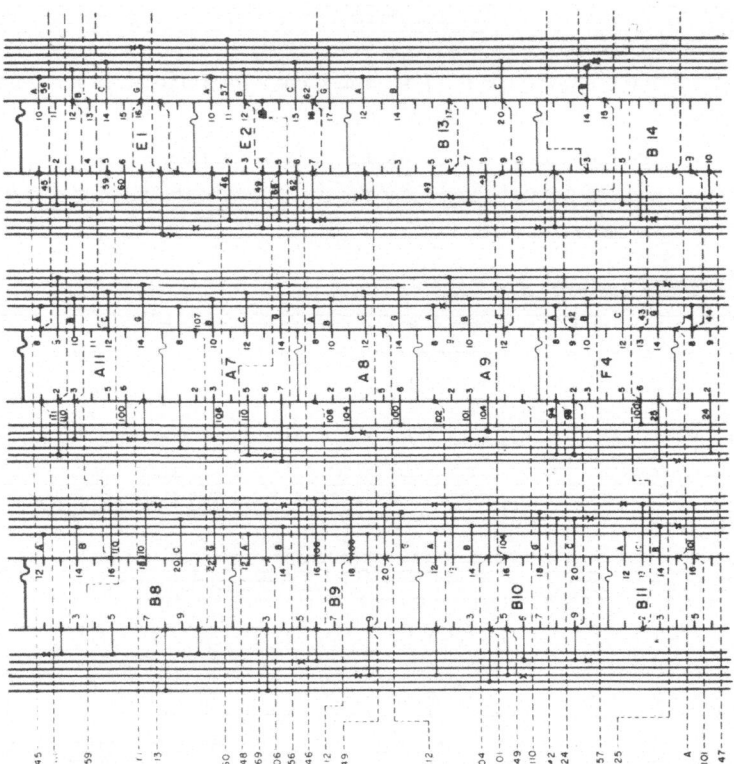

Fig. 14

Fig. 15

processing system for the National Aeronautics and Space Administration's
Orbiting Astronomical Observatory (OAO). It was stated that the digital data
processing system would be the most sophisticated yet attempted for nonmilitary
satellite use.

The OAO will be the first satellite designed to probe the far reaches of space
by analyzing ultraviolet radiation from stars and planets masked from terrestrial

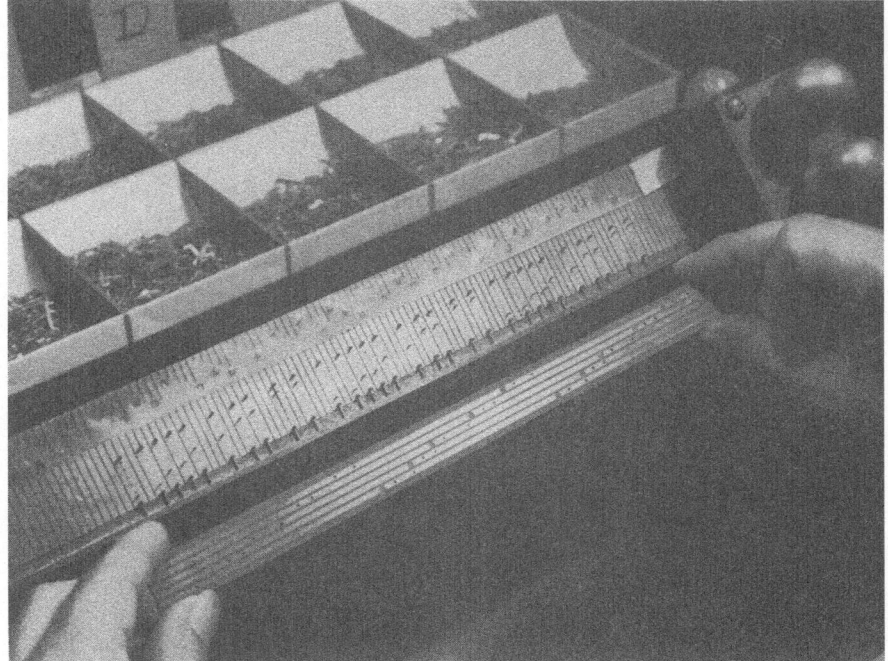

Fig. 16

Fig. 17

observatories by the atmosphere. This will give earth-bound scientists the ability to control or change the experiments while the vehicle is in orbit.

To afford the system and OAO assured reliability, the IBM system will use quadruple redundancy at the component level and triple redundancy at the circuit or module level.

To achieve high component density, a welded cordwood-type construction will be employed in the plug-in modules that make up the system (Fig. 17). The IBM circuit in Fig. 17 is illustrated prior to assembly into a MECA cell and also after

Fig. 18

Fig. 19

the circuit has been assembled into the cell and input-output connections have been welded to the cell contacts. Two individual ribbon leads are welded to component leads and input-output contacts for wiring interconnections.

In utilizing the MECA system with its almost infinite interconnection capabilities between modules, long established manufacturing techniques can be used. As opposed to long nonmaintainable stick modules, the plug-in submodule provides fast, economical access for replacement of the module in the event a change of circuit is required. This also provides test capabilities at the functional circuit level during assembly of the system. The same MECA interconnection philosophy is applied to the memory portion of the system.

The circuit used in Fig. 17 is no longer assembled into the cells in the manner shown. The welded cordwood package is still used, but it is inserted into the cell so that all components are aligned perpendicular to the contacts of the cell. This permits easier wiring and potting without decreasing the density of the package.

An Air Bearing Gyro Stabilized Platform Navigation System is being developed by Astro Space Labs in Huntsville, Alabama. MECA proved to be the most effective way to package the entire Servo Amplification System. Adoption of a modified 0.200-grid MECA system makes it practical to package 25 modules, the time-delay relay systems, and conventional AN plugs in a 12 by 12 by 3 in. housing fastened adjacent to the unit containing the Gyros. This facilitates the packaging of 135 2½ by 1 in. circuit boards which contain various types of components such as power transistors, ¼-w resistors, conventional transformers and capacitors as large as 4.0 µf at 50 v. Utilization of the MECA system eliminates 25 intermodule connectors. Shown is one module of a servo amplifier section where power-type transistors (2N1483) are used (Fig. 18). This type of transistor dictates the use of conventional heat sinking which has been adapted to MECA in this application (Fig. 19). Aluminum transistor mounts are mounted on an aluminum plate which serves as a sink and replaces the bottom of the MECA cell. In the module, this leaves space for an additional component circuit board, transformer mounting board, amplifier interwiring and the input-output connections to the cell contacts. Two small screws are used to attach the module with its aluminum base to the chassis base plate. This insures intimate contact for very good thermal conductivity.

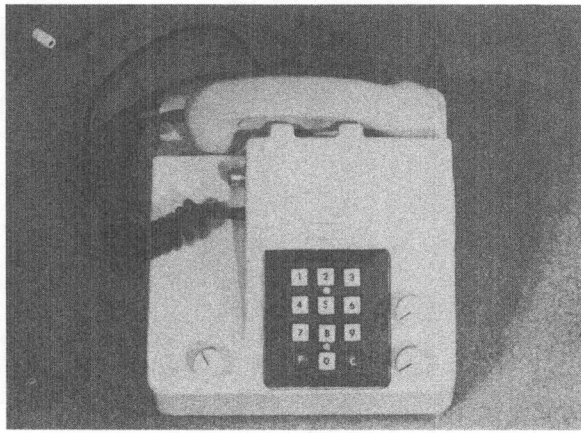

Fig. 20

Astro Space Labs has indicated that an important reason for choosing MECA is the ease with which it conformed to MIL specifications. It was important to Astro Space to have the breadboards and the prototype units as similar to production units as possible, and allow for easy trouble shooting and changes while in development. Because of the plug-in ability of the module, a great deal of GI-proofing is designed in from the beginning, and the prototype itself is a rugged piece that will stand up well during system testing and evaluation. Astro Space plans to continue this type of basic design using the MECA technique on future similar projects.

North Electric Company, Galion, Ohio, has developed a solid-state switching system which has incorporated the use of MECA for the packaging and interconnection of its circuits in the telephone instrument (Fig. 20). The following information reveals the designer's reasoning for using MECA in his application:

The packaging of the circuit for the Electronic Telephone Instrument was undertaken with two major goals in view:

1. The package must be compact and light in weight.
2. It must be easily maintained by people who have a minimum of experience both in locating and correcting trouble.

In addition to these major goals, the package should also be able to withstand a great variation of operational environment, such as temperature, humidity, shock, and vibration and still maintain constant operational levels.

To meet these goals, a program was instigated to get the smallest components available which would meet the circuit requirements and to find the best possible package for the components. The first solution considered was the use of the printed circuit board containing the entire circuit. This provided a small

Fig. 21

package, but it required specific knowledge and tooling for maintenance. It was also doubtful that the circuit would pass environmental tests as the miniature components were not hermetically sealed. The answer to the maintenance problem was smaller separate circuit boards to shorten the time to isolate troubles. This did not eliminate the necessity for a soldering iron or improve environmental quality.

The next step was the investigation of a module-type construction of solid or lattice encapsulated modules. This again resulted in smaller circuit sections or modules, further reducing the trouble-isolation time; but it did not reduce the need for soldering-iron maintenance. The unsoldering of multilead modules was considered unsuitable for unskilled maintenance personnel. It would not offer immunity requirements needed. The next consideration was to provide a plug-in module. This module should take up a minimum of space, offer the greatest flexibility, and provide low contact resistance and high retention. All these requirements were found in the MECA system (Fig. 21). The module cell and its receptacle connector provide the most efficient packaging densities for use with

Fig. 22

Fig. 23

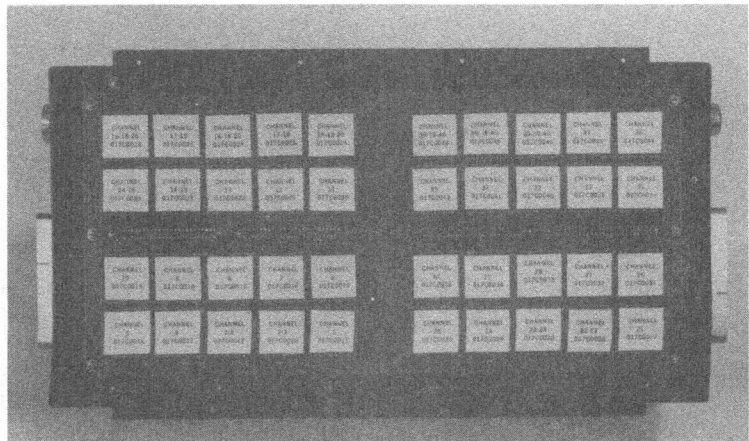

Fig. 24

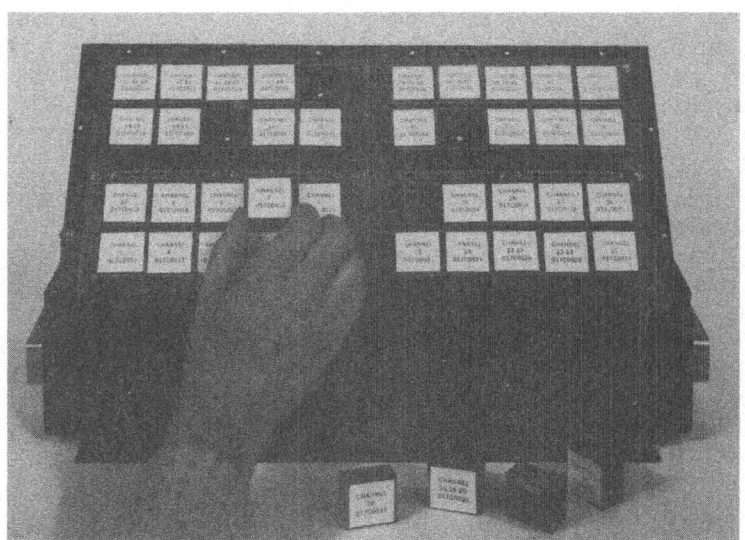

Fig. 25

a reliable, plug-in module. The contact pressures developed between the blade and four-point spring contact insured low contact resistance and high retention.

The circuit was then packaged in the following manner to fulfill all design goals: it was first sectioned into basic functional units, such as transmit amplifier, tone oscillators, ring detectors, etc., and then split into modules as required.

Components were mounted on small printed circuit boards, and these subassemblies were placed in the cells and encapsulated. The modules were then arranged in rows to provide maximum use of the side rail circuits for interconnection between modules. This usually placed all modules of a functional section in a row. These rows were located on a master printed circuit board. Connections between the rows and to the external connector were arranged for a minimum of artwork layout. Therefore we have a total of 78 large components in 21.25 in. This package contains a flat 40-db gain amplifier, voice frequency range, dual tone dialing, and 95 db of dual tone ranging from 180 Mw of power. The package is also able to satisfy rigid shock, vibration, temperature range, and humidity requirements.

The MECA program was selected for use in the AFCAN (Analog Factor Calculations Network) unit of the MADREC (Malfunction Detection and Recording) System at Lockheed Electronics Company, Military Systems/Stavid Division, and Lockheed Aircraft Corp., Marietta, Ga., for several reasons.

The MADREC airborne test equipment must always be prepared to monitor various electronic systems in a given aircraft. Thus, any malfunction in the AFCAN must be quickly located and corrected.

Electrically, the circuitry can be subdivided into small, integral functions which, when packaged in MECA cells, make for easy fault location and quick replacement.

Mechanically, the MECA cell provides a secure environment for all components in the system, ensuring protection at all times during manufacture, through shipment and storage to final use in the AFCAN unit. Figures 22, 23, 24, and 25 illustrate this application.

INTERCONNECTING OF WAFER DEVICES

Philco Corp., Research Division, and Philco Western Laboratory are considering MECA for interconnecting advanced designed circuits. The first application considered was for use with a solid-state diode logic commutator. The solid-state commutators differ from mechanically driven types in that they contain no moving elements, are synchronized electronically, and can be operated at fast speeds.

The primary function of such a system is to sequentially sample and channel a number of input signals (24, in this particular case) into one output, or vice versa. This particular function is applicable to an earth satellite communications system.

Fig. 26

A solid-state diode commutator was developed at Philco Research which consisted of 96 silicon diodes and 16 resistors. Solid-state diffusion was utilized to process six silicon wafers ($^3/_8$ in.^{2}) (Fig. 26). Each wafer contained 16 unilaterally constructed diodes and 4 silicon resistors. The schematic for one such wafer is shown in Fig. 27. Evaporated conductors were employed to connect these components into a functional block. Each block was mounted on a ceramic substrate which contained prefired metal terminations along the outer edges of the ceramic wafer. Soldered contacts were provided to these areas by using 20-mil-diameter nickel wires. The entire assembly was encapsulated in a potting material and contained within a MECA cell. The MECA interconnection system is expected to be used to interconnect circuit blocks of wafers into operating systems of the future.

A system which is intended for use with the Texas Instrument semiconductor network circuit is in advanced development (Fig. 28). Each of twelve semiconductor network circuits has been assembled within wafer-type connectors using ten four-point contacts located on two sides of the wafer. The input and output contacts and interconnecting conductor buses of the conductor circuit rails are spaced on 0.050-in. centers. Additional development is under way to determine if a higher density interconnection connector system is practical.

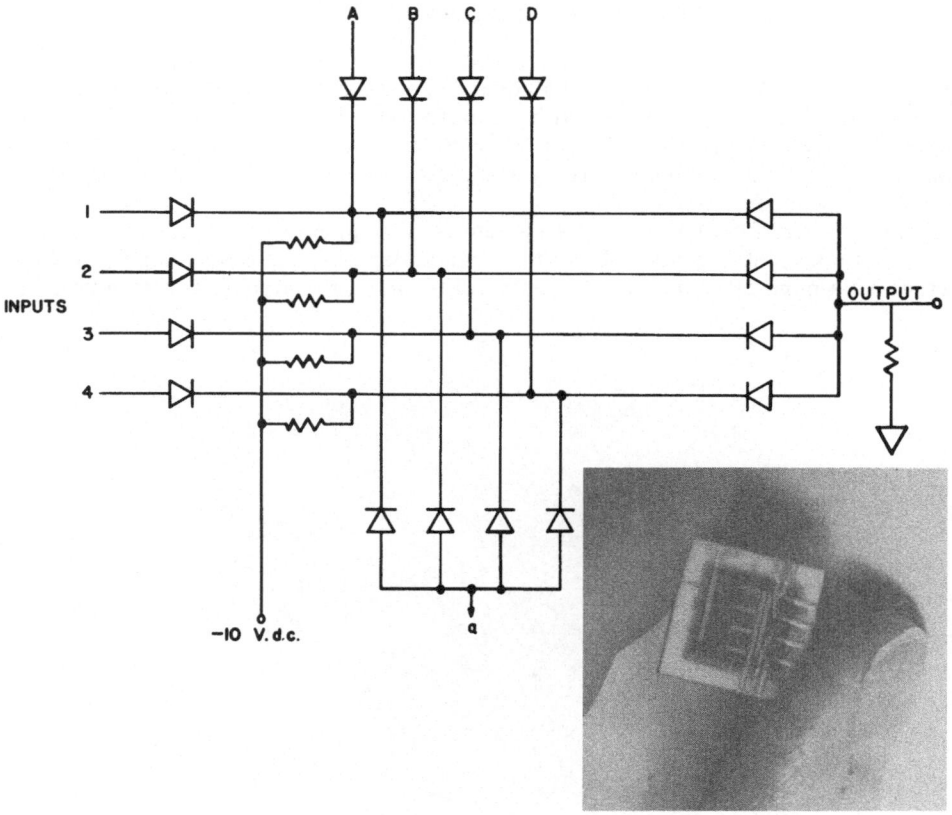

Fig. 27. Philco diode commutator—one of six sections.

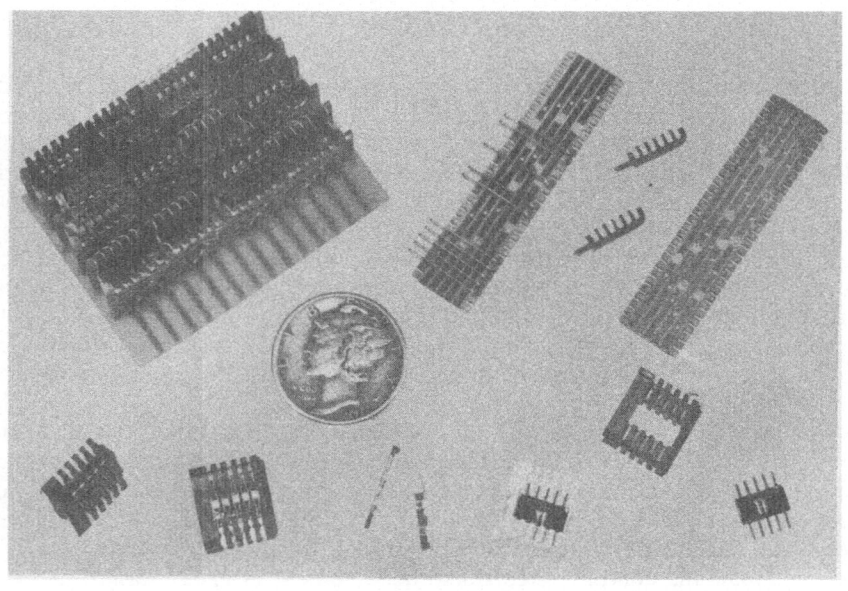

Fig. 28

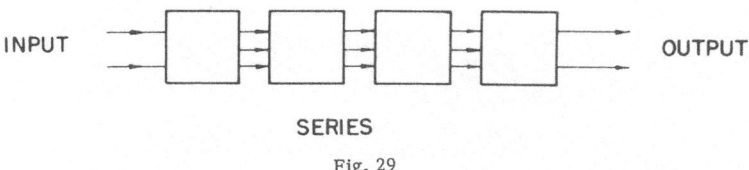

INPUT   OUTPUT

SERIES

Fig. 29

Fig. 30

Developments which utilize welding techniques for the interconnection of thin film circuits are also being evaluated.

· The MECA system described is generally used with complex-parallel or series-parallel wiring. A modified system, which lends itself to series interconnecting as used in transmit-and-receive-type applications (Fig. 29), employs much of the same type of packaging philosophy. Cell circuits are interconnected by a blade contact connector which serves as a shorting bar when engaged between cell contacts of two opposed cells. Figure 30 illustrates the MECA application of a 5.5-Mc I.F. strip.

Figure 31 is the new all-transistorized FM transceiver built by Selna Industries Incorporated of Orlando, Florida, for the commercial two-way radio market. The plug-in maintainable construction feature is employed with color coding of cells for production ease and simple maintenance. This approach minimizes down time and reduces service-stock parts to only 12 cells. The manufacturer feels that he is setting a precedent in price and performance using this unique packaging technique in the first FM transceiver of its type in the mobile communications market.

CONCLUSIONS

1. The packaging design objectives, which were established during our survey of the electronics industry, have been successfully accomplished with the new MECA system.

The first of these has been realized in the design of a reliable incremental plug-in module which employs contacts, each having a redundancy factor of four.

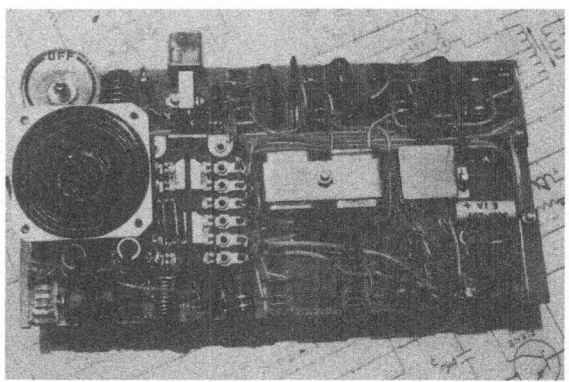

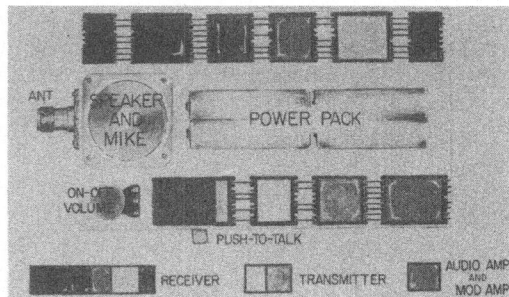

Fig. 31

The second of these objectives has been met in the use of a new receptacle design that incorporates additional conductor paths. It may be emphasized here that this new system of module interconnection has proven more than capable of accommodating the most complex computer logic and power interconnections.

2. The mechanical design of the system enables the designer to vary the concept to suit his particular needs. It was not intended that the MECA system be limited in use to a few specific equipment types. The preceding applications described point out that the system's flexibility is limited only by the imagination of the equipment designer.

The author wishes to extend his sincere appreciation to John Pratt, International Business Machine Company; Edward Quinn and Robert C. Martin, Astro Space Labs, Subsidiary of Belock Instrument Corporation; Dean Trump, North Electric Company; George Curtis, Lockheed Aircraft Corporation; Robert Fliedner, Lockheed Electronics Company, Military Systems/Stavid Division; Vincent Ramono, Thomas Sikina, and Steven Idzik, Philco Corporation; Thomas Nations and Frank Sela, Selna Industries Inc.; and the numerous other individuals who have been so generous with contributions to this paper.

REFERENCES

[1] R. C. Swengel and W. R. Evans, "A New Mechanical Approach for the Construction of Modular Electronic Equipments." ASME, Aviation Conference, Los Angeles, Calif. (March, 1961), No. 61—AV—3.

DISCUSSION

Question: Martin Camen, Bendix Corporation. In your approach you exhibit a great deal of flexibility and you seem to have incorporated the philosophy that you will modify your system to meet our needs, which I think is very good. The problem that I would like to know about is, what tooling expense, if any, can we anticipate due to these modifications of your system to meet our needs?

Answer: Much will depend upon the degree of modification that you require. We have at least 14 different cell sizes of the one-tenth-inch grid available today. We have two basic cell contact designs that are available; one lends itself to soldering techniques, while the other is more adaptable to welding. Our tooling is designed to enable us to vary the rib contact die and the cell contact die to heights other than those we have seen in the 0.700 high system. To incorporate a new cell design would require a cell-mold cost of approximately $900.00. We had hoped to hold this figure down to $650.00. As yet, we have not been able to reduce to that cost. We have been absorbing the additional cost up until now. I don't know whether to say we will continue to do so or if we will impose upon you this additional $300.00 charge. Can you be any more specific about the type of modification you would require?

Question: (Mr. Camen) We have an approach that is similar at the present time. It is a module approach, that is, a plug-in package, which could be easily adopted to the MECA. What we would gain in this sense is a third plane of interconnection, that plane being perpendicular to the normal board. The only problem that I see offhand would be the tooling expense to modify the standard MECA cells to meet our particular requirements.

Answer: I would say for the most part—something less than a thousand dollars would probably cover the expense of your application. I would be delighted to get together with you and find out a little more specifically what your problems are and give you a firm quotation of any costs involved.

Question: Do you have any specific experience or general experience or thinking relating to the application of this form of interconnection to say, modules operating at frequencies between 50 and maybe 5000 Mc?

Answer: At this time, we do not have any specific applications that run much beyond 5 Mc. We are exploring the possibilities of making a completely shielded system. If you have an application and you feel the basic philosophy of the MECA system can be adapted, assuming now that you can overcome the problem of high-frequency operations, we would be very glad to get together with you and explore this particular aspect.

Question: R. A. Dahlberg, Automatic Electric Laboratories. Would you care to comment on the long-term reliability of your connections?

Answer: I would like to, but unfortunately, we do not have an abundance of testing information as far as long-term reliability. We have conducted a series of accelerated tests which are available to you, and if you would see me at the conclusion of the session, I would be happy to take your name and send you the complete test results on the system.

Question: Boeing Company. In all your photographs, you show the modules just plugged in; there is no form of device to hold them in. Does this mean that they ought to stay in position under all forms of vibration?

Answer: No. I think you can anticipate what has to be done under some conditions. Frequently, you may not require too many contacts for a given circuit. Perhaps, in some instances you only require six contacts for input-output connections for the module. In other cases, you may require as many as 10 or 20 contacts in a circuit. If you are going to subject the over-all system to severe shock and vibration, it would be desirable to provide a locking bar or positive hold down which would assure the retention of the six-contact module. There are several ways of approaching this particular problem, depending upon how you are going to package the system. I would be glad to point out to you how we can adapt a simple bar that can be extended across the entire length of all of the modules in a string to serve as a hold down. In some instances, you may be sandwiching boards together such as in IBM's OAO program. In this instance, no special hold down was necessary. I think we can give you some very satisfactory methods for positive retention which should be considered in specific situations.

Question: Jim Siebert, Magnavox Company, Fort Wayne. In looking at this system, I don't notice any system for keying these blocks so that one given circuit could not be plugged into a position which required a different circuit of identical size.

Answer: We have looked at this condition for some time. Most people have not considered this to be much of a problem since they have been confronted with it with tube type applications and many other similar situations. We do have the module polarized so that you can only insert it in one particular way. Should you feel that in your particular system, you are going to design more than one module of a given size, for performance of different operating functions, we can provide a means of keying. By loading or closing off a cell slot location and keying rib contacts in the receptacle connector, you can eliminate the possibility of interchanging like-sized modules in the system.

Question: Jack Peterson, Ampex. What is the response from the military as far as your connector is concerned?

Answer: We haven't made too many approaches directly to the military. We did make a direct contact with the Air Force. We found that they were more inclined to suggest that we work with the equipment manufacturers. They felt that they did not want to dictate to you the way in which you should package your system. They felt that you should be responsible for deciding how the system should be packaged. I might say that we have had an awful lot of interest and inquiries from the various different military divisions requesting information on the system. We are also working with RCA and the Signal Corps on a package that can be adaptable to the RCA micromodule. So long as we meet the reliability requirements for the system, there doesn't seem to be much resistance from the services. In fact, I think they are probably one of the primary groups that have emphasized, in many instances, need for reliable connectors for use with plug-in modules. They are endeavoring to reduce or eliminate the need for specialized techniques for replacing these modules in the field.

Question: W. Jeromin, ACF Electronics. Do you have the vibration data for these modules as of now?

Answer: We should have them within the next 30 days. We have started out on a program to release to the field all the available data on testing, shock vibration, humidity, and other pertinent information. To date, we have released all of the results except for the data on vibration, which is in its final stages. Our initial vibration experiences have indicated that long strings of modules should be adequately supported. Support of the module card along two edges of the card only is not usually adequate if the system is to be subjected to critical shock and vibration conditions. There are several preferred methods for mounting. One is to support the three-dimensional string and base plane board at both ends of the assembly. Another method would be to mount the base plane board around the entire periphery rather than just at two opposite ends of the board. Still another method of mounting that is in practice is the multistacking of numerous boards sandwiched together and properly secured to the frame assembly. This results in a rigid cube-type construction. The IBM OAO application utilizes this philosophy. From the results we have obtained thus far during vibration, we have not had any degree of failure in the contacts or any significant increase in resistance or noise in the contact connection.

Question: R. C. Wildman, Martin Co. I'm just curious. You mentioned that you are running tests on environments, including humidity. What techniques are you using for moisture-proofness and humidity?

Answer: Let me point out here—as far as the moisture and the salt spray test—the system discussed today is not intended to be an environmentally sealed connector. We do have the printed results of test data that will give you the complete procedure for each test. It also indicates how the system was mounted, how the test was conducted, the degree of humidity, and so forth. I will be glad to supply to any or all of you a copy of these test results if you request it.

Question: Charles Mason, Raytheon Co. Referring back to the Philco diode assembly, how do you make contact from the cell contact to the flat printed board?

Answer: The cell contact tines project through an additional set of holes along either edge of the board. All input-output conductors are run to individual cell contact tine pads.

Question: Do they bend over?

Answer: This really depends on your preference. In the first application, they were bent over.

Question: Tom Mahoney, Boeing. On your smallest board, what are the dimensions of the bus line, and what are the dimensions of the tines? How much edge of copper do you have between the hole and the edge of the bus?

Answer: Are you referring to the little 0.050 grid that's in development now?

Comment: Yes.

Answer: Are you considering the system that we saw here?

Comment: I believe that you do have a subminiature system.

Answer: I don't remember the specific dimensions on this unit. It was a development project. The conductors are spaced on the 0.050-in. centers. We can modify this dimension, since at the present time, we do not have any production tooling established for the stripping of the board. No production dies have been developed for the rib contacts so we can still consider changes. If you have a standard that requires a specific conductor pad dimension around a hole, we would be happy to consider it. At this point, we could still modify the system to accommodate your needs. I'd like to point out, gentlemen, that all the stripping of the conductor rails is done mechanically. It is not an etched process. A one-sided clad board is stripped leaving only a predetermined number of conductor buses. Circuit interruptions are also handled mechanically. I'd like to point out that you do not have to buy the system completely programed. This system can be made available to you on a parts basis, where you would actually do the programing of the circuits and the assembly of the contact yourself. We recognize that many changes occur, particularly during the development stages and that you might not be able to purchase completely fabricated assemblies. In this instance, we would supply you with component parts and the procedure for their programing and assembly by you. Once your design is more fixed, you may then prefore to have us supply completed assemblies.

Question: Don Clifton, Bendix Corporation. You may have just answered my question, but I wondered if we could purchase these components for our prototype work?

Answer: Yes, you can. In fact, you are probably aware that your Radio Division is now offering to industry their sandwich-constructed modules packed in the MECA system. I think the intent here is to initially supply the module packaged in cells. You would supply the complete system into which these modules would be plugged.

Question: K. A. Allebach, Nortronics. Do you have complete literature describing this system?

Answer: Your copy of my paper describes specifically, and in more detail, how the system works and how it can be made adaptable to various types of applications.

ELECTRONIC PACKAGING FOR 5000 g SURVIVAL[*]

Wayne F. Miller

Seismological Laboratory, California Institute of Technology, Pasadena, California[†]

INTRODUCTION

One objective of NASA's "Ranger" program is to make the first attempt by the United States to set operating instrumentation on the moon. The specifications as originally set forth called for a semihard lunar landing of the order of 5000 g. The landing package will contain a seismometer for the measurement of moonquakes and meteorite impacts. The seismometer electronics consist of a test-mark generating circuit, and although the Ranger specifications have been relaxed to a g level of 3000 g or less the design objective of 7000 g for the seismometer has been met.

Early in 1962 geophysicists will be analyzing new and exciting seismic data obtained from a seismometer placed on the moon. This experiment will be one of the first performed in NASA's many scheduled planetary explorations.

The seismometer will be contained within a capsule launched from the Ranger spacecraft in the vicinity of 20 miles from the moon and slowed by a retrorocket to a semihard landing of a maximum of 350 ft/sec at the moon's surface.

Why send a seismometer to the moon? Weight limitations on this initial shot allow for only one experiment to be performed and a single instrument most likely to survive and give the greatest amount of data concerning the moon's structure and origin is a seismometer which will measure moonquakes and meteorite impacts.

Figure 1 shows a cross section of the lunar capsule survival sphere containing the seismometer, electronics, and battery pack. A detailed account of the design of this sphere has appeared elsewhere [1] and it will be the purpose of this report to discuss only the ruggedization of the electronics to withstand the landing impact. The expected average acceleration is of the order of 2000 g for a 5 msec duration. Studies of high-g effects on telemeters and amplifiers have appeared in the literature [6]. Therefore, emphasis will be placed upon the ruggedization of the system test-pulse-generating circuit which presented some unique problems.

Earlier work had shown [7] that certain transistors were capable of withstanding very high accelerations. Accelerations of 200,000 g and above [9] have been successfully experienced by removing the metallic case and potting the internal structure with epoxy resin. It was our intention, however, to avoid this approach if possible and to produce a rugged package using readily available standard components, and more or less standard techniques.

TEST PULSE CIRCUIT

The operation of the lunar seismometer is shown schematically in Fig. 2. A pot magnet is loosely suspended from the instrument case by spider springs.

[*]Contribution No. 1041, Division of Geological Sciences, California Institute of Technology.
[†]The work disclosed in this report was performed by the California Institute of Technology Seismological Laboratory and the Columbia University Lamont Geological Observatory in cooperation with the Jet Propulsion Laboratory under a joint effort for the National Aeronautics and Space Administration.

The transducer coil is attached to the case which is solidly coupled to the moon's surface by the survival sphere. The suspension is designed such that the mass has a natural period of 1 sec. Relative motion between the magnet and the frame produces a voltage in the coil.

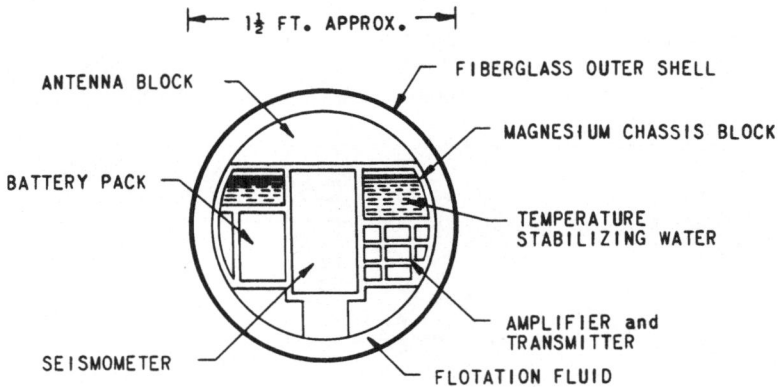

Fig. 1. Lunar capsule survival sphere.

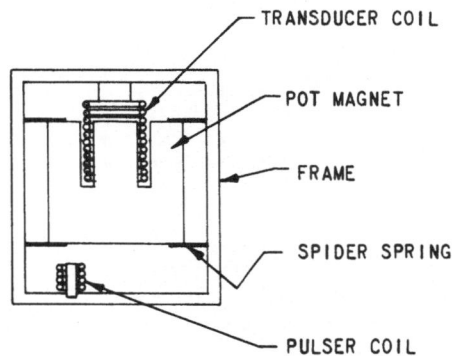

Fig. 2. Schematic illustration of lunar seismometer construction.

Another coil is wound on a phenolic form and placed near the magnet. When a current step is injected into this coil, the magnet is displaced a distance of approximately $10^{-3}\,\mu$, causing a voltage output which is the differential of the mass displacement. When the current is removed, the mass returns to its original position and the negative of the previous output is produced. Note that this is a truly generated output from which the proper operation of the entire lunar systems may be ascertained. For the experiment to be a success the survival sphere must erect itself so that the antenna points earthward. This ensures that the seismometer is vertical. The polarity of the generated test pulses will polarize the transmission and receiving systems and indicate the phase of ground waves arriving at the seismometer.

It was desired that these marks be placed on the record at as widely spaced intervals as could be obtained electronically within the limitations of the space and power available and be easily recognized against a seismic background. The circuit developed to perform this function is shown in Fig. 3. It consists of two asymmetrical free-running multivibrators plus an "and gate." The total periods of the multivibrators are slightly different so that they come into synchronism only once in ten cycles. The "and gate" selects the synchronized pulses resulting in a period multiplication of ten to one. As will be pointed out later in this paper, a maximum reliable RC time constant of $1\frac{1}{2}$ min was chosen which results in a period of 15 min between test pulses with a pulse duration of 4 sec. A typical record showing test marks and a local earthquake is shown in Fig. 4.

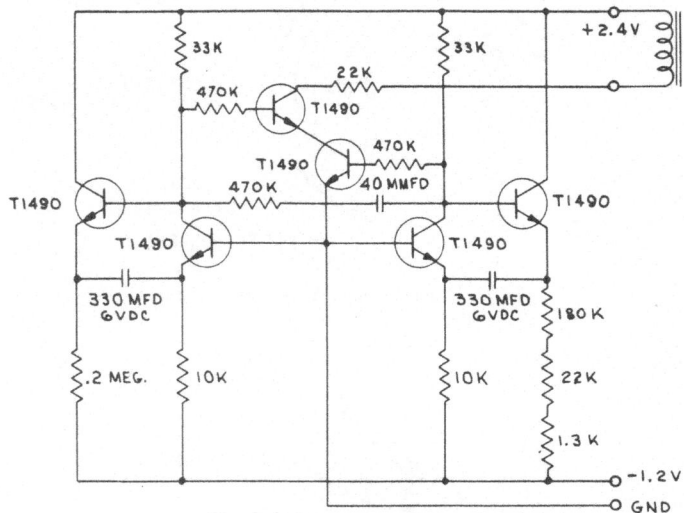

Fig. 3. Pulser-circuit schematic.

Fig. 4. Seismogram showing microseisms, minute marks,
pulser-generated output, and a local earthquake.

The design of a rugged seismometer began during the Ranger feasibility study and although the final landing specifications indicate that g level previously mentioned, the design objective set forth by Caltech's Seismological Laboratory of 5000 g for the instrument mechanics had been met. The test pulse electronics are physically located within the seismometer case, which was the motivation for ruggedizing the circuitry to the high g level, and thus have an integrated design.

STEPS IN HIGH-g DESIGN

There are four steps in the design of an electronic package to withstand severe shock.

1. Avoid fragile components in the electronic circuit design.
2. Test and evaluate components, selecting the most rugged types applicable.
3. Design a rugged mechanical package, taking into account component mounting and orientation relative to the expected direction of acceleration.
4. Test the package under actual or simulated conditions.

Note from (1) above, the problem of electronic packaging starts with the circuit designer. It is not unusual that he make sacrifices in his design in order to gain ruggedness.

Steps (2) and (4) above constitute the bulk of the package designer's work. When manufacturers were approached with the problem of supplying components to survive 5000 g, the first question asked was always, "How do you test it?" This is a good question, for testing is the key to successful high-g package design. For this reason our testing equipment and techniques will be discussed.

TESTING

A simple method of high-g testing which is easily implemented in the laboratory consists of a mallet and block. A piezoelectric accelerometer is attached to the mallet to obtain a dynamic oscillographic recording of the g amplitude. A practice swing may be taken to "calibrate" the arm before the unit for testing is mounted on the mallet. The block material is selected to give the desired acceleration waveform, amplitude, and duration. In this fashion accelerations of 3000 or 4000 g's for $\frac{1}{2}$ msec may be obtained with small items such as transistors, resistors, etc.

For more controlled testing at higher g levels a tower-drop tester was constructed at the Jet Propulsion Laboratory. Figure 5 illustrates a normal test setup. Basically, it consists of a steel-faced concrete impact block at ground level, a pair of taut vertical guide wires attached to the block, and at the top of a 50-ft support beam, a specimen carriage guided by the wires, a hoist, and a release mechanism. The drop height is continuously variable up to 45 ft, giving impact velocities up to 50 ft/sec. The test acceleration is determined primarily by the combination of the penetrating tool, target material, and total carriage mass. The optical system is for measuring impact velocity and triggering an oscilloscope which records the accelerometer output. A typical waveform is shown in Fig. 6. The area under this acceleration time curve is equal to the velocity at impact. That is

$$\int a\, dt = v \tag{1}$$

for no rebound. This equation shows that a trade-off must occur between pulse amplitude and duration regardless of the means employed to stop the carriage. It also permits a good check on measured accelerations by comparing the area under the a - t curve calculated by Simpson's rule, with the measured impact velocity.

Another useful expression obtained by equating the work done at impact to the change in kinetic energy of the carriage and specimen is

$$\int a\, ds = \frac{1}{2}v^2 \tag{2}$$

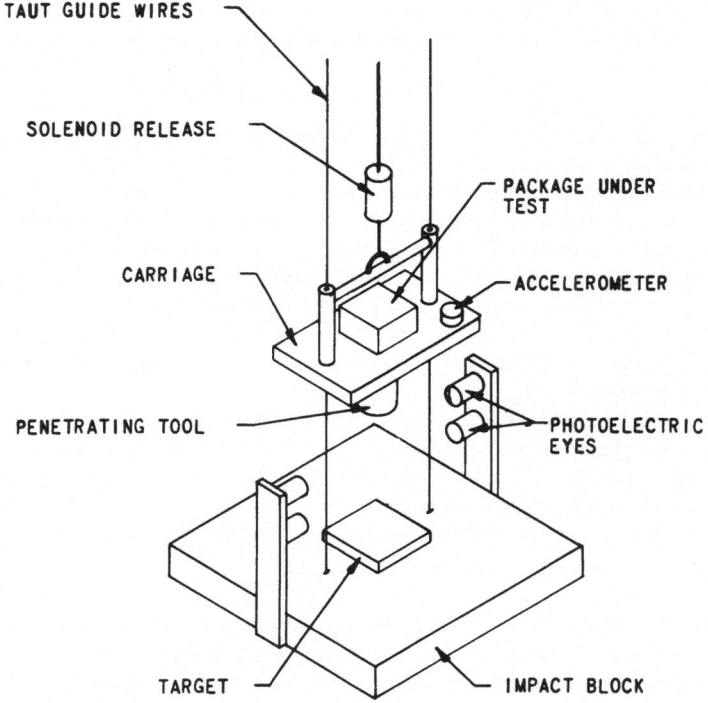

Fig. 5. Tower-drop tester.

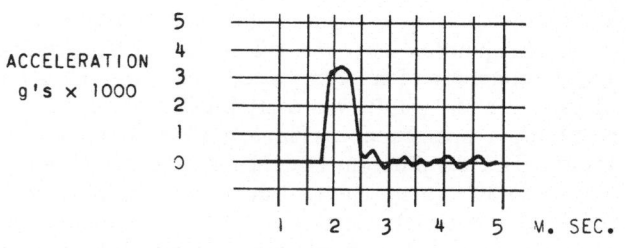

ACCELERATION VS TIME

Fig. 6. Accelerometer waveform for tower-drop test, recorded from
an oscilloscope preceded by a 4-kc low-pass filter.

where s is the carriage displacement during impact. This equation says that no matter what method is employed to stop the carriage there is a relationship between acceleration, displacement, and impact velocity. Various penetrating tools, target materials, and drop heights have been used to obtain accelerations from 800 g to 10,000 g and pulse durations up to 2 msec.

For greater impact velocities a horizontal test facility at the Jet Propulsion Laboratory is also used. It is similar in principle to the vertical unit except that the taut wires are replaced by a pair of 18-ft I-beam guide rails, and a $^3/_4$-in.-diameter elastic bungee cord is used for propulsion. The carriage is designed to ride along the rails and a winch is used to pull it back to a release point. With this "sling-shot" arrangement, impact velocities up to 200 ft/sec and accelerations up to 30,000 g have been obtained.

Another machine for testing at high g levels is the pneumatic cannon, whose name describes its operation. This writer has no actual experience with such equipment; however, it is understood that very high impact velocities with high peak accelerations may be obtained [4].

A final method of testing was developed to give higher impact velocities and to simulate the actual lunar landing conditions. This consisted of putting the units to be tested aboard a bomb-shaped vehicle shown in Fig. 7 and dropping this from a helicopter onto selected surfaces. Although it was not possible to monitor the dynamic acceleration, the peak acceleration could be obtained from a statically calibrated steel-ball-and-lead-disk accelerometer contained within the bomb. Under the simplifying assumption that the impact acceleration is a constant, the average acceleration may be calculated from equation (2). The acceleration in g's is simply the ratio of the drop height to the depth of penetration. An altitude of 2000 ft was set for most of the drop tests, for this gave the estimated maximum landing velocity to be encountered on the moon (fall time of the bomb indicated negligible retardation due to air resistance). This method has resulted in accelerations of approximately 12,000 g peak and 7000 g average with a 5 msec duration.

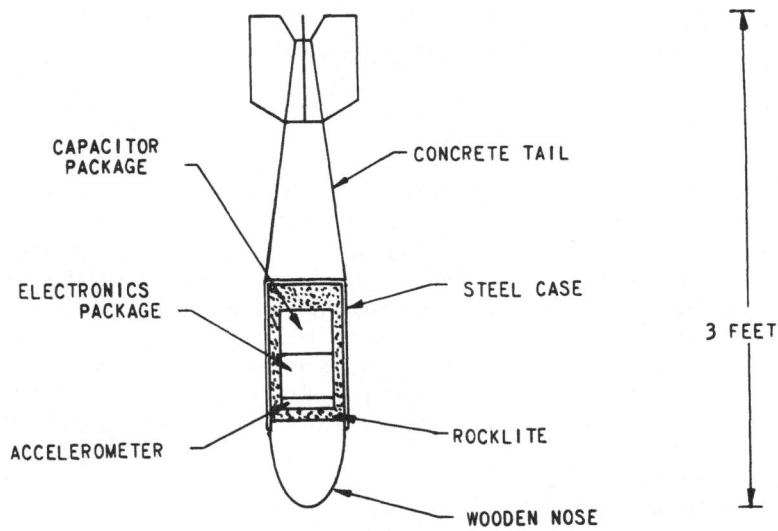

Fig. 7. Drop-test vehicle.

COMPONENT SELECTION

Component evaluation and selection is one of the major steps in high-g design. Considerable testing can be avoided by choosing only those components which are inherently rugged. Figures 8 and 9 show the internal construction of several typical transistors. In Fig. 8 are shown a medium power transistor, a small signal, low noise transistor, and a switching transistor in a T0-18 case. All of the devices shown are unsuited for high-g design. The large wafers and supporting structures and the cantilevered construction make them relatively fragile.

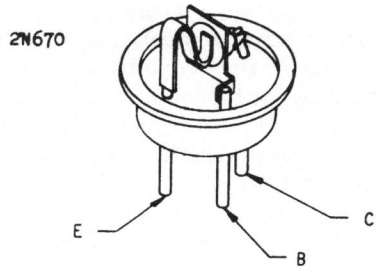

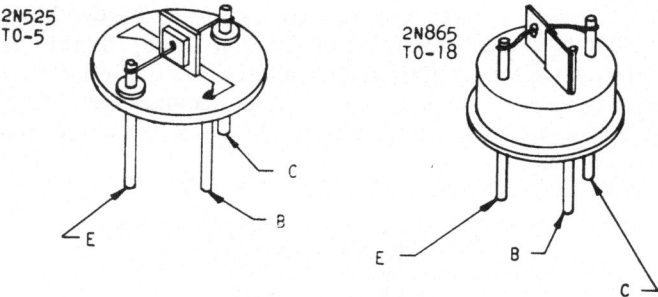

Fig. 8. Transistor internal construction.

A great improvement in ruggedness can be found in the mesa or fixed-bed type of construction shown in Fig. 9. Here the crystal is attached to the bed of the case rather than to any supporting structure. In general, we found the mesa transistors constructed in the TO-18 case to be more rugged than those in the TO-5. The reason appears to be due to the smaller size of the elements, i.e., less mass and, therefore, smaller forces exerted during the shock. This was demonstrated in tests performed on the transistor shown in Fig. 9D which is similar in construction to those shown in Fig. 8 except that the wafer is considerably smaller. Note the size of the wafer relative to the supporting wires which are the same as those in the other transistors shown. This transistor was the first one found to be sufficiently rugged for the seismometer design objectives. It was tested in all orientations at g levels of 5000 and above for a 5 msec duration with no impairment in circuit performance, and some were repeatedly dropped with not one catastrophic failure.

The transistor finally selected for the pulse circuit is shown in Fig. 9C. This device retains the mesa construction as well as the small elements, and has resulted in no failures at g levels above 12,000 peak or 5000 average. It also is of the silicon microenergy type, that is, high beta at low collector current, which is desirable for low-power-drain circuitry.

One item worth mention is that transistors which contain a getter should be avoided. This getter may be resting freely on the wafer or supported by a retaining ring at the top of the case. High-g shock often results in the getter disk disintegrating, causing considerable damage to the internal construction.

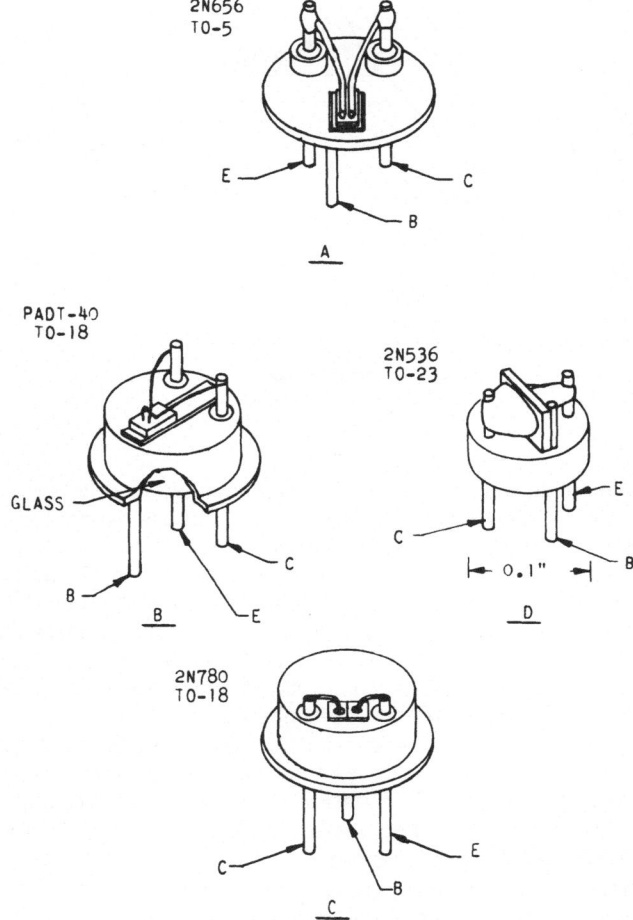

Fig. 9. Transistor internal construction.

Before leaving the discussion of transistor construction, some note should be made of the relatively new microtransistors, although at this writing the testing of these devices is only beginning. Their construction seems to be ideal for withstanding severe physical shock. In essence, the situation here is that the problem of mounting has been brought outside the case and falls upon the package designer. These devices, it is hoped, will be as rugged as a transistor whose metallic case has been removed and the internal structure potted with epoxy resin [4], a procedure which was avoided in the test pulse circuit design. These transistors have another desirable feature in that unlike some attempts at miniaturization the devices can keep up with the advances of transistor development and thus offer the designer the newest transistor in a rugged package.

In Fig. 10 the typical construction of point-contact or gold-bonded diodes is compared to the newer diffused junction microdiode. In the point-contact diode the molybdenum S-shaped whisker is fused to the germanium slice to form the active junction. In the gold-bonded diode a gold-gallium doped wire replaces the whisker and a short pulse of current fuses the gold wire to the germanium. This

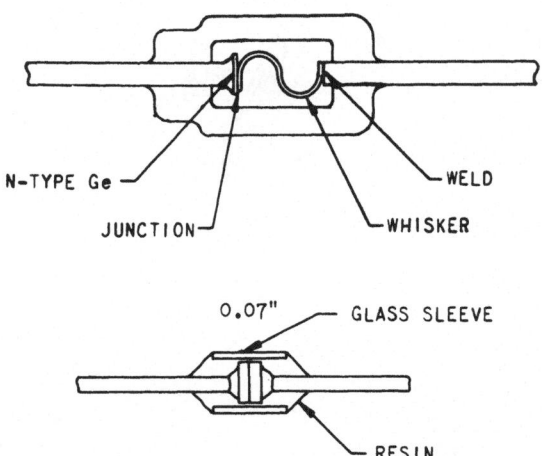

Fig. 10. Diode internal construction.

junction is somewhat more brittle than the corresponding alloy formed in the
point-contact diode. Germanium point-contact diodes were found to survive g
levels of 6000 or 7000. Postshock evaluation of encapsulated electronics is made
very difficult because the polyester potting material is virtually impossible to
dissolve. Some early problems in reliability were attributed to diode failure
although it was never conclusively proven. A circuit was developed which required no diodes and their further testing was abandoned. The desirable features
previously mentioned with respect to the microtransistor also apply to the microdiode, and it should prove to be a rugged device.

Figure 11 shows the typical construction of a sintered-anode tantalum capacitor. The magnified section illustrates the principle of operation, the capacitance
being formed between the tantalum metal and the manganese dioxide with tantalum
pentoxide as the dielectric. The capacitance and especially the dc leakage is
dependent upon the integrity of this film.

As the operation of the test pulse circuit indicates, the greater the period
of the individual multivibrators, the greater will be the period between output

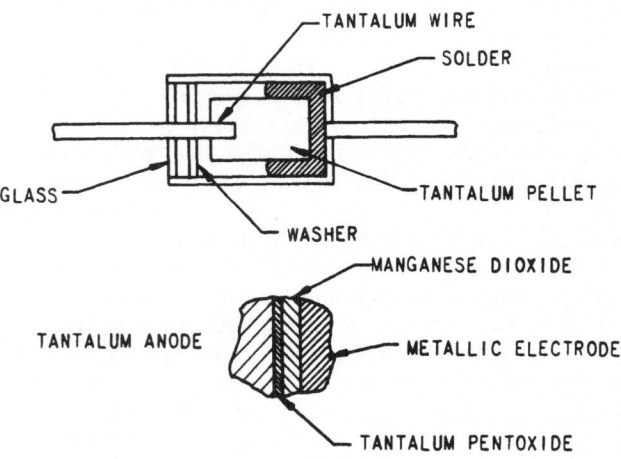

Fig. 11. Sintered-anode tantalum capacitor internal construction.

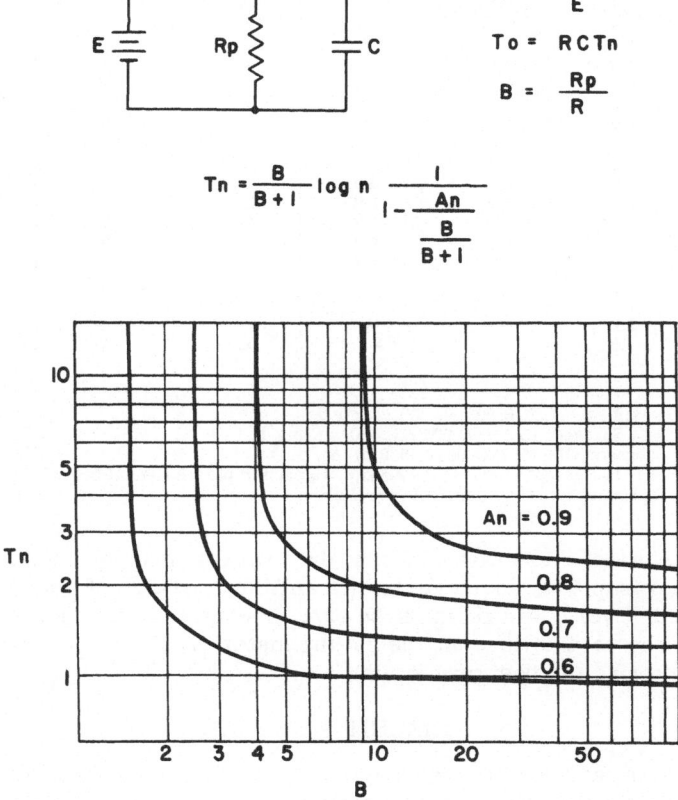

$$An = \frac{Vn}{E}$$

$$To = RCTn$$

$$B = \frac{Rp}{R}$$

$$Tn = \frac{B}{B+1} \log n \frac{1}{1 - \frac{An}{\frac{B}{B+1}}}$$

Fig. 12. Curves relating effective time constant (T_n) to the capacitor leakage resistance (R_p).

pulses. Figure 12 indicates the effective time constant of an RC network as a function of the capacitor leakage. Note that capacitor leakage sets a limit upon the maximum reliable time constant and that increased leakage can lead to an infinite period, i.e., the multivibrator stops operating.

A study of the effects of high acceleration on tantalum-capacitor leakage was in order. A typical leakage specification given by the manufacturer is 0.01 μa/μf-v. This indicated that the maximum time constant obtainable was about 4 sec if no increase in leakage occurred. It was discovered, however, that capacitors could easily be selected with leakages an order of magnitude below that specified. Selected capacitors were potted with polyester resins of different resilience into wells drilled into a rolled epoxy-impregnated fiberglass disk and tested by the methods previously discussed.

Acceleration levels up to 10,000 g peak and 5000 g average for 5 msec were experienced. Circuit requirements specified that the maximum postshock leakage be less than 24 μa. Table I shows capacitor leakage before and after the high-g tests. As is indicated, some of the capacitors were dropped twice. Note that in some cases the leakage was found to have decreased after 48 hr, indicating self-healing of the tantalum-pentoxide film.

Table I. Tantalum Capacitor Leakage Before and After High-g Shocks

Cap. No.	Predrop leakage	6000 g 0.3 msec	7000 g 5 msec	+48 hr	Orientation	Potting material
4	3.0 μa	8.5 μa	4.4 μa	— μa	V	Helix 465 Pot
6	0.5	0.6	100.0	60.0	V	Flex
7	1.6	2.3	13.0	10.0	V	465 Pot
15	6.9	7.9	8.2	6.9	V	465 Pot
18	0.9	0.9	0.9	—	V	Flex
22	1.8	2.3	20.0	9.6	V	465 Pot
26	0.0	0.0	0.0	—	V	Flex
27	0.6	1.0	1.0	1.0	V	465 Pot
28	3.2	—	6.2	4.4	H	Castipast 535
29	3.6	—	4.0	3.5	V	3M-235
30	2.8	—	4.2	8.9	H	3M-235
31	4.4	—	50.0	32	V	Eporost 15E-2pt 9010A1 -1pt
19	1.2	—	—	300	V	Flex
21	1.8	—	—	Open	V	Flex

Identification:
  Nos. 4, 6, 7 — T. I., SCM337HP006A4, 330 μf, 6 v dc
  Nos. 15, 18, 22 — Sprague, 150D337X0006S2, 330 μf, 6 v dc
  Nos. 26, 27 — Sprague, 109D567X0006T0, 560 μf, 6 v dc
  Nos. 28, 29, 30, 31 — Sprague, 150D337X0006S2, Two each in parallel—total 660 μf
  Nos. 19, 21 — Sprague, 150D337X0006S2 in parallel

The results of these tests indicate that, with the proper selection of a potting material for support, an increase in capacitor leakage of only 1 or 2 μ a could be expected. Every failure in these experiments was attributed to a greater or lesser degree to a failure in the potting material, a hard setting but nonbrittle material offering the best protection.

CIRCUIT PACKAGING

Results of acceleration shocks on the order of 2000 g indicated that such items as rf crystals, variable capacitors, coil forms, and other hardware associated with the telemeter transmitter, when properly selected could survive with little or no protection [10]. In fact, a large mass of potting material could actually be a detriment by producing stresses on the potted components.

The type of construction used for the seismometer amplifier and transmitter is shown in Fig. 13. The circuitry was broken into a number of small sections and fitted into a honeycomb structure within the magnesium casting. The smaller components are mounted between terminals on an epoxy fiber board which is supported from the periphery. Larger components are mounted beneath the board and given additional support with a light foam. After final checkout a light film of epoxy resin is applied to the top of the board. This form of construction lends itself to ease of quality control and servicing while still maintaining sufficient ruggedness for the g level anticipated in the lunar landing.

The design of the test pulse package was more difficult. Basically the problem was to design a package which was no more fragile than its individual components.

Figure 14 shows the steps in the evolution of the pulser package. The first approach was the conventional one of mounting the components on a printed board and potting the unit into a block of polyester resin. In this type of construction almost all the support must be supplied by the potting material. There are many potting materials with suitable electrical and physical characteristics for high-g construction. The selection of a material is primarily based upon the

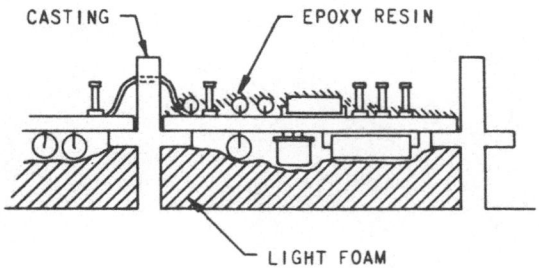

Fig. 13. Seismometer amplifier and transmitter packaging.

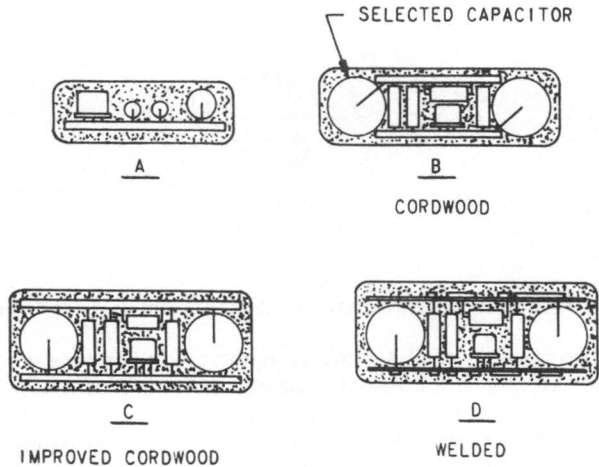

Fig. 14. Evolution of the seismometer pulser packaging.

designer's experience and familiarity with certain compounds as far as ease of handling and quality control. Of course, it must not have a large coefficient of thermal expansion or high shrinkage on cure. In this case the selection of the potting material was subjected to further restraints by the Ranger sterilization requirements in which the unit must be free of all viable microorganisms. Sterilization would consist of a heat soak at 125°C for 36 hr. In addition the resin must be inert when immersed in n-heptane caging fluid contained within the seismometer for protection during impact.

The reaction of potting materials to high-g acceleration is not straightforward. Many materials were found to be too brittle and would crack, damaging components in the plane of the crack, or some would shatter completely. Other materials would allow large, heavy components such as tantalum capacitors to shift position, causing distortion of the case, resulting in increased leakage.

In an attempt to eliminate some potting problems it was decided to make the package more rugged before potting and to use as little potting material as was necessary. This resulted in the cordwood type of construction shown in Fig. 14B. Here two printed circuit boards are used with the components closely stacked between them. The tantalum capacitors had to be selected for each circuit and so they were added externally to the cordwood construction. Shock testing of the unit, however, revealed cracks forming in the resin between the capacitors and

Fig. 15. Photograph of cordwood package before potting (scale in inches).

the boards. In an improved version, provision was made for inserting the selected capacitors between the boards after circuit checkout and before potting.

This type of construction, using components which have survived the design objective g levels, has resulted in no failures at accelerations up to 12,500 g peak and 7000 g average for 5 msec. It would be interesting to see just how high a g level the units would withstand; however, time does not allow us to pursue this effort.

Cordwood construction has some undesirable characteristics as far as quality control. The units are very prone to catching and holding fine bits of dirt, solder, and wire and so must be scrupulosly cleaned before potting. Also, rework of the circuits is virtually impossible, resulting in the necessity for extra meticulous care in assembling and testing.

One final design was attempted. By simply replacing the epoxy board with a piece of Mylar and the printed circuit with a nickel ribbon, the identical component layout was made into a welded construction. The advantages of welded construction have been elaborated on in the literature [8]. Testing of this design is being performed now; however, preliminary results indicate that the weld, when properly made, can withstand very high accelerations. We are optimistic in the hope of improving our assembly techniques.

After assembly, the pulser is tested, its period adjusted and the unit encapsulated. The seismometer case is machined out of rolled epoxy-impregnated fiberglass stock and a well is included in the case into which the pulser unit is potted. At first it was decided to pot the pulser into the case with Silastic silicone rubber so that in the event of a malfunction it could be easily removed. Shock tests indicated that this technique resulted in no sacrifice in ruggedness. Unfortunately, however, Silastic proved to be incompatible with n-heptane, and vibration tests revealed that the Silastic did not restrain the unit sufficiently. Therefore, the Silastic was replaced with firm polyester.

CONCLUSION

An attempt has been made to describe the design approach, test procedures, and results culminating in success for a particular objective. It is hoped that this may contribute to the foundation for further study in this area.

There is certainly room for improvement. The approach so far has been one of trial and error. Improvements in high-g testing and instrumentation will lead to improved high-g ruggedization.

But what lies ahead in the way of uses for high-g packaging? Last year at the First International Electronics Packaging Symposium at this University, Mr. Letarte and Mr. Moir speculated on the packaging of electronics for a hard landing on the moon. This is already a possibility for the near future and plans are presently being formulated for parachute landings on Mars and Venus. In these landings it is even harder to anticipate the acceleration forces that may be encountered. More emphasis upon high-g level ruggedization will be demanded.

ACKNOWLEDGMENTS

The author is indebted to the following persons whose work contributed to this paper: Mr. F. E. Lehner, Mr. E. O. Witt, and Mr. R. D. Gurney of the Seismological Laboratory, who designed the lunar seismometer and the drop-bomb vehicle; Mr. J. O. Lonberg and Mr. D. L. Daigle of the Jet Propulsion Laboratory, who designed the horizontal and vertical shock test facilities; and Mr. S. N. Thanos of the Lamont Geophysical Observatory, who suggested the principle of using asynchronous multivibrators for period multiplication.

REFERENCES

[1] "Design Study for a Lunar Capsule," Aeronutronic Publication No. U-870 (April 15, 1960).
[2] George H. Didinger, "Solid Tantalum Capacitors in Timing Circuits," Electronic Industries (May, 1960).
[3] James O. Lonberg, "Impact Test Facilities," JPL Report (April 10, 1961).
[4] M. Letarte and L. E. Moir, "A High-g Telemetry System for Gun and Rocket Firing," CARDE Technical Memorandum 351/60 (Aug., 1960).
[5] Wayne F. Miller, "Results of the JPL Tower and Sante Fe Dam Drops on Seismometer Electronics," Seismological Laboratory Report (Sept., 1960).
[6] G. M. Rosenberg, J. C. Tomasello, and W. L. Gieseler, "DOFL High-g Telemetry System," DOFL Report No. TR-841 (May 27, 1960).
[7] Willis Dworzak, "High Acceleration Impact and Spin Evaluation of Semiconductor Devices," USASRDL Tech. Report No. 2027 (April 15, 1959).
[8] C. G. Clark, "Potting Embedment, and Encapsulation of Welded Electronic Circuits," STL Report No. TR-60-0000-19354 (Nov., 1960).
[9] Raymond Young, "Packaging of a Telemeter to Withstand Impulse Accelerations of 500,000 g," Electronic Circuit Packaging Symposium (Aug. 16-18, 1961).
[10] F. E. Lehner, E. O. Witt, and R. D. Gurney, "Notes on Joint CIT Seismological Laboratory and JPL Drop Test Experiments at Goldstone, Cal." Seismological Laboratory Report (Jan. 26, 1960).

DISCUSSION

Question: Joseph Ritter, Electronic Modules Corp., Pasadena, California. You said when you first potted the unit you got very bad results and then you later showed the cord wood welded-type packages. Were these packages potted also?

Answer: Yes they were. I am not saying the potting is bad. In fact you have to pot these units to protect against the shock. The picture was of an unpotted unit and the one I have here is potted. Our problems in potting were mainly due to our inexperience. Everything, including phase of the moon, seems to affect the potting material. Basically, the material we settled on was a mixture of our own which we had been using to pot transitor coils, a polyester resin. The welded units are being constructed now by a local company. One reason for farming out the welded construction was so that we could get rid of the potting problem.

Question: Gorden Short, International Telephone and Telegraph. Do you have any experience in your recent adventures in high-g relating to what happens to a transformer under the high-g load? For instance, does the core dig into the windings and cause shorts and that sort of thing, or just what happens?

Answer: I do not know of any transformers that were tested at all, either small signal or power transformers. I don't think there is a transformer in the system.

Question: G. H. Pettibone, General Electric. Can you tell me first how much acceleration the actual components themselves see, and secondly is your entire module supported solidly to your case or does it have an elastic support?

Answer: As far as the acceleration the components themselves see, since the unit is solidly potted and we are discussing a long pulse duration, all the components see the same acceleration over the pulse duration, except possibly internal parts like the internal leads of a transistor. As to using something like this elastic material for isolation, it doesn't help at all. During the shock any soft material becomes very rigid, and, in essence, you are directly coupled to the case. I think there are 60 solder joints in the cord wood construction and in that test over 30 of them broke. The material was soft enough, however, so that we were able to go into the material and resolder all the broken solder points, after which it worked. Although this type of potting does not protect it, mechanically, the components were not affected.

Question: Frank Jarvia, Raytheon Company. Along what axis of this cord wood assembly was your loading applied? was it perpendicular to the components or longitudinal?

Answer: If you remember the photograph, I think it's the same as in the paper; the direction of the acceleration was vertical, that is up the leads as shown in the paper. Actually we tested the unit in all configurations and found no difference. This was not true of batteries. In testing batteries we found that we did have to position them for the shock.

Question: Anton Oswald, Kearfott, Clifton, New Jersey. You mentioned in your cord wood module packaging that you used two different techniques, soldering and welding. In your drop test did you drop each of these? If so, which stood up better?

Answer: Yes, we did drop both. I could find no difference. I think the welding is a better approach from a quality-control standpoint.

Question: (Mr. Oswald). My next question is why you make the statement that welding is better than soldering?

Answer: Quality control. We found that the weld, if properly made, was just as good as the soldered connection.

PACKAGING OF A TELEMETER TO WITHSTAND
IMPULSE ACCELERATIONS OF 500,000 g

Raymond Young

Von Karman Gas Dynamics Facility, ARO, Inc.*

INTRODUCTION

Aerodynamic data can be obtained from small-scale, free-flight models launched from guns in aeroballistic ranges. A large, variable-density range is now being constructed in the von Karman Gas Dynamics Facility (VKF) of the U.S. Air Force's Arnold Engineering Development Center (AEDC). The range will be 1000 ft in length and 10 ft in diameter and is to simulate testing conditions at altitudes to 70 miles and free-flight velocities beyond 20,000 ft/sec.

To obtain pressure data, heat-transfer rates, and other aerodynamic information from the scale models, a radio telemetry system is required. Two stringent packaging requirements must be met to provide a telemetry transmitter for use in aeroballistic ranges. The major requirement is that the packaging of the required electronic circuitry shall survive the extremely high, short duration, launching accelerations. For the range now being constructed it is expected that the launching accelerations to which the telemeters might be subjected will be approximately 10^6 g. The second requirement restricts the physical size to a cylindrical space nominally 1 in. in diameter and 1 in. long. The size limitation is necessary so that high velocities can be attained in the guns presently used. As gun development progresses, the maximum permissible size of the telemetry package might be increased.

A telemeter transmitter consisting of a simple rf oscillator (FM, 150 Mc) incorporating a variable-capacitance pressure transducer has been developed. Standard, miniature, electronic components were selected for use in the rf oscillator. At present, this telemeter meets the size requirements as stated and has successfully transmitted free-flight stagnation-pressure data after being launched at peak accelerations of 125,000 g. Accuracy of the data has varied from 4 to 15%. Another similar telemeter is being developed to survive peak accelerations of 200,000 g. In early testing, this telemeter produced promising results, but errors in transmitted data have been intolerable. The errors encountered were found to be directly related to the packaging of the telemeters. Static and dynamic tests were devised to study packaging techniques with respect to component orientation so that errors in telemetered data could be minimized. With the pressure transducer replaced by an electrically equivalent capacitor, oscillators were constructed to check the results of these two tests.

RANGES AND INSTRUMENTATION

Cold-Gas Gun Range. The cold-gas gun range (Fig. 1) is specially constructed for telemetry development. The gun uses high-pressure cold gas as a propellant and has a short barrel length (approximately 30 in.) so that high initial accelerations can be imposed on a model without necessitating excessive muzzle ve-

*Contract operator of Arnold Engineering Development Center, Air Force Systems Command, U.S. Air Force.

Fig. 1. Cold-gas gun range.

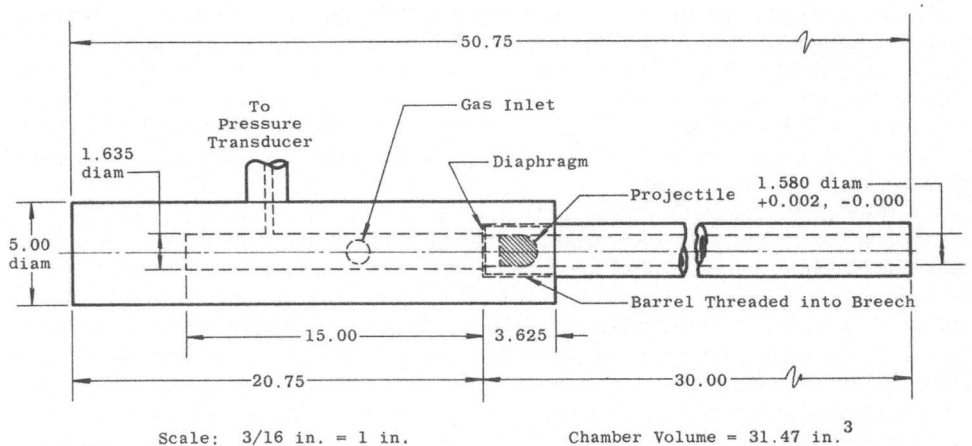

Scale: 3/16 in. = 1 in.   Chamber Volume = 31.47 in.^{3}

Fig. 2. Schematic diagram of cold-gas gun.

locities. Low velocities increase the data-recovery time and permit recovery of the projectile. Figure 2 is a schematic of the cold-gas gun. The propellant gas is introduced into the chamber from a high-pressure reservoir by a solenoid valve. When the pressure in the chamber is sufficient, the diaphragm separating the chamber from the barrel ruptures, and a pressure approximately equal to the chamber pressure is applied at the base of the projectile. From a record of chamber pressure at the time the diaphragm ruptures, the initial acceleration can be calculated by the equation:

$$a_i = \frac{PA}{W} (V_1/V_2)^\gamma g$$

where a_i is the initial acceleration, A is the gun-bore area, and W is the model weight. This expression is an approximation because it assumes adiabatic ex-

Fig. 3. Cold-gas gun range recovery box.

Fig. 4. Range instrumentation.

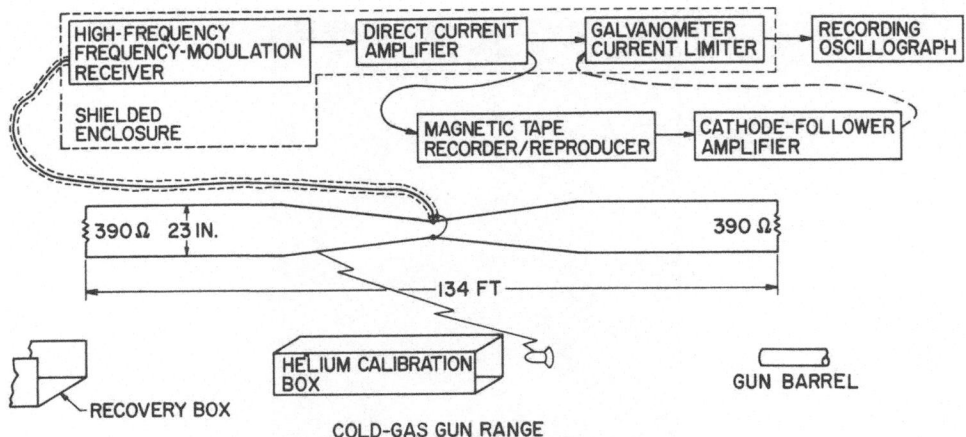

Fig. 5. Telemetered data recording system.

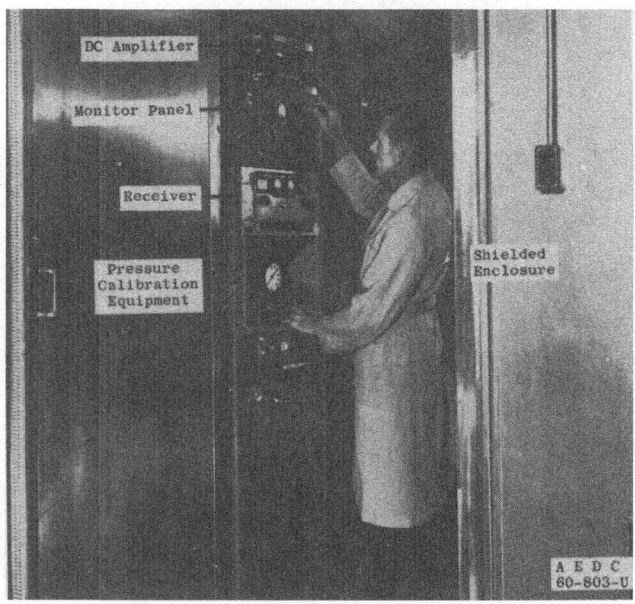

Fig. 6. Telemetered data equipment.

pansion of the propellant gas into the space $(V_2 - V_1)$ between the model and the diaphragm. V_1 is the chamber volume, and V_2 is the volume between the model and the diaphragm plus the chamber volume. The value of specific heat (γ) is taken as 1.4 for air.

At low velocities (below 1500 ft/sec) projectiles can be recovered with little damage in a recovery box at the end of the 135-ft range; thus, some telemetry models can be fired several times. The recovery box, which is 30 in. wide, 36 in. high, and 16 ft long, is filled with wood shavings to minimize damage to the projectile (Fig. 3).

Range Instrumentation and Data Acquisition System. A variety of data were taken during the telemetry test launchings. The instrumentation system used is described in [1] and [2]. Photographic data were available from optical shadow-graphs, Fastax cameras, a Beckman & Whitley high-speed framing camera, and spark X-ray equipment.

Additional range instrumentation was used to record gun chamber pressure, position of the projectile, and the telemetered signal [1]. The gun chamber pressure was sensed by a 60,000-psi, strain-gauge transducer. A 20-kc carrier system was used with the pressure transducer, and the output was recorded by an oscillograph and a magnetic tape recorder. Light screen units detected the projectiles at known positions and recorded time-index pulses on the two recording devices previously mentioned; thus, the average velocity was established between light screen units. Figure 4 is a photograph of the range instrumentation equipment.

Frequency changes of the oscillators were measured by recording the discriminator output of a Nems-Clarke 1510A receiver. An oscillograph and the magnetic tape system simultaneously recorded those data. A block diagram of the entire telemeter data recording system is shown in Fig. 5. A photograph of the equipment within a shielded enclosure is shown in Fig. 6. The readout equipment was located immediately outside this shielded enclosure. With this system and a long transmission-line antenna (Fig. 5), the frequency of the oscillator could be monitored throughout its flight in the cold-gas gun range.

TELEMETRY CIRCUITRY AND COMPONENTS

Two main packaging requirements are placed on the telemetry transmitter: miniature size and structural stability. The guns used in VKF limit the telemeter envelope to a cylindrical shape about 1 in. in diameter by 1.2 in. in length. Standard miniature components were used to design telemeters to meet these requirements with only the transistors being modified. Typical miniature components with an assembled rf oscillator are shown in Fig. 7.

The telemeters now incorporate a variable-capacitance pressure transducer (Fig. 8) in a simple rf oscillator. The stagnation pressure at the nose of the model produced changes in center frequency of the oscillator. The accuracy of measured data becomes a function of frequency stability of the oscillator during launching and flight. An analysis of this effect is found in [3]. An rf oscillator with the pressure transducer replaced by a capacitor which was electrically equivalent to the transducer was constructed to determine the effect of the launching acceleration on the oscillator frequency. Figure 9 is a photograph and schematic diagram of the variable capacitance telemeter, and Fig. 10 is a photograph and schematic diagram of the simple rf oscillator. Several types of construction employing various component orientations are represented. The component values indicated are not necessarily those used in the various types of oscillators.

Telemeters incorporating thin-film, variable-resistance, temperature transducers are also under development. Very little work on this type of telemeter has been done at the AEDC. A temperature-measuring telemeter is being developed at the Canadian Armament Research and Development Establishment (CARDE), and this work is reported in [4] and [5].

All of the components used in telemetry circuitry have been selected because they were capable of surviving high accelerations with only small changes in their quiescent values. A static compression test in conjunction with a dynamic test determined the orientation, with respect to the direction of the acceleration vector, in which the components would function best. These tests are described herein.

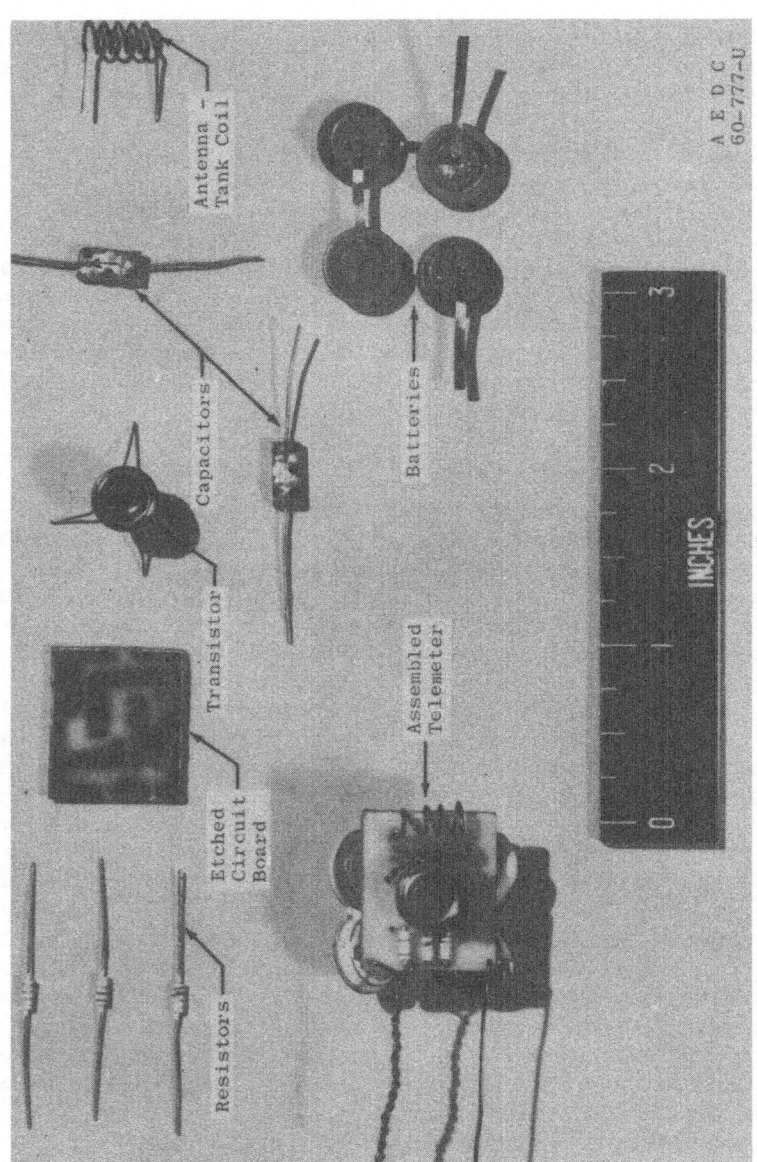

Fig. 7. RF oscillator and components.

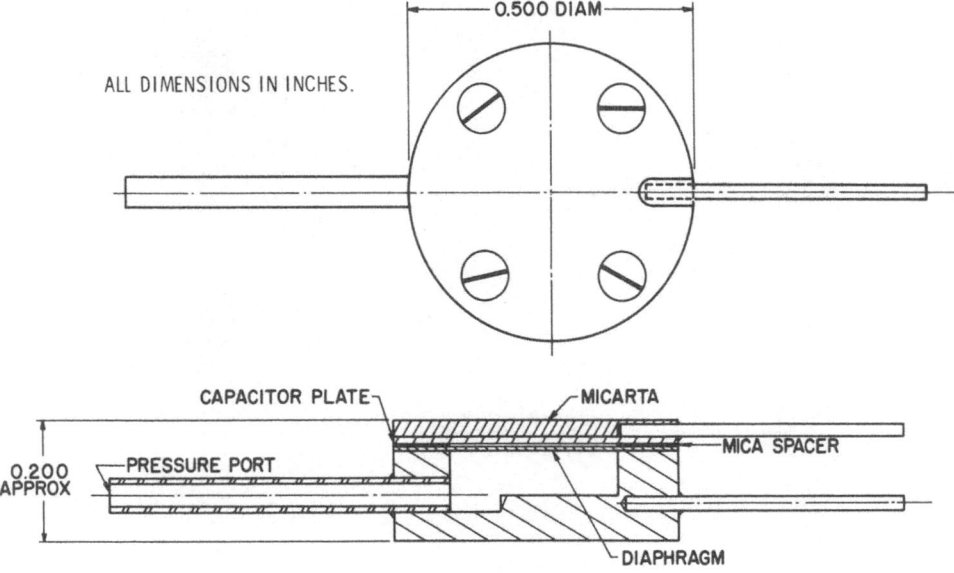

Fig. 8. Capacitance transducer.

Fig. 9. Variable-capacitance pressure telemeters.

Corning Cy-10 glass capacitors with values from 5 $\mu\mu$f to 220 $\mu\mu$f were used in the oscillator circuit. Ohmite Little Devil 0.1-w resistors have also proved satisfactory. The antenna-tank coil consisted of four turns of No. 24 wire with 0.1-in. spacing and an outside diameter of 0.25 in. Four Mallory RM 400R mercury cells connected in series were used as the power supply. Smaller batteries have been tried with little success.

The components mentioned in the preceding paragraph were used as received from the manufacturer, but the use of unmodified transistors has proved un-

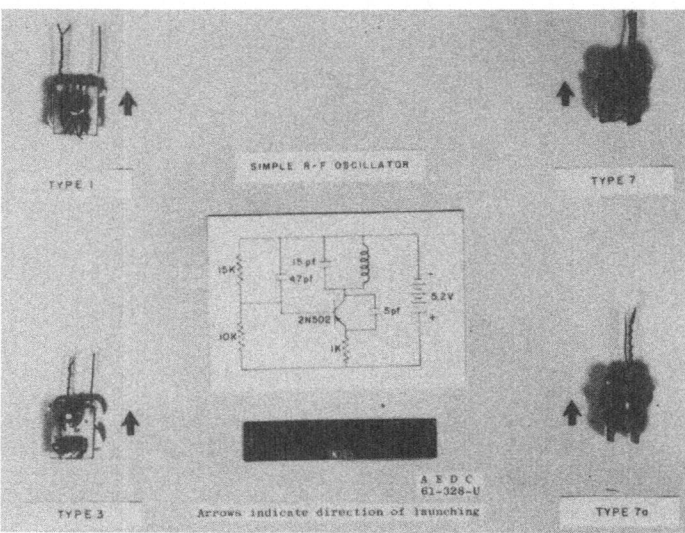

Fig. 10. RF oscillator-construction types.

satisfactory. The transistors were modified by removing the tops of the cases and filling them with an epoxy resin. For some of the epoxies used, a protective coating was required prior to potting. Philco 2N502 transistors, potted with Armstrong C-7, have been used predominantly because they are inexpensive and give good results. Motorola 2N700 transistors have also been used in the oscillator circuit with good results. Unpotted Texas Instruments T1451 transistors have survived launchings at peak accelerations up to 200,000 g without modification. The cost of this transistor, however, restricts its use.

COMPONENT TESTS

Static-Compression Test. It was found that the center frequency of the rf oscillators changed as a result of the acceleration experienced in gun launching. Static and dynamic tests were devised to determine the cause of this shift in frequency. The purpose of the static-compression test was to determine the effect on the different components of removing a compressive load, and to establish the effect of mounting orientation with respect to the direction of the force vector. A means of mechanically loading electronic components (similar but not identical to the loading experienced during launching) was the application of a static, compressive load with a hydraulic press. The load was transmitted to the components by an encapsulating material. Normally, two test components were mounted in a cylindrical slug of epoxy resin having a cross-sectional area of 1 in.² One component was connected to an unstrained rf oscillator, and the other was connected to an appropriate measuring device to monitor its electrical value. The apparatus used for this test is shown in Fig. 11. The center frequency of the rf oscillator was adjusted, and the electrical value of the second component was recorded. A stress of 1000 psi was placed on the slug, and the deviations of frequency and component value were recorded. The stress was then removed and the same reading again taken. This procedure was followed beyond a 1000-psi stress level for incremental steps of 500 psi until the specimen failed.

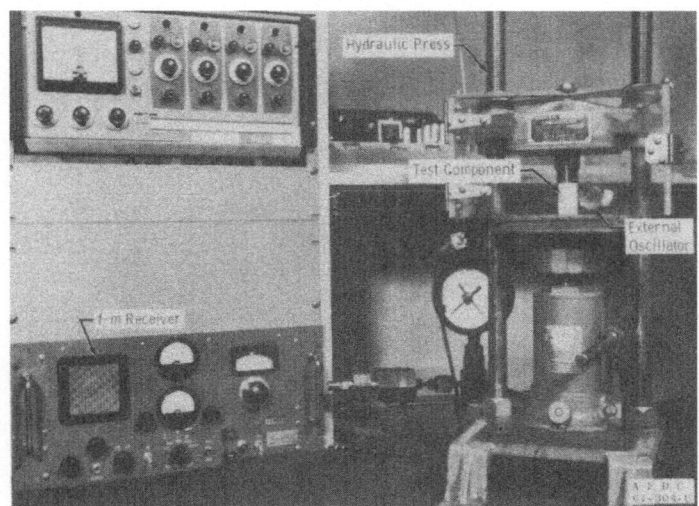

Fig. 11. Static-test apparatus.

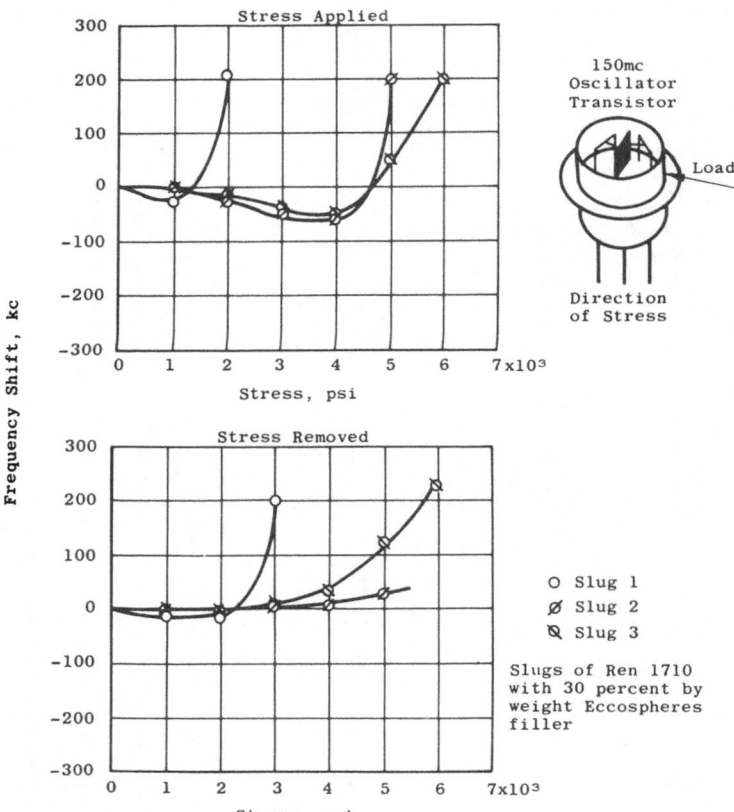

Fig. 12a. Frequency shift vs. stress, 2N502 transistor. Plane 1.

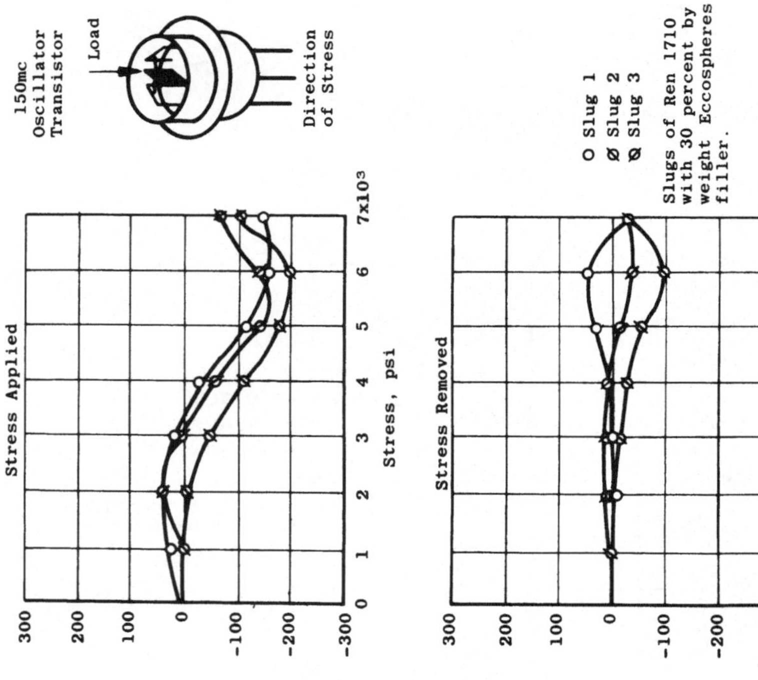

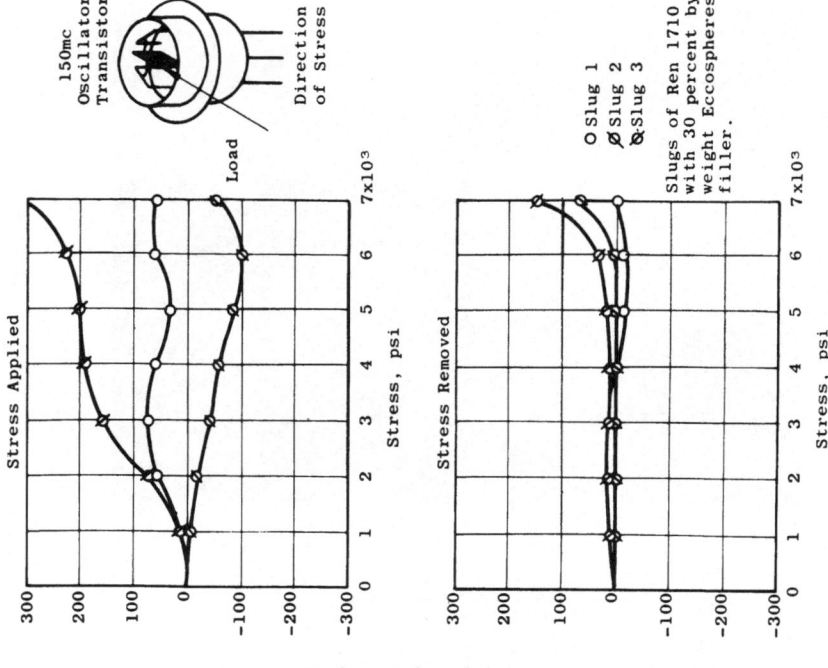

Fig. 12c. Frequency shift vs. stress, 2N502 transistor. Plane 3.

Fig. 12b. Frequency shift vs. stress, 2N502 transistor. Plane 2.

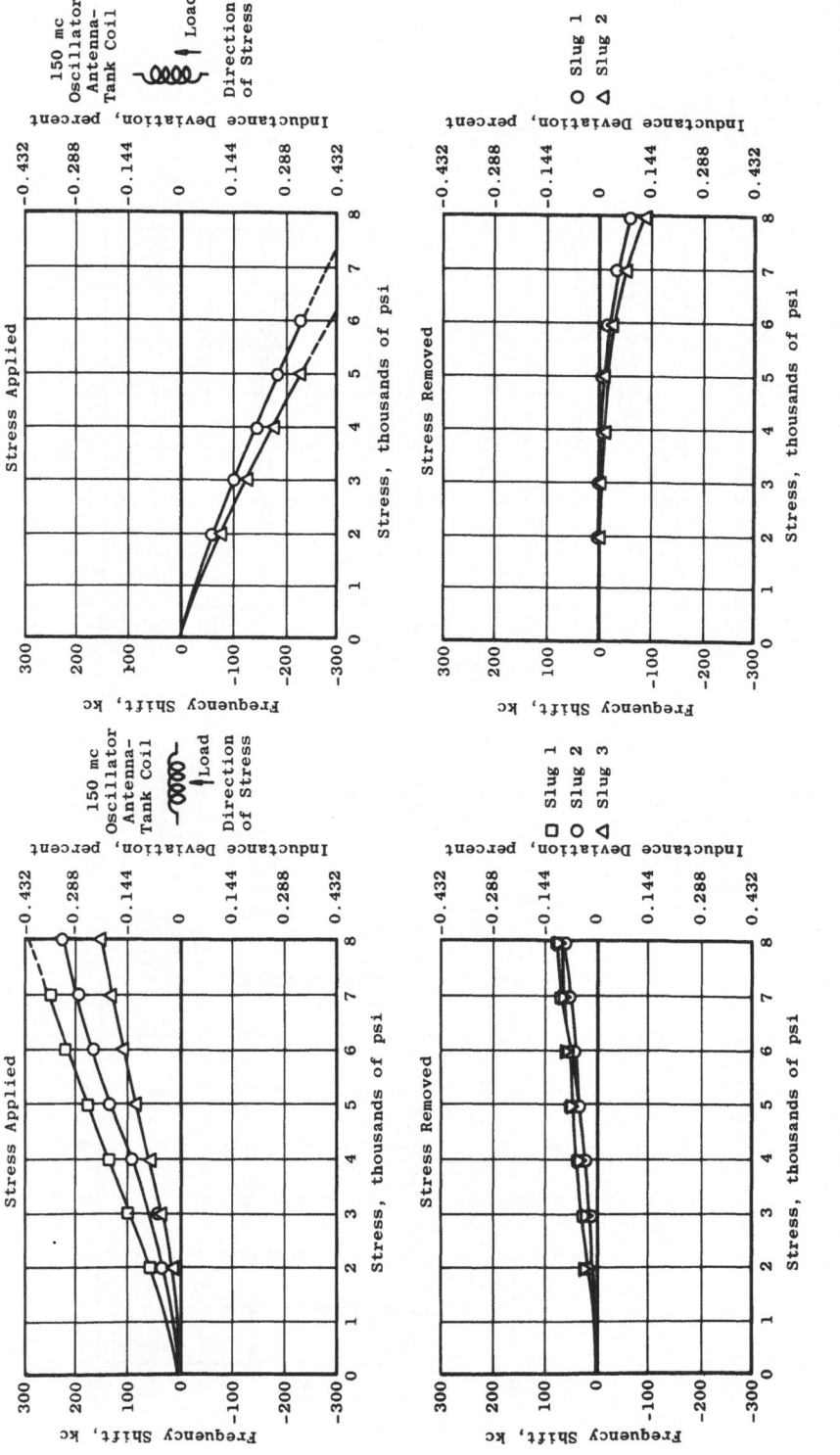

Fig. 13. Frequency shift vs stress, tank coil, a) Plane 1; b) plane 2.

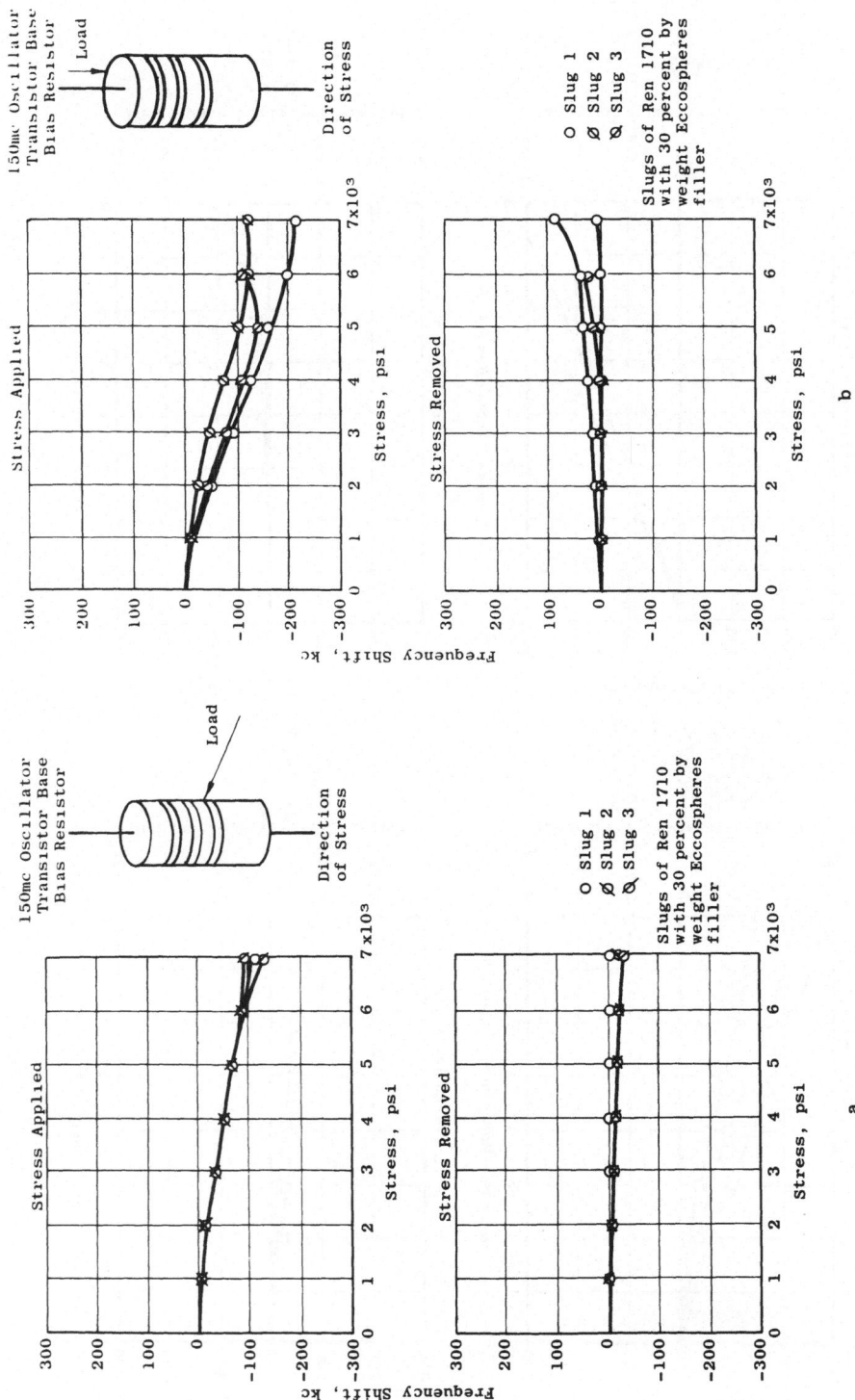

Fig. 14. Frequency shift vs. stress, 15-K resistors. a) Plane 1; b) plane 2.

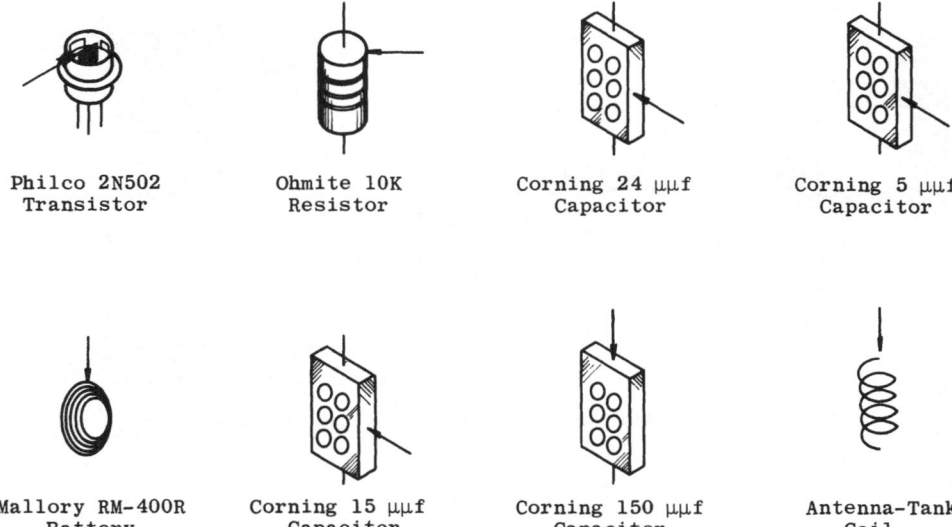

Philco 2N502   Ohmite 10K   Corning 24 μμf   Corning 5 μμf
Transistor   Resistor   Capacitor   Capacitor

Mallory RM-400R   Corning 15 μμf   Corning 150 μμf   Antenna-Tank
Battery   Capacitor   Capacitor   Coil

Fig. 15. Most favorable component mounting orientation as determined by static-compression tests. (Arrow indicates most favorable orientation with respect to acceleration vector.)

Static-compression tests of nearly all of the components used in the rf oscillator have been conducted. Typical data from the static tests are represented in Figs. 12, 13, and 14. Figure 15 is a summary of the results of the static tests and represents the proper mounting orientation for each component used in the oscillator circuit.

The results of this investigation indicated that most of the electronic components, if subjected to high static stresses, would cause the operating frequency of the oscillator to change. In most cases, however, if the component were mounted in its most favorable orientation, little frequency shift would be retained when the applied stress was removed. An exception to this was the Philco 2N502 Transistor. Proper orientation of this transistor will minimize the frequency shift retained after load removal but will not eliminate the shift completely.

The static-compression tests were not intended to simulate completely the actual mechanical loading on components launched from hypervelocity guns. The yield and ultimate strengths of the materials of which the components are composed may be much higher when compressed at very high strain rates. Also, the internal parts of the components are subjected to equally distributed loads in the static tests, but during launching inertial forces cause unequal stress distributions. This unequal stress distribution cannot be simulated by static-compression tests. The static-compression tests served to provide comparisons of component orientations and transistor potting materials, and helped to predict the behavior of launched telemeters.

Dynamic Tests. The dynamic tests were used to correlate the results of the static tests. They consisted of launching individual components or complete oscillators from the cold-gas gun.

The first test consisted of potting several components in a one-caliber Scotchply cup (Fig. 16) in different orientations with respect to the acceleration vector. Resistors incorporated in the rf oscillator were the only components evaluated in this manner. Static tests indicated that the small changes in the electrical

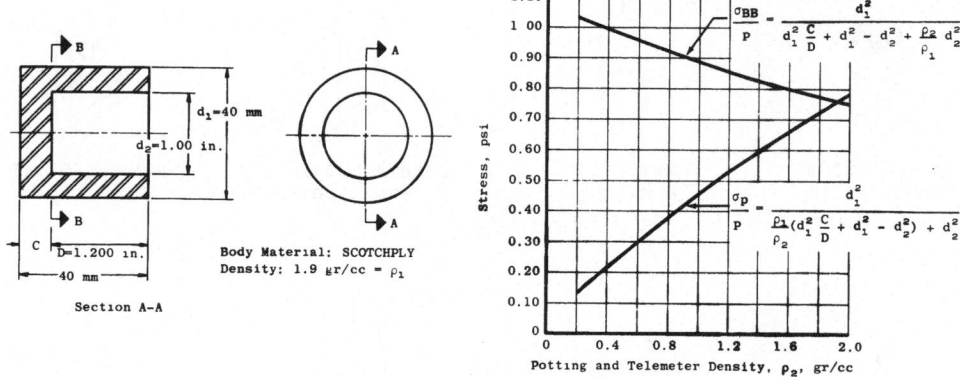

Fig. 16. Body and potting stress curves, one-caliber Scotchply body shell.

TABLE I

a. OHMITE LITTLE DEVIL RESISTORS				
ACCELERATION, g	RESISTANCE* ohms		PERCENT CHANGE	REMARKS
	BEFORE	AFTER		
Acceleration Vector Parallel to Leads				
256, 000	1701	1702	0.05	Normal Shot
	14, 900	14, 900	0	
378, 000**	1701	1698	0.18	Hit Range Hardware (Light Screen); Some of Potting Lost from Cup; Deceleration Unknown
	14, 900	14, 900	0	
Acceleration Vector Perpendicular to Leads				
297, 000	1708	1708	0	Normal Shot
	1754	1754	0	
429, 000**	1705	1706	0.059	Gun Malfunction
	1750	1751	0.057	

b. GLASS ENCASED RESISTORS				
ACCELERATION, g	RESISTOR	RESISTANCE* ohms	PERCENT CHANGE	
		BEFORE	AFTER	
Acceleration Vector Parallel to Leads				
256, 000	1	1472	1473	0.07	Normal Shot
	2	13810	13970	1.16	
378, 000**	1	1473	Open	100	Hit Range Hardware and Removed Some of the Potting Material; Deceleration Not Known
	2	13930	14830	7.2	
Acceleration Vector Perpendicular to Leads				
297, 000	1	1473	1473	0	Normal Shot
	2	1473	1473	0	
429, 000**	1	1473	1474	0.07	Gun Malfunction
	2	1473	1474	0.07	

*A Wheatstone Bridge (ratio tolerance of ±0.05%) was used for measuring resistance.

**The acceleration here may be slightly higher than that actually attained. The gun was precharged, and a leak caused the projectile to move down the barrel before the final pressure was reached.

values of the inductor and capacitors sufficient to change the frequency of the oscillator would be negligible as compared to the stray capacitance deviations encountered in connecting the test component to an rf oscillator before and after launching. The transistor characteristics (β and α) which could be measured did not appear to be functionally related to the operating frequency of the oscillator; thus, transistors were not tested in this dynamic test.

The results obtained from the resistor-dynamic tests were as predicted by static testing. No changes in resistances of the Ohmite Little Devil resistors could be detected regardless of their orientation after launchings at 250,000 g (Table Ia). These resistors, properly oriented as determined by static-test results, have survived launchings at peak accelerations of 300,000 g without changing resistances. Tests of glass-encased resistors, not as well suited for this application, have shown that the orientation data obtained by static tests correlate well with dynamic-test data (Table Ib).

Resistor orientation as determined by static tests was considered to be of greater value than the results obtained in the dynamic tests because it is known that, given enough time, the resistors will return to their prestressed values. In the dynamic tests on the resistors, data could not be taken until the projectile was recovered from the catcher box, and in this time the slight changes in resistances could have disappeared.

The second set of dynamic tests consisted of launching complete rf oscillators which differed from one another only in the orientation of their components. These dynamic tests were paralleled by the static tests; consequently, many different component orientations were studied. Several construction types are represented in Fig. 10. The tests were also used to evaluate different types of transistors, potting materials, and protective coatings for transistor junctions.

INTERNAL TRANSISTOR AND CIRCUITRY ENCAPSULATION

Internal Transistor Encapsulation. Experience has shown that commercially available transistors will not survive accelerations much higher than 200,000 g without modification. The technique in this test consisted, generally, of removing the top of the transistor case and filling the remaining case with an epoxy resin. When using some epoxy-hardener systems, a protective coating was required on the transistor junction before encapsulation. The general requirements of the material used for encapsulating the junction of a transistor are as follows:

1. Low viscosity
2. Low temperature cure
3. High compressive yield strength
4. High compressive modulus
5. Purely elastic action
6. Low density
7. No detrimental effects on the transistor-performance characteristics

Low viscosity was a requirement because of the pouring problem. The case diameter of the transistors used in high-g telemetry is generally less than 0.3 in., and high-viscosity fluids will not flow into this small an opening. The air bubbles entrapped in the pouring process were more easily removed from low-viscosity fluids. Epoxies with a viscosity below 15,000 cp at 25°C will meet the viscosity requirement; however, the higher-viscosity resins must be heated before pouring.

There was generally a maximum temperature to which a transistor could be subjected and still operate after cooling to room temperature. For the Philco 2N502 transistor, this temperature was considered to be 150°F. Further testing has indicated that this temperature could be increased to 200°F. However, when the transistor was subjected to a temperature over 200°F, the emitter lead became disconnected at the junction and became inoperable. Since there was a maximum temperature to which this transistor could be subjected, the cure temperature of the potting material was limited. In general, the exothermic heat produced when curing the epoxy resins could be neglected because of the small quantities used.

High compressive strength and modulus are required to hold the transistor leads in place during gun launchings. Any relative motion between the leads and base wafer can cause disconnection of the leads. A high compressive modulus was therefore needed because the mass-to-support area ratio for the leads differs from that of the base wafer. The low modulus of vulcanizing silicon rubber at room temperature was the cause of failure at relatively low accelerations of telemeters which contained transistors potted in this material. A high compressive yield strength* is necessary, as shown by static tests, because any permanent strain remaining in the potting material will appear as a pressure on the germanium wafer and will change the transistor characteristics.

The material must also be capable of purely elastic action. Elastic action occurs if the strain which accompanies a stress vanishes upon removal of the stress (i.e., strain is not a function of time after load removal).

Epoxy resins with amine curing agents have fulfilled most of the requirements stated in the preceding paragraphs, and have been used extensively. Until recently, Armstrong epoxy C-7 and 8 phr activator A (triethylene tetramine— TETA) have been employed for potting transistor junctions. Because this activator has a detrimental effect on the transistor performance, a protective coating of Broma Plastic Preservative has been used. Another coating, Sealbrite, has protected the transistor from the potting material but has been undesirable because oscillators using transistors coated with this material have had large center-frequency shifts after launching.

Because little data describing mechanical properties of coating materials were available, and difficulties were encountered in obtaining consistent coating thicknesses, materials which did not harmfully affect the transistor characteristics were sought. Several epoxy systems were found which could be used without the need for a protective transistor coating. Three of these materials are Armstrong C-7 with 8 phr Activator E, Epon 828 with 7 phr dimethylamino propylamine (DIMAPA), and Exp-35, an experimental epoxy formulated by Hermitage Plastics, Inc.

Circuitry Potting Material and Body Shell. The general requirements of the circuitry potting material very closely resemble those of the internal transistor encapsulation described previously. The most important of these properties are high compressive modulus and yield strength, low density, purely elastic action, and low-temperature cure. Since the mass of the material was relatively high, care had to be taken in selecting a system which did not produce a high temperature as a result of exothermic heating.

A single laminating epoxy resin has been used for the potting material because of the necessity for comparing results during the dynamic testing. Evaluating the different model constructions would have been very difficult had the potting materials been variable. The compressive yield strength of Ren-1710 when mixed with Eccospheres† is relatively low as compared to other epoxy resin

---

*The compressive yield strength is defined as the stress at which plastic deformation begins.
†Emerson and Cuming trade name for hollow glass spheres, 30 to 300 μ in diameter having a density of 0.4 g/cc.

systems of comparable densities; however, it appears high enough for launchings to 300,000 g.

A low-density epoxy resin, Stycast 1090* gave very promising results when tested at very low strain rates. Test launchings with rf oscillators, however, always gave poor results. Generally, the center-frequency shifts were excessively large and varied as the projectile traveled down the gun range. The reason for the discrepancies in the data was found to be in the epoxy—catalyst ratio. The manufacturer's suggested ratio was always used. The amount cast for static testing was large when compared to the amount used for circuitry potting, and the gel temperature was much higher. In small quantities the potting would not cure properly because the gel temperature was not sufficiently high. As this epoxy system used a catalyst hardener, it was possible to greatly increase the percentage of hardener without appreciably changing the cured-system properties. Sixty-six percent excess of hardener was used to increase the exothermic heat and gel temperature. Only one rf oscillator potted in this material has been launched using excess hardener. The results of the launching were very promising, and more testing of this material is in progress.

A one-caliber body shell was designed to minimize the compressive load imposed upon the components by the potting material [1]. This was done by making the ratio of the weight of the body to that of the telemeter package within it as large as possible for a given package weight. The effect of varying this ratio by decreasing the density of the internal package is shown in Fig. 16. Launching the cup with the open end forward provided another means of lessening the load on the components. The advantage of designing the cup to carry the major loading caused by acceleration was that the cup could be fabricated from much stronger, high-temperature cured, epoxy systems. Scotchply 1002, unidirectional, single-ply tape, manufactured by Minnesota Mining and Manufacturing Co., was used for the fabrication of this cup. The strength-to-weight ratio of this material in compression is 5.7 to 1.

Cups fabricated from Scotchply tape were very elastic and did not retain a permanent strain after load removal. They also restricted the radial strain of the internal potting material. These were proved to be necessary properties of the material used for the construction of the cup because other materials not having these properties gave results similar to those obtained when the uncured Stycast 1090 was used [1].

TEST RESULTS

The results of the static and dynamic tests are summarized in Table II. At first glance, it appeared that the type-1 rf oscillators (Fig. 10) had performed better than any other type tested. There were two reasons for this: (1) the transistors used in these oscillators were given extremely long cure times; and (2) the most sensitive component in the circuit, the transistor, was oriented properly. A close examination of the static-test results revealed the possibility of frequency shifts caused by one component cancelling those caused by other components.

Type-3 oscillators were similar to the type-1 oscillators, but the transistor was given a different orientation (Fig. 10). The resulting frequency shifts of launched models were always greater than could be tolerated. This correlated with the data concerning transistor orientation which were obtained from the

*Emerson and Cuming trade name for low-weight epoxy casting material containing eccospheres.

TABLE II

TYPE CONSTRUCTION	TELEMETER NUMBER	ACCELERATION, g	FREQUENCY SHIFT, kc	TYPE TRANSISTOR	PROTECTIVE COATING	POTTING MATERIAL
1	111	224,000	Negligible	2N502	Broma	C-7 A
1	114	216,000	Negligible	2N502	Broma	C-7 A
1	121	309,000	-7	2N502	Broma	C-7 A
1	192	213,000	+8	2N502	Broma	C-7 A
3	199	214,000	-43	2N502	Broma	C-7 A
3	200 No. 1*	199,000	+185	2N502	Broma	C-7 A
	No. 2*	208,000	-(35 - 113)			
7	201 No. 1*	209,000	-12	2N502	Broma	C-7 A
	No. 2*		-19			
7	202 No. 1*	198,000	-85	2N502	Broma	C-7 A
	No. 2*	194,000	-112			
	No. 3*	261,000	Off Scale			
7	208 No. 1*	215,000	-48	2N502	Broma	C-7 A
	No. 2*		+8			
7	209 No. 1*	126,000	+4	TI-451	None	None
	No. 2*	172,000	+8			
	No. 3*	232,000	+5			
7	225 No. 1*	201,000	+22	TI-451	None	None
	No. 2*	200,000	+30	TI-451	None	None
	No. 3*	306,000	+75	TI-451	None	None
7	226 No. 1*	207,000	+4	TI-451	None	None
	No. 2*	333,000	-80	TI-451	None	None
7a	203	208,000	27	2N502	Sealbrite	C-7 A
7a	204 No. 1*	210,000	-28	2N502	Sealbrite	C-7 A
	No. 2*	238,000	+17			
7a	205 No. 1*	226,000	+18	2N502	Broma	C-7 A
	No. 2*	235,000	-(70 - 140)			
7a	207 No. 1*	190,000	+10	2N502	Broma	C-7 A
	No. 2*	211,000	-43			
7a	211 No. 1*	217,000	+13	2N700	Sealbrite	C-7 A
	No. 2*	196,000	+50			
7a	212	210,000	-165	2N700	Sealbrite	C-7 A
7a	213	211,000	+27	2N502	Broma	C-7 A
7a	214	201,000	-50	2N502	Broma	C-7 A
7a	215	197,000	+32	2N502	Broma	C-7 A

*Indicates multiple firings of single telemeter package.

static testing. From the static tests, it was indicated that models of this type should operate within the band pass of the receivers after being launched at very high accelerations. Models of this type, potted in partially cured Stycast 1090, have survived successive launchings of over 400,000 g. The only oscillator launched at an acceleration exceeding 500,000 g was of this type. The frequency shift during flight was approximately 100 kc. The results of tests using rf oscillators potted in Stycast 1090 are not listed in Table II because the epoxy was only partially cured, and the data were considered inconclusive.

The type-7 oscillator (Fig. 10) was the result of successive modifications of the type-3 oscillator. The transistor and coil were given the same orientation as in the type-3 model, but the other components were all properly oriented with respect to the results obtained from static tests. Also, the more sensitive components, i.e., transistor and inductor, were located near the front of the projectile in an attempt to reduce any compressive loading attributable to the potting material. The frequency shifts during launchings were generally much less than those obtained with the type-3 oscillators but still were not within required tolerances. Results obtained from launching type-3 and type-7 models indicated, however, that the premise of cancelling effects stated about the type-1 oscillators was plausible.

Unpotted TI-451 transistors were also tested in models of the type-7 configuration. Only three have been launched to date. Two of the three performed

TABLE II (continued)

TYPE CONSTRUCTION	TELEMETER NUMBER	ACCELERATION, g	FREQUENCY SHIFT, kc	TYPE TRANSISTOR	PROTECTIVE COATING	POTTING MATERIAL
7a	216 No. 1*	235,000	0	2N700	Sealbrite	C-7 A
	No. 2*	255,000	-152			
7a	217 No. 1*	200,000	+16	2N700	Sealbrite	C-7 A
	No. 2*	210,000	TM did not function			
7a	218 No. 1*	198,000	+58	2N700	Sealbrite	C-7 A
	No. 2*	226,000	0			
7a	219	213,000	+82	2N502	Sealbrite	C-7 A
7a	220	209,000	-88	2N502	Sealbrite	C-7 A
7a	221	206,000	-16	2N502	Sealbrite	C-7 A
7a	229	160,000	-70 to -58	2N502	Broma	E-4640
7a	230 No. 1*	141,000	-18 to -12	2N502	Broma	E-4640***
	No. 2*	216,000	+46 to +115	2N502	Broma	E-4640
7a	233	204,000	-240	2N502	Broma	E-4640
7a	234	193,000	Did not function	2N502	Broma	E-4640
7a**	241 No. 1*	170,000	-7	2N502	Broma	C-7 A
	No. 2*	275,000	+385	2N502	Broma	C-7 A
7b**	242 No. 1*	230,000	Negligible	2N502	Broma	C-7 A
	No. 2*	223,000	+65	2N502	Broma	C-7 A
7b**	243	222,000	-46	2N502	Broma	C-7 A
7b**	244 No. 1*	223,000	+28	2N502	Broma	C-7 A
	No. 2*	225,000	-100	2N502	Broma	C-7 A
7b**	245 No. 1*	231,000	-15	2N502	Broma	C-7 A
	No. 2*	223,000	Did not function	2N502	Broma	C-7 A
7b**	246	260,000	+20	2N502	Broma	C-7 A
7b**	259	188,000	-77	2N502	None	C-7 E
7b**	260 No. 1*	184,000	-10	2N502	None	C-7 E
	No. 2*	170,000	+70	2N502	None	C-7 E
7b**	261	182,000	-22	2N502	None	C-7 E
7b**	265	187,000	+14	2N502	None	C-7 A
7b**	266	177,000	Negligible	2N502	None	C-7 A

*Indicates multiple firings of single telemeter package.
**Transistors cured for at least two months.
***Emerson and Cuming Eccosil 4640, an RTV silicon low density potting material.

satisfactorily after launchings near 200,000 g , but attempts at 300,000 g produced intolerable frequency shifts. This transistor operated well when unpotted and will probably operate better when potted. (Models with potted TI-451 transistors are being constructed for testing at accelerations above 500,000 g.)

The next variable considered was the orientation of the transistor. Static tests had revealed that, even though transistors would fail at lower acceleration values when mounted as in the type-7a model, the frequency shifts at these accelerations would be much less than if the transistor were oriented as in the type-7 model. The first two models launched using 2N502 transistors, protective-coated with Broma and potted in C-7, functioned well, but the next three gave poor results. The poorer results were attributed to the internal transistor potting. At the time these models were constructed, the rate of transistor consumption dictated a curing period of approximately one week as compared to a normal curing time of two months. The last model built in this configuration incorporated a potted transistor which had been given a two-month cure. The little frequency shift which resulted indicated that the longer cure cycle was necessary. This model configuration was also used to check the results of the static tests with respect to the protective coating used on the transistor junction. Again the dynamic test confirmed the results obtained during the static tests. The Sealbrite coating did not perform nearly as well as the Broma coating; the importance of the proper selection of the coating material is indicated. Motorola 2N700 transistors were tried, but since their junctions were coated with Sealbrite, the data were not considered to be representative of the performance of this transistor.

In static testing of the coil, it was found that some improvement should result if the coil were reoriented in the type-7a model. This new construction type was designated type-7b. Results to date have revealed no improvement with respect to frequency shift, but as the 7b was much easier to fit into the Scotchply cup, it was considered an improvement over the 7a model. Coil orientation changes did not improve the operation of the telemeters because the coil was located very near the front of the projectile and received only minor loading at the accelerations tested. At high accelerations, the effect of coil orientation will probably become apparent.

Some of the erratic results from dynamic tests were believed to be caused by variations among the coatings applied to the transistor junction. Transistors potted in Armstrong's C-7 with 8 phr activator E were not adversely affected in their electrical characteristics. However, launched models using these transistors had excessive frequency shifts. A spectrographic analysis revealed that the activator used had deteriorated because of aging and would seriously affect the quality of the cure. Large samples of this material were poured, and it was found that the strength of the cured systems was extremely low. Two other models using Armstrong C-7 with 8 phr activator A were launched, and the results were very promising. However, only a fraction of transistors potted with this material can be used because of detrimental effects of the potting materials on transistor characteristics. Models using uncoated transistors and other epoxy systems which do not affect the transistor characteristics are being built, but none have been launched to date.

CONCLUSIONS

A simple rf oscillator for use in conjunction with a variable-capacitance pressure transducer has been evaluated with regard to frequency instability caused by gun launching acceleration. It was determined that frequency stability was directly related to the method of packaging this oscillator. Because each potted 2N502 transistor responds differently to loading, a frequency shift of ± 20 kc cannot presently be eliminated from 150 Mc oscillators launched at $200,000\,g$. Stronger, more elastic plastics could possibly improve frequency stability of oscillators containing this transistor. More rugged and expensive transistors can be incorporated into the same oscillator circuit to give a decrease in frequency shift.

If a pressure gauge produces a shift in oscillator frequency of one Mc, the error introduced by using the Philco 2N502 transistor can be neglected at low accelerations. It is doubtful that this transistor will furnish accurate results when $500,000\,g$ accelerations are imposed. It will probably then become necessary to use a more rugged transistor having a mesa or planar structure.

REFERENCES

[1] M. K. Kingery, R. H. Choate, and R. P. Young, "Progress Report on Telemetry for a Hypervelocity Range," AEDC-TN-60-214 (Dec., 1960).
[2] P. L. Clemens and M. K. Kingery, "Development of Instrumentation for a Hypervelocity Range," AEDC-TN-60-230 (Dec., 1960).
[3] M. K. Kingery and P. L. Clemens, "Progress in the Development of a Radio Telemetry System," paper presented at the National Telemetering Conference, Chicago, Ill. (May, 1961).

[4] M. Letarte, "A Miniaturized Telemetry System for Guns and Rocket Firing Status Progress," CARDE Technical Memorandum 255/59 (July, 1959).

[5] M. Letarte and L. E. Moir, "A High-g Telemetry System for Gun and Rocket Firing," CARDE Technical Memorandum 351/60 (Aug., 1960).

DISCUSSION

Question: H. L. Roberts, EDN. Have you tried the PICO transistors?

Answer: No, the PICO transistors which are commercially available at the present time are switching transistors and have too low a cut-off frequency. The telemeters launched in the pilot range must operate at frequencies above 140 megacycles. This is because we excite the tank as a waveguide and use a quarter-wavelength stub antenna.

Question: Mark Hurowitz, Sylvania. Mr. Miller mentioned that he used mesa transistors with some good effects. Have you tried those at all?

Answer: Yes. The TI-451 (2N850) and the 2N700 are mesa transistors. They have both worked quite well.

Question: (Mr. Hurowitz) Are you encapsulating these?

Answer: The only transistor that has been launched at high accelerations and worked without being internally potted, was the TI-451 (2N850).

Question: Bob Rooney, RCA, Burlington, Mass. If I understand you correctly, you said that if your potting compound was basic, this would affect the transistors, whereas if it were slightly acidic it would have only a slight effect on the transistor characteristics?

Answer: The acidity of the compound was determined with p Hydrion paper and I am not sure how it works. I asked a chemist if it was permissible to use this paper to determine the acidity of organic materials and he said it was. If this is true, the compounds which were slightly acidic did, in almost every case, have little effect on the germanium transistors.

Question: This is the transistor with its case, right?

Answer: This was the transistor with the top of the case removed to allow pouring around the transistor junction.

Question: One other question: On the 200,000 g and the 500,000 g, do you have a time duration to go with this? You said it was on the order of microseconds.

Answer: We cannot measure the time the maximum acceleration is applied to the projectile. The timing lines on the oscillograph trace are millisecond spacing. It is very difficult to determine just how fast the chamber pressure decreases.

Question: Jerry Skaug, Raytheon, Santa Barbara. Did you make any test during your loading of capacitors and resistors to see if bending of the leads in proximity to the body had any effect in the change in either capacitance or resistance?

Answer: We have never tried a test of this kind. We kept the leads parallel and in the same plane so that when the slugs were loaded, the leads would not move with respect to each other.

Question: (Mr. Skaug) Do you have any feeling for whether or not this would cause a change in the capacitance or resistance?

Answer: I don't believe that bending the leads would cause a change in the value of the components because of the way the leads are attached. If the leads of the capacitors were moved relative to each other, it would definitely have an effect because of the low values of capacitance.

Question: Wayne Miller, Cal Tech. Do you purchase your transistors with the cases removed? If not, how do you remove them without damaging them?

Answer: We use a Sears Roebuck lathe which we purchased for $300 new. The 2N500 transistors are chucked in a three-jaw chuck and the tops are removed with a pointed lathe tool. The 2N700 transistors pose a little problem because of the lip on the back of the case. A special collet was made for this transistor case. It is necessary to fill the transistors directly after removing the top. The humidity in the air seems to have more effect on the transistors than the potting material.

Question: Bill Hayes, Martin Company. You say the potting materials change the characteristics of the transistors for about three days after potting. Couldn't you possibly use a material which has a curing time of three days?

Answer: We use a potting material that is supposed to cure in four hours in small amounts, but I don't believe it does. We have found that if a projectile is not allowed to set approximately three weeks, a frequency shift similar to that obtained with telemeters equipped with pressure transducers results. This is due to the relaxation effects of epoxy potting materials which are not completely cured.

Question: (Mr. Hayes) Then you say it's a curing matter instead of a characteristic flex and aging period of the compound?

Answer: I believe it has been generally accepted that epoxy potting materials keep curing after they have solidified. In fact, I believe this is the general reason for a post cure of an epoxy system. We cannot postcure for short times at high temperatures and therefore must allow the long times at low temperatures.

THE WELDED-WIRE MATRIX—AN IMPORTANT STEP
BEYOND PRINTED WIRING

Thomas Telfer
General Electric Company, Utica, New York

Since the Industrial Revolution, man has dreamed of a society in which machines would do most of his work. Until the computer, however, this dream included only his physical labor. Today the ideal in industry is to automate all phases of production from design to the finished product.

In some cases this goal has been achieved with remarkable success. In others, it has been elusive. One of the elusive areas is electronic circuit packaging. The early packages—point-to-point wiring, chassis construction, terminal boards—all had to be designed and built by hand. Printed wiring boards were an improvement in that after they had been designed for a given circuit, subsequent steps in making the boards, and in placing and soldering the components could be mechanized. But the art work for printed wiring-board layouts requires considerable time and effort. Rough layouts have been attempted on a computer; however, because the computer has to concern itself with position and curvature of runs to avoid crossover points on the same side of the board, the results have never been completely satisfactory. Furthermore, no method has been found to mechanize economically the preparation of printed wiring-board artwork.

The welded-wire matrix, on the other hand, is a circuit-packaging technique which lends itself to automation. The matrix is built by assembling wires side by side and spacing them at a standard grid increment, which may be as small as 0.050 in. A sheet of insulation is next placed on top of the wires, and a second set of wires, at right angles to the first set, is then added over the insulation. Projection welding is used to connect the intersecting wires through the insulation wherever a connection is required. The final step in forming the basic circuit is cutting away excess wire. In simple terms this is the welded-wire matrix as it is shown in Fig. 1. Its simplicity and orderliness make it highly adaptable to machine fabrication. It can also be built in any length and in any width.

If we trace its design and fabrication, it is obvious how tape-controlled machines can turn out the great variety of assemblies required for different circuits. Assuming that the electrical engineer designs his own circuit, the first input for the computer is a circuit diagram. Ideally, we would like to lay the schematic under a computer reader and let the computer lay it out in matrix form. We do not yet have such a technique at General Electric's Light Military Electronics Department (LMED), but we do have a computer program that will process information which has been coded from the schematic drawing onto punched paper tape. The computer then lays out the matrix, making allowances for the size of components, their lead location, required electrical connections, and critical nodes. This program, for example, directs the computer to analyze how components may be paired off to take advantage of the component-form factor in fitting components more closely together. To accomplish this and other tasks, information on the size, shape, and lead-location of components is first coded onto the input tape.

The computer's work in laying out matrices is greatly simplified because the welded-wire matrix is on a standard grid, and all wires are straight lines at right

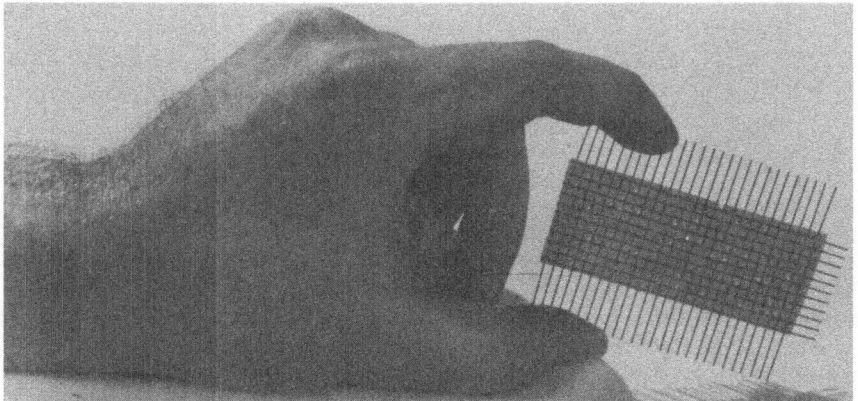

Fig. 1. The basic welded-wire matrix.

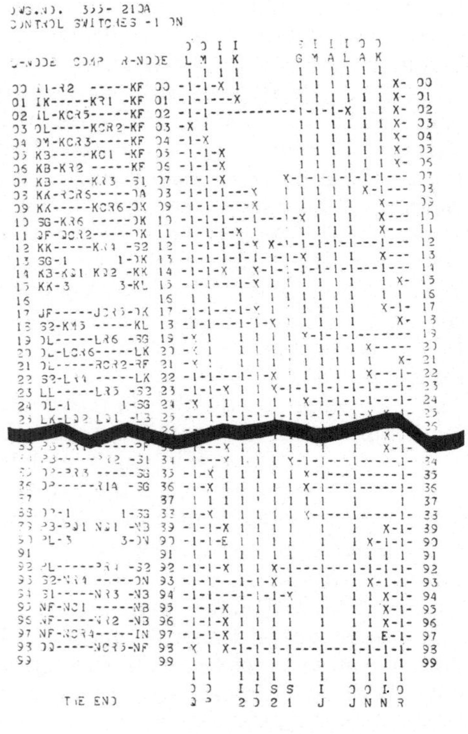

```
DWG.NO.  355-213A
CONTROL SWITCHES -1 ON
```

Fig. 2. Computer sheet of a welded-wire matrix
layout of 79 components.

angles to each other with unlimited crossover possibilities. The output of the computer is a printed sheet and a punched paper tape. As an indication of its speed of
operation, an electronic circuit of 79 components can be reduced to matrix form
in nine minutes of computer time. Of the nine minutes, one minute is required to
feed the tape into the computer, three minutes to lay out and wire up the matrix,
and five minutes to punch the output tape.

Fig. 3. Automatic fabrication machine for the welded-wire matrix.

Fig. 4. Automatic testing machine capable of running
330 checks per minute.

The output tape is finally fed to the printer, which in 15 min produces a sheet showing the location of components in proper, staggered positions together with the interconnection pattern for the matrix. A magnetic-tape system and a high-speed printer should reduce the time for print-out to less than five minutes.

An example of the printed sheet is shown in Fig. 2. It permits the engineer requesting the matrix layout to inspect the computer's work and may also be reproduced photographically in drawing format to satisfy customer requirements. A repetition of dashes represents transverse wires, repeated figure "1" represents longitudinal wires. The character "x" designates weld points. Components and node symbols are listed at the left of the sheet; component designations are staggered in the positions in which the components are to be placed on the matrix. If the engineer desires some change in the layout after he inspects the printed sheet, an editorial routine is available to instruct the computer to make the change.

The paper-tape output of the computer is fed into a translator to produce a number of specialized tapes required by other machines. One of these, the automatic fabrication machine shown in Fig. 3, is capable of turning out 50 ft of matrix ribbon per hour in any desired width up to the maximum of 18 wires, and in any length up to a maximum set only by the amount of wire on its spools. Rails are also welded into the assembly and later in the process formed into slotted terminals. Because it is tape-controlled, the machine can be used for breadboard matrices as well as for production quantities. With the computer and fabrication machine working together, it is possible to produce either a breadboard or a production matrix in hours, as compared to several days for printed-wiring breadboards.

Another paper tape controls an automatic testing machine shown in Fig. 4. This machine, which tests for continuity and for possible shorts, can run through 330 checks in 1 min. Machines of this type are also available at LMED for checking out complete circuits after the components have been added.

Slotted terminals are provided to accept the component leads so that accurate positioning is required only in the direction along the length of the matrix assembly. The terminals themselves provide convenient indexing registry.

Thus, we have the possibility of a circuit being machine-fabricated all the way from the schematic to the finished and tested circuit. In addition, tape-controlled drafting machines, available commercially, can be used to produce the drawings required by the customer. This is automation of both the mental and the physical portions of the task.

Fig. 5. Welded-wire matrix in a clear encapsulant.

Fig. 6. A repairable matrix assembly.

Fig. 7. Size comparison between 0.050-in. grid-matrix assembly (left) and cordwood subminiature module (right).

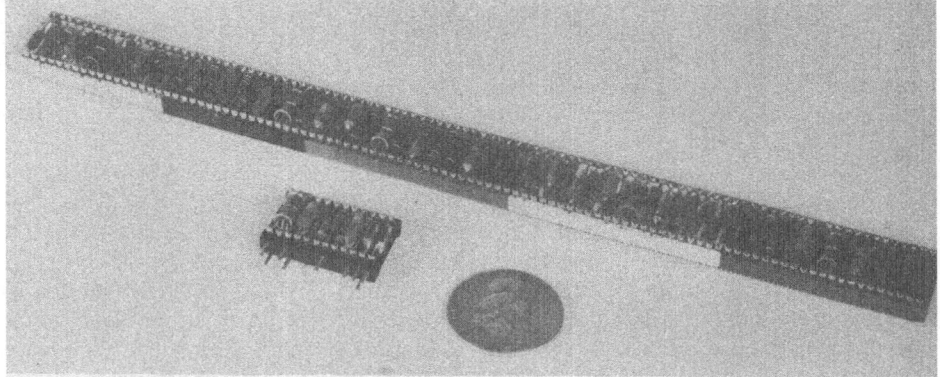

Fig. 8. Individual functional module compared with functional modules in a continuous strip.

We have discussed the features of the matrix which make it amenable to automation; now let us consider the additional features which make it attractive to design engineers. One of these features is its versatility. Until a certain point, the construction of the matrix is the same regardless of the type of assembly for which it is intended. At the Light Military Electronics Department, the matrix is made in "ribbon form" of any length and to a maximum width of 14 or 27 longitudinal wires, depending on the grid selected.

From this point on, the matrix may be tailored to suit a number of different applications. If it is desirable to solder on the components, slotted terminals are welded to the transverse wires of the matrix in a continuous form; reinforcing insulation is applied to maintain wire and terminal positions; and then the continuous form is separated to make individual terminals. The reinforcing insulation may be rigid for straight assemblies, or flexible so that the matrix may be formed into the odd shapes and configurations necessary, for example, for high-density packaging in the circular cross section of a small missile body.

Final encapsulation of the assemblies may be performed either after or before the addition of components to the matrix. When the final encapsulation is made after the addition of components, the encapsulant can be either a filled epoxy for strength and heat conductivity or a foam material for lightness. Figure 5 is an assembly in a clear encapsulant.

If the matrix is encapsulated before the addition of components, the repairable assembly which results is a terminal board in which the wiring between terminals is enclosed within the board. Because all connections in the matrix are made by welding, components may be unsoldered and resoldered without fear of lifting runs or damaging other connections. A repairable assembly is shown in Fig. 6.

If it is desirable to weld components directly to the basic matrix, the terminals may be left off the assembly. Next, either the rigid or flexible reinforcing insulation is applied, and the component leads are welded directly to the transverse wires of the matrix. To make the welded-wire matrix more versatile, pins for external connections can be extended from the assemblies at one or both ends, one or both sides, or from the bottom.

Another feature which interests the design engineer is the size reductions possible when substituting the welded-wire matrix for printed wiring boards. Reductions of two or three to one are common when using the 0.1-in. grid-matrix assemblies, and reductions of six to one have been achieved by using a 0.050 in. grid-matrix assembly. In fact, 0.050 in. grid-matrix assemblies are competitive in size with cordwood style, subminiature modules when both use the same size components. A comparison is shown in Fig. 7. When the assemblies are joined in longer lengths, as shown in Fig. 8, the matrix assembly becomes smaller than an equivalent number of interconnected cordwood assemblies.

The illustrations point out another consideration—repairability. The matrix assembly is repairable; the encapsulated cordwood style is not. There has been a great deal of discussion, pro and con, about repairability. Two kinds of repairability must be considered—repairability during manufacture and repairability in the field. No one disputes the fact that repairability in the factory can save a great deal of money and time. Repairability in the field, especially on smaller modules, is open to some question. One school maintains that such modules should be thrown away; another claims that they should be repaired. If the modules can be made repairable without a penalty in size and weight, it is best to give the user the option of either repairing or throwing away.

The design engineer is also concerned about reliability. We have a continuing test in which welded-wire matrices are subjected to various environments. As of the end of June, 1961, we had a total of 44,480,592 welded-joint hours without any failures. In fact, we have never had a welded-joint failure in-service or on-test in any of our assemblies.

Matrix assemblies have been checked for operating characteristics in high-frequency applications. For example, a transistorized 11-Mc, video-amplifier circuit was built both on a matrix and on a breadboard. Great care was taken in the construction of the breadboard layout for high-frequency operation so that the breadboard assembly could be used as a basis for comparison (see Fig. 9). When

Fig. 9. Breadboard assembly compared with a matrix assembly of the same circuit.

tested, the matrix assembly compared favorably with the breadboard. Other amplifier circuits in matrix form operated satisfactorily at 60 Mc. A number of other types of circuits have also been evaluated for electrical operation when packaged in matrix form. No degradation of operation occurred in any of the tests. The matrix thus supplies the design engineer with an electronic circuit-packaging technique which he can use for a large variety of circuits.

The welded-wire matrix is a going technique at LMED. It is in production, and equipments containing matrix assemblies have been shipped to the customer. As our design engineers have become more experienced in the use of matrix assemblies, they are finding novel methods of using them to solve their circuit-packaging problems. They are also finding that with the matrix, less lead time is needed to design and build equipment and to make circuit changes as the design evolves. For these reasons and because of its repairability and its amenability to automation, the welded-wire matrix is truly an important step beyond printed wiring.

DISCUSSION

Question: Do you use a roller-type welding device on the matrix?

Answer: No, we don't use the roller-type device; we use ordinary electrodes.

Question: You mean you use an individual electrode for each one of your welds in the matrix in an automatic system?

Answer: No, one set of electrodes is repositioned to do all of the welding on a matrix assembly.

Question: You also made a statement that you have 45 million welds and no failures.

Answer: We have had no failures in 45 million welded joint-hours. I think we can attribute this to the fact that we selected our materials and configuration very closely so that we had an optimum condition. This is a little bit different from when you try to weld component leads, for you hit all diameters, all coatings, and so forth.

Question: Martin Camen, Bendix Corporation. I have three questions. In your comparison of the cordwood construction vs. the matrix, you pointed out that the matrix was both smaller and lighter due to the absence of the encapsulant.

Answer: I think that we said that the matrix was lighter at the first stage of assembly. It was not smaller.

Question: Just taking the lighter part then. This is due mainly to the absence of the encapsulant. Right?

Answer: Yes, that's correct. The matrix assembly was repairable; it was not completely encapsulated.

Question: Does the absence of this encapsulant give you any problems in environmental conditions?

Answer: It would depend, of course, on what the environmental conditions were. We tested this under humid conditions and it proved satisfactory. If you had an extreme vibration or extreme shock, of course, it would be possible to make either of them fail. In cases of extreme vibration we have put a conformal coating on the matrix assembly. This is a very thin film that bonds the components to the matrix assembly. But we can still overpower this if we wish to replace the component.

Question: Second question: I notice that all the matrices that you showed were of small size. Have you attempted to make any large-sized matrices comparable to some of the large-sized printed circuit boards that are available?

Answer: When we say large-sized, a 79-component printed wiring board would be within typical size range for computers.

Question: I was thinking somewhere on the area of a 10 by 10 matrix.

Answer: We make it in ribbon form in any length, depending on the amount of material on the roll. The width we have set a standard for is 14 longitudinal wires. If the need arose, we could make it any width that would be desired.

Question: Do you run into any problems or any special fixturing to maintain a tolerance between conductors on long runs?

Answer: No. The reason for this lack of trouble is that we have the tooling, the fixturing, and the machines to meet our specialized requirements.

Question: Third question: When you were speaking about the matrix, you stated that circuit design by the computer was simpler than programing the same thing in terms of a printed circuit board. Could you clarify this? Why is the printed circuit layout more difficult for the computer than is the welded matrix?

Answer: For example, on the printed wiring board the computer has to concern itself with crossover points. If you have a doublesided board you can, of course, have some crossover points. It also has to concern itself with placing the components and physical locations across the board. Curvature of runs is another problem. After the computer has laid the board out, you have a straight-line designation of what connects to what. Beyond that you have to give it to someone to make up the art work, so that you can make the boards. The example of the matrix layout shown in the paper is such that an engineer could look at it and tell exactly what layout he has. The components are listed. We even list the node points. We don't need any further art work to make the matrix assembly. The tape that comes out of the computer goes into a translator which makes up a special tape in the language of the automatic fabrication machine. To summarize, the computer has a lot more trouble in placing the components and considering the curvatures of the runs and crossovers with printed wiring boards than it does with the welded-wire matrix.

Question: What I had in mind was some of the assumptions Dr. Kodres had made. This would probably eliminate a lot of the components and runs problems which you cited.

Answer: We have done some work at the Light Military Electronic Department of G.E. in optimum placement of modules on a "mother" board, and we find that this is a lot more difficult for the computer than is the layout of the welded-wire matrix.

Question: Joe Ritter, Electronic Modules. It seems to me that in comparing your matrix with the other modules you took the best examples of your matrix and the worst examples of the other modules. Most of these things would compare favorably if you took the same type of construction. In fact, I think you might find some results reversed. But the question I have is: do you hand-solder or dip-solder these components when you put them on this matrix?

Answer: The assemblies that you see on display here have been hand-soldered into the matrix assembly. This may raise some eyebrows, but we find that we can hand-solder these on in about the same time you can put components into the printed wiring board and dip them. If you will consider it a little bit, most of the time in assembling components on printed wiring boards is spent in taking the components with the previously bent leads and sticking them into two holes in the printed wiring board. In short, most of the time is spent in placing the component on the board or in placing it on the matrix. We find it is very simple to place the component, to put the iron on it, and solder it in place. Thus, there is not too much difference between hand-soldering components into the matrix assembly as compared with placing components on the board and then dip-soldering.

Question: I don't entirely agree with that, but the point is that you declare on the reliability of welded matrix and yet the welded matrix is potted and particularly protected from the environment. At the same time you expose all your soldered joints. Obviously, you think soldering is more reliable than welding.

Answer: No, that wasn't my point, nor do I believe that.

Question: I am sorry for putting it that way, but the point I am trying to make is the fact that you have got supposedly reliable joints completely protected from the environment and your supposedly unreliable joints vulnerable to the environment.

Answer: The words "supposedly unreliable joints" is your phrase. But even at that, we put the reliable assembly inside and put the less reliable components on the outside where you can get at them to replace them if you have to.

Question: Wes Sargent, Hoffman Electronics. I would like a little clarification on your potting again as opposed to a welded module that is potted. Are you saying that your vibration and shock resistance are the same as the potted welded module?

Answer: No. We find that in welded assemblies, in some cases, you get embrittlement on these welded joints using a nickel ribbon welding to a Kovar lead. There is some tendency for embrittlement, and you will find that the welded joints, unless they are supported, are not too satisfactory in extreme vibration. In general, we have adopted a policy of encapsulating welded subminiature modules.

Question: What techniques do you use to put these pins into the matrix? Is this done by the machine, or is this done at a later point?

Answer: This is done at a later point. We have a fixture which accurately places the pins and the wires. A template is placed on top of the fixture, and the proper welding is done.

Question: Jack Peterson, Ampex. You made an estimate of laying-out a matrix containing 80 components as such in nine minutes on the computer, as compared to some given time for a draftsman. How much time has it taken to prepare the program for the computer and what is the cost of the programer plus the cost of the computer?

Answer: I assume that you want to know how much time we spent in coding this information for input to the computer. In that particular case it took two hours to code it. We may be able to speed that up somewhat, but it is still an appreciable amount of time.

Question: Beck, Bendix Research. When you are using the welding technique in welding the transverse matrix wire to the components, do you use components with nickel leads or do you weld to copper or tinned copper?

Answer: We can weld to the component leads as they occur, and we find that in some cases we have to have preparation of the transverse wire of the matrix.

Question: What is the preparation of the matrix wire?

Answer: In some cases this consists merely of preflattening. This round wire will weld very well to materials like dumet. When we get into materials like soldered-coated copper, we have to preflatten the matrix wire in order to get the proper heat balance.

Question: When you use solder lugs on the transverse matrix wires, do you find that the cost is less than the ordinary PT board when you have to add the solder lug to the matrix wires?

Answer: No, the solder lug, when the matrix is machine-made, is put on as part of the process. This doesn't add anything to the complexity or cost of the assembly.

Question: Is this a V-shaped solder lug that your component lead drops into?

Answer: No, it is more of a U-shaped slot.

Question: When you're welding the matrix wires to your crossover points or your interconnection point, the pressure of the welding electrode pierces the separating plastic material before the weld is made? Is that right?

Answer: The wire is in contact before the welding current is applied.

Question: And you get 100% efficiency in that operation?

Answer: We do. We find that it works out very well. In manufacturing these matrix assemblies there is a further precaution that after every so many welds, a weld coupon is made and pulled to see if there is any gradual degradation of the welded joints. We find that they are amazingly consistent.

Question: Is it, in effect then, a cold flow piercing operation?

Answer: You might term it that.

Question: Does the automatic drafting machine you're talking about work only on a matrix system of boards or can you apply it to any type of drafting?

Answer: We have been testing out these automatic drafting machines for other purposes as well. For example, for sheet metal work there are a number of different machines that appear capable of doing the drafting job. Right now we are considering the question of economics—whether it is cheaper to pay for the machine or have it done by draftsmen.

Question: Ahmet Erdogan, Engineered Electronics Co., Santa Ana, Calif. When you were talking about your matrix, what is the thickness of the insulator film between those wires? Do you prepunch that insulator film?

Answer: To answer the first question, the material we are using right now is about 10 mils thick. No, we do not prepunch the material.

Question: You said that after several tests, there is 100% accuracy for your welds. Do you have any technical data? Do you have any technical record that those have not failed? What kind of test methods do you use for your weld inspection?

Answer: Let me answer the last question first. We find that the only reliable test that we can put on the welded joint is a pull test. We can tell something about it by how much the material has been set down, how it appears, whether it cracks. But the final analysis is the destructive tests, and this is done by pulling a sample coupon.

Question: This pull test is enough for your record to be able to say this is a good weld or bad weld or do you go further, perhaps for some metallurgical analysis?

Answer: We use a number of methods when we are getting the machine set up. We have monitored the welding machines and we have metallurgical microphotographs; we did a number of things in setting the process up. Once we found that the welds were consistent, we used the pull test on sample welds to check on any gradual degradation that may be occurring because of malfunctions in the welding equipment.

Comment: Jim Kuller, Bendix Radio. I would like to make a little comment here with reference to the exposed solder joints. It does seem reasonable to me that they should be in an exposed area. Carrying a welding machine around a field may be a bit difficult, and therefore, the provision of maintainable solder connections for repair, I think, is an excellent idea.

Answer: I thank you for that comment.

Question: With regard to cost, I would like to ask you what you are comparing the welded matrix to when you state it is cheaper. Are you comparing it to old hand-soldering processes or are you comparing it to something else?

Answer: I am comparing it with printed wiring board assemblies.

WELD IMPROVEMENT STUDY INDICATING FUTURE
ELECTRONICS PACKAGING APPLICATIONS

R. M. Steigerwald
The Sippican Corporation

The fact that more than 60 electronics firms, military representations, and university research groups were present at the recent Welded Electronics Packaging Association conference gives an indication of the growth of welding as a technique for production of electronics equipment. In addition, the published reports and statements of some of the country's largest electronics-production sources indicate the ever-increasing application of welding as a controllable and reproducible production technique for use in ground and airborne equipment, satellite, missile, and space vehicle developments.

Sippican's principal welding contributions to these developments have been obtained through the perfection of the welded-wire matrix principle, the programing of component lead materials for weld production-line operations, the development of weld schedules and quality control procedures, and the development of welding equipment.

When we first began welding, in September of 1958, our governing criteria for acceptable welds were:

1. No splashing.
2. No more than 50% embedment.
3. No "burned" welds.
4. Highest possible pull strength.

We found that our rules on splashing had to be treated fairly leniently, and sometimes we had to relax the 50% embedment law.

In order to establish some reliable methods for setting margins on welds, we undertook an extensive investigation of analytical tools and procedures, including radiography, ultrasonic investigation, metallography, and physical testing. From this we determined that metallography was the only sure way to evaluate a weld, and that physical tests could be employed as check tools, but not used as the sole criteria for establishing margins.

WELD SCHEDULES

Significant work has been done in improving weld schedules, one of the most important aspects of welding for production of electronics. As standard practice, Sippican designs and fabricates a prototype package or subassembly and provides the production interest with the weld schedules necessary to duplicate the equipment via assembly line methods.

At Sippican, weld schedules are determined experimentally and are constantly checked and upgraded. The tendency toward standardization of lead materials and interconnecting wiring has also contributed to this work. More time is available for quality control on the materials in use and, because of the lesser number of combinations to contend with, for research into new materials.

The margins for good welds are also directly contingent upon equipment, as well as upon a reduction of the number of combinations of materials to consider and the corresponding increase in attention given to the materials in use. We find that with research and actual use in production, the weld margins grow

wider. Of course, the wider the acceptable margins are, the greater is the reliability factor in the package.

Tests have been run at one of the university laboratories where welds were
deliberately made just outside of the allowable margins. These welds were then
encapsulated under the worst possible encapsulating conditions and the potted
assemblies were temperature cycled every 30 min. using dry ice and acetone,
to +225°C for 48 hr. This was done with approximately 500 welds and no incidence
of failure was recorded.

With the materials that are usually acceptable in the electronics industry,
past experience offers little toward defining metallurgical joints. The major
portion of this information is derived from experimenting and recording the
data obtained.

In reviewing "schmoo diagrams" to determine acceptable limits, the chief
governing criteria are those which are listed as standard acceptable practice
by such organizations as the Resistance Welder Manufacturer's Organization,
the American Welding Society, and the American Society for Metals. Consistent
with these practices, we try to establish settings which would give the best bond
with the least number of defects such as cross-sectional deformation and porosity.

After a sufficient "library" of information is gathered, it can be used to
establish the limits of a "schmoo." These hand-picked limits are always thoroughly researched before being published as acceptable margins.

MATERIALS

If we hold to the fusion theory of welding (which Sippican does), we find that
we can weld any combination of materials which do not, when in company with
one another, form a detrimental alloy. Dissimilarity of materials, contrary to
some opinion, does not necessarily imply a lack of compatibility. Nickel and
copper, for example, alloy over their complete range. The problem of forming
these alloys is mechanical rather than metallurgical. Furthermore, we are
learning how to control the volume of a melt in a joint and are thereby opening
the way for using combinations of materials which alloy only in fixed percentages.

WELDED-WIRE MATRICES

Sippican has pioneered the development of welded-wire matrices for circuit
and wiring module applications. We feel that the accuracy and ease of construction of the matrix is superior to other component connection and back-panel
wiring schemes with respect to the majority of system package and component
module applications. All of Sippican's system prototypes and experimental
assemblies have utilized the matrix principle with very successful results. It
might also be noted that new equipment designs permit the efficient production
of welded-wire matrices.

EQUIPMENT

The design of welding equipment has advanced with the state of the packaging
art. The most modern of today's weld-equipment designs are based on production
demands, and eliminate all of the weld station setup operations of previous
equipment. Most of the previous welding equipment in use was originally designed for industries other than electronics. In short, the new equipment developments offer "process control for production interests." Besides Sippican, Hughes
Aircraft, Unitek, and Raytheon are also building welding equipment.

Some of the larger firms actively engaged in or entering into welded package and subassembly production include Lockheed, Boeing, General Dynamics, Hughes, IBM, Sprague, Sylvania, Raytheon, AC Sparkplug, Rheem Semiconductor, Mallory, GE, Kearfott, Arma, Unitek, STL, WEMS, Burroughs Research Center, The Alloyd Corp., Electro Mechanical Research, etc. Many of these firms have conducted extensive research studies in order to evaluate welding for electronics and to improve welding as a connection process.

The current welding activities of just a few of these firms include—

1. Redesign and redesign recommendations for a number of missile projects to accommodate welding techniques. Some of the missile projects involved are Titan, Atlas, Minuteman, Sidewinder, Ranger, Polaris, Terrier, Tartar, Redeye, and Mauler.

2. Replacement of etched circuitry with welded modules.

3. Application of preprototype welding techniques to missile ground-support systems.

4. Welded modules for magnetic drum-memory systems, with packaging densities of 200,000 components per cubic foot.

5. Developmental infrared units.

6. Use of welded modules in programs such as Advent, Nimbus, the Mark VI Reentry Vehicle, Skybolt, etc.

At The Sippican Corporation, we have been chiefly concerned with the design and prototype fabrication of welded electronic packages, the improvement of welding techniques, and the development and manufacture of welding equipment for production-line operations.

We have been particularly involved in the design and prototype fabrication of Missile Guidance Computers, Airborne Computer Logic and Digital Control Systems, an Analog Flight Control System, and a Multipurpose Ground—Air

Table I. Failure Rate of Welded Joints in Polaris FBM Guidance Computer (1X Developmental Models as of May 1, 1961)

Computer design and operating mode	No. of welds	Hours operated	Weld operating hours	No. of weld failures
1X-1 bench model (logic sticks) (unpotted, controlled, ambient environment, life test)	16,800	17,800	$299 \cdot 10^6$	1 (See Note 1)
EDU test model (potted, uncontrolled ambient life test)	4,900	12,360	$61 \cdot 10^6$	0
1X-3 flight test model (potted, temperature cycling test)	31,500	520	$16 \cdot 10^6$	2 (See Note 2)
1X-4 flight test model (potted, cycling life test, uncontrolled ambient environment, on 8 hr, off 16 hr, followed by 24 hr/day operation since Jan. 6	31,500	3,000	$94 \cdot 10^6$	0 (See Note 3)
		Totals	$470 \cdot 10^6$	3

$$F_R = \frac{3}{470 \cdot 10} = 0.006 \cdot 10^{-6} \text{ failures/operating hour}$$

Note 1: Resistor failed at 10,857 hr (due to overheating during welded fabrication using current transformer type of welder).

Note 2: Opens in end connector under temperature cycling. Probable cause was missed welds. It is also noted that all welds were made prior to institution of metallographic weld analysis program in Fall, 1959.

Note 3: The 1X-4 flight test model has had no failures of any kind. Power was first turned on May 10, 1960.

Digital Package. Our claims for higher reliability, reductions in size and weight, efficiency in production, reduction of fabrication errors, etc., have been supported by extensive testing by our prime contractors. Sippican has published reliability data for such programs as the Polaris 1X-4 Guidance Computer (see Table I).

Welding will be used with many of the electronics techniques still in the research and development stage. It is Sippican's opinion that welding will not only continue to develop for various modular electronic concepts, but will be used extensively in conjunction with thin film and molecular electronic concepts. Sippican is already working with two other firms on the development of reliable welded connections between vacuum-deposited circuits and between wafer-electronic circuits. Based on a prediction of our own future efforts and an analysis of the present nature and scope of welding activities, we believe that welding will become the leading electrical-connection mechanism for military equipment, and that it will find many new markets in industrial areas.

* * *

To summarize, welding has very positive prospects as the leading electronics-connection technique for future electronics applications. Our opinion is based on the facts that—

1. Suitable production welding equipment is now available to industry.
2. Accurate welding schedules are available to industry.
3. The associated techniques for production welding operations have been developed and made available for use. These techniques include weld metallurgy, quality control, matrix fabrication, and use of materials.
4. Prototype welded system packages for ground, airborne, missile, and satellite programs have been designed, fabricated, and successfully tested. Production applications are already underway in these areas.

DISCUSSION

Question: Bill Hayes, Martin Company. Do you use prepunched Mylar or insulator between your matrix?

Answer: We draft our film photo, reduce it, and print in on photosensitive Mylar. We then prepunch the Mylar and weld through the holes.

Question: (Mr. Hayes). Could you explain to me the difference between your method and Mr. Telfer's. He did not say they weld through the insulator, did he?

Answer: Yes, they do. When they make a welded joint, they burn through the insulating material.

Question: (Mr. Hayes). Because of the insulating material, doesn't the voltage rise and endanger components?

Answer: When a matrix is welded, there are no components attached so the voltage has no effect.

Question: This welding technique, is this what is called the resistance weld using a capacitor discharge?

Answer: Yes, we use a capacitor discharge welder to form our resistance weld.

Question: A technique I have heard about for monitoring the welding is the actual recording of the resistance of the discharge itself? Have you utilized this method?

Answer: We have done considerable work in trying to monitor the output of the capacitor discharge welder across the electrodes. Wendell Hess has been very successful with a small shunt and an air toroid current meter. To use this system, however, we would have to modify our standard machine because there is no available conductor around which to put the air toroid. As far as monitoring the weld pulse itself, we have some ideas about how to use deviations in the wave shape to give us a go—no-go signal.

Question: George Jones, Minneapolis-Honeywell, St. Petersburg, Florida. What precisely do you mean by the statement that you furnish weld schedules to other people?

Answer: We are a prototype fabricating firm of consulting engineers and when we undertake a job we are compelled to furnish tooling, process sheets, schedules, etc. We've had a good bit of trouble in the past making out weld schedules with the equipment we used at Sippican so that the schedules could be used elsewhere in the country. This was primarily due to equipment vibrations. With the equipment we are now marketing, we have tried desperately to eliminate the variables in the equipment, that is, induction losses in the cables, etc. We now feel, therefore, that we can issue weld schedules usable at Sippican, then turn the job, the process sheets, and the list of weld schedules over to the customer. If necessary, we can even deliver preset weld heads to be used at the customer's facility. We have already done this on a limited basis.

Question: (Mr. Jones). In other words, you describe a cartel of sorts whereby you sell the welding machines as well as the weld schedule.

Answer: We have tried using other equipment to furnish weld schedules but it hasn't worked out too satisfactorily. The equipment is not entirely to blame. Some people prefer, for example, to orient their heads differently; therefore, they have a greater or lesser run of cable to the power supply. Some people prefer using fixed buses of varying dimensions. Some people strap cables together. Some people separate cables by two or three feet. It is, therefore, very difficult to have information which we have developed on one configuration at our shop transferred to another facility unless the configurations are identical. Even then, severe variations in temperature and humidity conditions will influence the weld schedule.

Question: Jim Kuller, Bendix Radio. What equivalent basis of comparison do you have to state that the matrix system is cheaper? Cheaper than what?

Answer: Again, I must use the same frame of reference that Mr. Telfer used and this is a component count. If you are trying to achieve high density packaging, then I would say that at 50 components or above the actual subassembly cost per module, that is, from the time the components come into the shop until the time the element or module is encapsulated, tested, and sent out is less on a matrix assembly than it is on a point-to-point assembly. On short runs of modules containing 10 to 15 components, point-to-point wiring is probably less expensive than going to a matrix technique where you incur the engineering cost of matrix layout, photo reduction costs, etc. .

Question: (Mr. Kuller). When you say point-to-point, do you mean point-to-point welding?

Answer: Yes. Point-to-point wiring; welded connections.

Question: In other words, you are comparing one welding technique to another?

Answer: Yes. I can make no comparison on solder joints.

Question: Ken Plant, Minneapolis-Honeywell. In your outline you mentioned some tests that were run at the university laboratories on some joints. The joints were welded outside of the normal margins and then encapsulated under the "worst possible encapsulating conditions." Would you describe to me something of the configuration and what you mean by "worst possible encapsulating conditions?"

Answer: Yes, I can describe the test in more detail. The laboratory concerned chose weld settings which were just outside the allowable limits of heat and pressure; that is to say, that the strength of some of these welds would fall outside the ± 3 sigma limitation normally established as a process control. The materials chosen for this test were two which made up a narrow margin weld. The welds were then encapsulated in foaming epoxy with no regard to stressing of the joint or interconnecting wiring. The encapsulated assemblies were temperature-cycled from dry ice and acetone to plus 600°F. The temperature cycling was done as rapidly as possible in a period of 56 hours. The first failure was a component failure.

Question: Leo Fiderer, RCA, Van Nuys. Getting back to the cost comparison business. I have read in the literature of many welding advocates that welding, and welded matrices, are cheaper than other methods of assembly. How does this compare to dip soldered costs?

Answer: I have had no experience with dip solder and I am not qualified to answer your question. The only comment I can make is that we know of two concerns that have said that they are now converting from dip solder to welding at no cost reduction, but at a higher reliability level. I don't know what they are paying for their reliability, whether they have doubled their cost or are merely adding 10%.

Question: Abe Levine, Honeywell Aeor, St. Petersburg. You mentioned that Kovar welds would show a high pull-strength figure, yet are bad welds. In what way will this type of bad weld show up as a malfunction?

Answer: No one has completed a real evaluation of this yet. The joint efficiencies, pull test wise, are 100% with most kovar welds; this is to say that during the pull test, the wire breaks outside the welded or heat-affected zone. Add to this the fact that we have built three completely air-free systems which have been severely environmentally tested and have had no incidents of failure with the kovar welds. All this does not detract, however, from the fact that metallurgically the weld appears unsound. There are several continuing projects to further evaluate the condition which is measurable in the microsections but, for the time being, we are accepting Kovar as a weldable material.

Question: Dave Caldwell, General Electric, Phoenix. For a weld schedule on a particular component, how many leads do you normally use?

Answer: An investigation of materials about which we have some previous information will require about 150 component leads or an equivalent length of similar wire. About 12 of these go for metallurgical investigation and about 100 for establishing quality-control information. The remainder are used for writing inspection criteria.

Question: Ernie Alford, Sandia Corporation. I am interested in your activities in welding with vacuum-deposited circuits and if there is any public information available.

Answer: We have no published information yet and I can't say exactly when it will be prepared. My best guess is around Christmas or first of the year. We are sure that this will involve a welding technique different from the one we are currently using. Consequently, we are working on different electrode configurations, sizes, and possibly a different type of energy delivery. We will mail out this information as soon as we have it available.

Question: Joe Bachus, Medtronic Inc., Minn. Do you know if work has been done to evaluate joint components using silver conductible epoxy?

Answer: Yes, I know of a funded research project going on using this technique. I have very little knowledge of the results but understand that the information will be released for military use through one of the missile programs.

AUTOMATIC PACKAGING OF MINIATURIZED CIRCUITS

R. L. Gamblin, M. Q. Jacobs, and C. J. Tunis

IBM, General Products Division, Endicott, N. Y.

INTRODUCTION

This paper reviews the effort to utilize the advanced capabilities of semi-conductor device fabrication and component interconnection techniques in the manufacture of digital computers. By taking advantage of the characteristics of these new devices and interconnections, substantial savings can be realized in manufacturing cost. If other potential advantages of this packaging capability are to be gained, it is important that all aspects of machine design, production, and maintenance be considered as an integrated whole.

Technical literature today is replete with references to "microminiaturization," "integrated circuits," "micromodules," "functional circuits," and numerous trade names. So that there shall be no misunderstandings, we state here what technology we are assuming is available. This is the manufacture of circuits which perform a single Boolean function (AND, OR, AND-INVERT, OR-INVERT) in a "module" (or "block") whose dimensions are referred to in terms of tenths of inches rather than multiples of inches. It is true there are special, more complex, circuits (triggers, flip-flops) available now in similar modules, but these may be treated in a similar manner. This paper will deal with computer considerations from the device or module "upward," as it were, into the computer system.

At present, for the manufacturer of general-purpose, commercial, digital data processing machines, the appeal of these new devices does not lie directly with their size or weight. These physical characteristics will appear rather far down on a list of machine characteristics the devices could affect. An order of decreasing importance might be:

1. Manufacturing cost
2. Reliability
3. Serviceability
4. Speed
5. Power consumption
6. Size
7. Weight

All these factors are inextricably related. The physical size of a semi-conductor junction will affect the device speed. The machine speed will be affected by device speed, and to some extent now, by the interconnection length; i.e., the over-all machine size. On a system level, the serviceability requirement will affect the reliability possible, and both will affect manufacturing cost. Our comments in this paper will be mainly concerned with the first three items on the list presented.

An examination of the items involved in the manufacturing cost of the central processor of a digital computer would reveal the following major factors:

1. Circuit components

2. Power supplies
3. Interconnections
4. Mechanical hardware
5. Assembly

Miniaturized circuit module use will directly affect the amount of mechanical hardware required to house a given machine. In addition, the interconnections and assembly costs will be reduced. The cost of circuit components is also expected to be reduced due to standardized methods for the manufacture of circuit "modules."

The particular choice of circuit scheme for a miniaturized digital computer is, in general, compatible with the search for minimum-cost circuits. In general, however, more significance is now attached to the following:

1. Low power dissipation
2. Reduced number of interconnections
3. High logic power and flexibility
4. Efficient usage of any integrated units

Packing schemes developed in the past would not be directly applicable to miniaturized circuits because of the following limitations:

1. Pluggable levels of packaging would tend to be "pin limited." A "pin" is a pluggable connector.
2. Larger "pluggable levels" tend toward uniqueness and this presents a field servicing problem.
3. The interconnecting wiring itself requires space.

Factors 1 and 3 imply that (a) the interconnections (pluggable connectors, printed or deposited wiring) should be miniaturized as well, and (b) interconnections should be made "efficiently" as far as space occupied is involved. The latter would imply positioning of those devices to be interconnected and efficient wiring of nets.

Factor 2 poses a problem to the system designer. In general the larger a pluggable level (in terms of the number of modules in it) the less often it is used in a given machine. Can machines be designed so that they use relatively large pluggable levels repeatedly without too many redundant circuits? Thus far, the answer to this question is "Practically, no."

What then should be the size of the "pluggable level"—the smallest easily replaceable unit in the machine? The optimum size of replaceable level is that which maximizes the availability of a digital computer system to a user for the minimum cost.

The availability of the machine will depend on:

1. The number and reliability of the various types of modular interconnections
2. The maintenance strategy used

The over-all reliability of the machine will be affected by the replaceable level size because the smaller this level is, the greater will be the number of interlevel connections.

For larger replaceable levels, however, the time required for fault location will be lower and the maintenance costs beneficially affected. Large replaceable levels, however, will require a costly field inventory, both because of their intrinsic cost and their low frequency of usage in any given machine.

Thus, the determination of an optimum "replaceable unit" size is seen to involve many aspects of computer design, manufacture, and maintenance. No simple solution to the problem will be presented here, but the electronic packaging engineer must now be acutely aware of the compromises involved.

Once the size of the "pluggable level" has been determined, it is important that this level not be "pin limited"; for this would be wasting the space we are intending to utilize efficiently. Thus, techniques for selecting those sections of a machine to be packaged on a given level would be convenient. The first part of this paper, "Partitioning" of Computer Circuitry, proposes a technique for this purpose.

Once the modules for a given level are selected, they must be interconnected. The relative positions of the modules will affect the total "wire" length used and the "ease" of interconnection. A method of positioning a number of modules to simplify the interconnecting wiring will be presented, and finally, an approach to efficient interconnection of a number of circuit modules will be considered.

"PARTITIONING" OF COMPUTER CIRCUITRY

The circuitry of a digital computer system can, for the purposes of discussion, be simplified by describing it in terms of a nondirected graph consisting of nodes and branches. Each "block" in the circuit will be represented as a node (or vertex) in the graph. A "block" will be considered as any collection of circuit components which the engineer may for various reasons wish to regard as a single entity. With each node of the graph there is associated a "size" which is the number of basic circuit components represented by the node or corresponding "block." A line connecting two or more nodes of the graph is called a branch. Such nondirected graphs can be represented conveniently in a matrix form resembling the classical matrices of incidence [1].

Matrix Construction. The construction of the matrix is as follows: There is a one-to-one correspondence between the rows of the matrix and the nodes of its corresponding graph. There is a one-to-one correspondence between the branches of the graph and the columns of its corresponding matrix. Let $Q = ||q_{ij}||$ be the $m \times n$ matrix representing a given circuit configuration. Thus, the entries in this matrix are binary-valued, i.e.,

q_{ij} = 1 if the branch corresponding to column j is incident at the node corresponding to row i.

= 0 otherwise.

Henceforth such a matrix will be called a Q-matrix.

It will be convenient for the purposes of this discussion to represent the nodes as numeric characters, and to represent the branches as alphabetic characters (with the option of using numeric subscripts on the alphabetic characters).

$$\sum_i q_{ij} = k$$

implies that the branch corresponding to column j is connected to k different blocks. There is another systematic matrix representation of a network of points. This matrix is generally referred to as an "interconnection matrix." This $n \times n$ matrix is denoted by $C = ||c_{ij}||$, where n is the number of nodes in the graph under consideration. The node numbers are now the indices of the matrix $||c_{ij}||$ where

$$C_{ij} = \sum_k q_{ik} \qquad i = j$$
$$= m \qquad i \neq j$$

(1)

where m is the number of separate connections (branches) which exist between node i and node j. Whence if $C = ||c_{ij}||$ and $Q = ||q_{ij}||$, then we have the interesting relation $QQ' = C$, where Q' is the transpose of Q. Proof: Let the entry in the ith row and jth column of the product QQ' be z_{ij}. Then

$$z_{ij} = \sum_k q_{ik} q'_{kj} = \sum_k q_{ik} q_{jk}$$

But

$q_{ib} \, q_{jb} = 0$ if and only if nodes i and j are not connected by branch b.
$\quad\quad\quad = 1$ if and only if nodes i and j are connected by branch b.

And if $i = j$,

$$z_{ii} = \sum_k (q_{ik})^2 = \sum_k q_{ik}$$

Whence $z_{ij} = c_{ij}$, and $QQ' = C$.

If P is a permutation matrix and $T = PQ$, then $TT' = PCP'$ so that an interchange of rows in $Q = ||q_{ij}||$ corresponds to an interchange of both rows and columns in the "interconnection matrix." Because of this feature the Q'-matrix has been chosen for manipulation. With these prefatory remarks completed, an example will be given illustrating the notation:

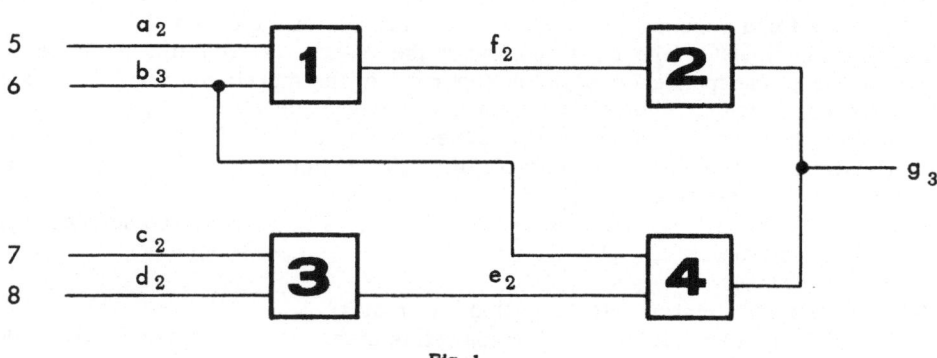

Fig. 1

Figure 2 is a graphical representation of Figure 1.

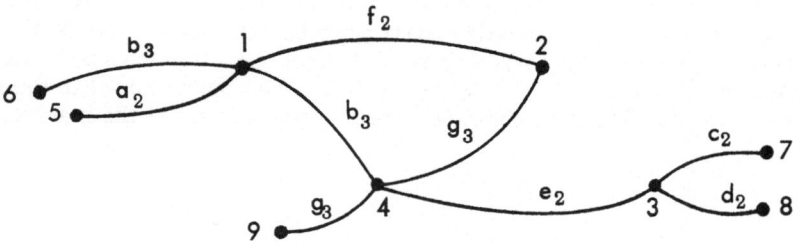

Fig. 2

The corresponding Q-matrix representation is:

$$
Q = \begin{array}{c} \\ 1 \\ 2 \\ 3 \\ 4 \\ 5 \\ 6 \\ 7 \\ 8 \\ 9 \end{array}
\begin{array}{c} a_2 \quad b_3 \quad c_2 \quad d_2 \quad e_2 \quad f_2 \quad g_3 \\
\left[\begin{array}{ccccccc}
1 & 1 & 0 & 0 & 0 & 1 & 0 \\
0 & 0 & 0 & 0 & 0 & 1 & 1 \\
0 & 0 & 1 & 1 & 1 & 0 & 0 \\
0 & 1 & 0 & 0 & 1 & 0 & 1 \\
1 & 0 & 0 & 0 & 0 & 0 & 0 \\
0 & 1 & 0 & 0 & 0 & 0 & 0 \\
0 & 0 & 1 & 0 & 0 & 0 & 0 \\
0 & 0 & 0 & 1 & 0 & 0 & 0 \\
0 & 0 & 0 & 0 & 0 & 0 & 1
\end{array} \right]
\end{array} \qquad (1)
$$

From the matrix, or from the logic diagram itself, the information needed to write each node as a vector may be obtained. The vector components will be the branches incident at the given node. From the matrix we write

$$
\overline{1} = (a_2, \ b_3, \ f_2)
$$

$$
\overline{2} = (f_2, \ g_3)
$$

$$
\overline{3} = (c_2, \ d_2, \ e_2)
$$

$$
\overline{4} = (b_3, \ e_2, \ g_3)
$$

The procedure for transforming this vector information to Q-matrix form is self-evident.

It will be shown how a local solution to the problem we have formulated can be obtained. No absolute optimal solutions are promised; in fact the only method presently available for obtaining an exact solution to this problem is the impractical method of exhaustion. Definition: Let $N(S)$ denote the "size" of the set S. Thus, by the conjunction of two sets, S and S', is meant $N(S \& S')$. By the disjunction of two sets, S and S', is meant $N[(\overline{S} \& S') \lor (S \& \overline{S}')]$.

Now let Q_i represent the ith row of a Q-matrix. Let $Q^{(i)}$ represent the ith column of a Q-matrix. Whence the disjunction, d_{ij}, of Q_i and Q_j is the number of nonzero elements in Q_i or Q_j which do not have the same column index. Clearly then,

$$
d_{ij} = \sigma_{ij} - 2c_{ij}, \text{ where } \sigma_{ij} = \sum_k q_{ik} + \sum_k q_{jk} \qquad (2)
$$

This can easily be demonstrated by realizing that for two mutually exclusive sets A and B, $N(A \lor B) = N(A) + N(B)$. It follows that $S \& S'$ and $S \& \overline{S}'$ are mutually exclusive sets and $(S \& S') \lor (S \& \overline{S}') = S$. Therefore, $N(S) = N(S \& S') + N(S \& \overline{S}')$ or $N(S \& \overline{S}') = N(S) - N(S \& S')$.

Applying this relation to the two mutually exclusive sets $\overline{S} \& S'$, $S \& \overline{S}'$ we have

$$
N\left[(\overline{S} \& S') \lor (S \& \overline{S}') \right] = N(S) + N(S') - 2N(S \& S')
$$

Equation (2) follows immediately.

The following algorithm is an effective means for realizing a close approximation to the partitioning problem formulated above.

Step 1. From a Q-matrix describing a circuit, the "interconnection matrix," $C \ (= QQ')$ is formed. As before, c_{ij} is the number of separate connections existing between nodes i and j.

Step 2. A starting node is defined in the Q-matrix. Let the index of this starting node in the interconnection matrix be 1 (one). Now we seek the column vector(s) $C^{(j)}$ with the property:

$$c_{1j} \geq c_{1k}, \quad k = 2, 3, \ldots, n; \quad j > 1,$$

where n is the number of nodes in the graph being segmented. If the set contains more than one column vector, a test for minimal disjunction is made among the nodes corresponding to column vectors in the above set. The minimally disjuncted node is selected to reside on the same card with 1 (one).

Step 3. Suppose p is the index of the first node chosen. A new row vector is formed, $Q(1, P) = Q_1 \vee Q_p$, where the symbol "vel," $\vee$, is a logical operation with the property

$$0 \vee 0 = 0$$
$$1 \vee 1 = 1$$
$$1 \vee 0 = 1$$

$Q(1, P)$ replaces Q_1 in the Q-matrix. Q_p is deleted from the Q-matrix. Steps 1, 2, and 3 are now successively repeated until either the pin limitations or node limitations for the card are reached.

Below is a table of disjunction and conjunction describing the circuit in Fig. 3. The upper entry $(r/m)_{ij}$ is the disjunction, n, of block i with block j; the lower entry, m, is the conjunction of block i with block j.

The analysis will begin with node 1 (in Fig. 3), and we will attempt to package "cards" of four "blocks" each. Column vectors obtained from step 2 correspond-

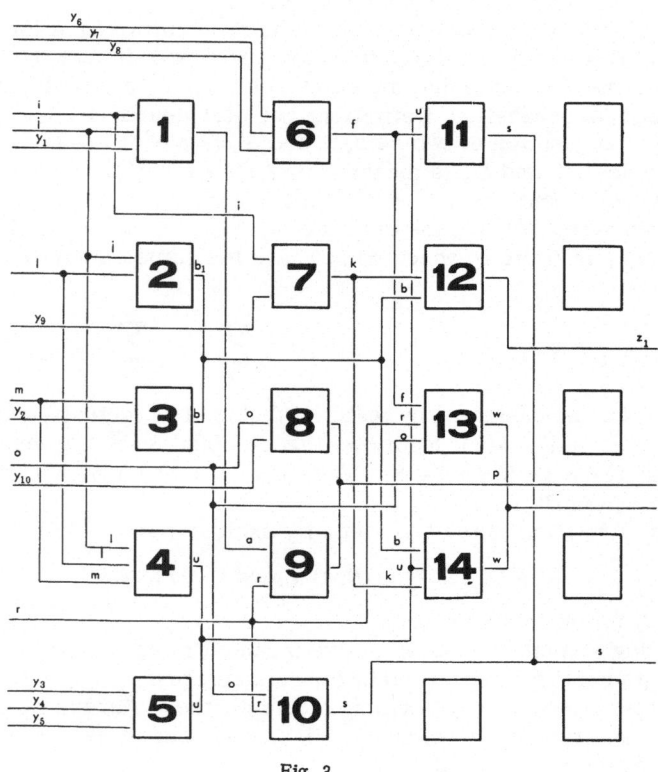

Fig. 3

TABLE I

	1	2	3	4	5	6	7	8	9	10	11	12	13	14
1	03	41	50	51	40	40	31	50	41	60	60	50	70	70
2	41	03	31	32	40	40	50	50	60	60	60	31	70	51
3	50	31	02	41	30	30	40	40	50	50	50	21	60	41
4	51	32	41	04	31	50	60	60	70	70	51	60	80	61
5	40	40	30	31	01	20	30	30	40	40	21	39	59	31
6	40	40	30	50	20	01	30	30	30	40	21	30	31	50
7	31	50	40	60	30	30	02	40	50	50	50	21	60	41
8	50	50	40	60	30	30	40	02	31	31	50	40	41	60
9	41	60	50	70	40	30	50	31	03	41	60	50	51	70
10	60	60	50	70	40	40	50	31	41	03	41	50	32	70
11	60	60	50	51	21	21	50	50	60	41	03	50	51	51
12	50	31	21	60	30	30	21	40	50	50	50	02	60	22
13	70	70	60	80	50	31	60	41	51	32	51	60	04	61
14	70	51	41	61	31	50	41	60	70	70	51	22	61	04

TABLE II

Total pins	Induced pins	Card number	Number of blocks
9	5	1	4
9	3	2	4
8	4	3	4
5	3	4	2

ing to $\bar{1}$ are $\bar{2}$, $\bar{4}$, 7, $\bar{9}$. The conjunctions are all equal, so 7 is chosen for its lesser disjunction. Thus, nodes connecting to $\bar{1} \vee 7$ are $\bar{2}$, $\bar{4}$, $\bar{9}$, $\overline{12}$, $\overline{14}$. These all have the same conjunction (one), whence $\overline{12}$ is chosen for its lesser disjunction. This procedure, which should be clear now, is continued until $(\bar{1}, 7, \overline{12}, \overline{14})$ reside on the first card, $(\bar{2}, \bar{3}, \bar{4}, \bar{5})$ on the second card, $(\bar{6}, \overline{10}, \overline{11}, \overline{13})$ on the third card, and $(\bar{8}, \bar{9})$ on the last card. The Q-matrix for the internal connections of each card is:

$$
\text{Card 1} = \begin{array}{c} \\ \bar{1} \\ 7 \\ \overline{12} \\ \overline{14} \end{array}
\begin{array}{ccccccc}
a_2 & b_4 & i_3 & i_4 & k_3 & u_4 & w_3 \\
\left[\begin{array}{ccccccc}
1 & 0 & 1 & 1 & 0 & 0 & 0 \\
0 & 0 & 1 & 0 & 1 & 0 & 0 \\
0 & 1 & 0 & 0 & 1 & 0 & 0 \\
0 & 1 & 0 & 0 & 1 & 1 & 1
\end{array}\right]
\end{array}
$$

$$
\text{Card 2} = \begin{array}{c} \\ \bar{2} \\ \bar{3} \\ \bar{4} \\ \bar{5} \end{array}
\begin{array}{ccccc}
b_4 & i_4 & l_3 & m_3 & u_4 \\
\left[\begin{array}{ccccc}
1 & 1 & 1 & 0 & 0 \\
1 & 0 & 0 & 1 & 0 \\
0 & 1 & 1 & 1 & 1 \\
0 & 0 & 0 & 0 & 1
\end{array}\right]
\end{array}
$$

$$
\text{Card 3} = \begin{array}{c} \\ \bar{6} \\ \overline{10} \\ \overline{11} \\ \overline{13} \end{array}
\begin{array}{cccccc}
f_3 & 0_4 & r_3 & s_3 & u_4 & w_3 \\
\left[\begin{array}{cccccc}
1 & 0 & 0 & 0 & 0 & 0 \\
0 & 1 & 1 & 1 & 0 & 0 \\
1 & 0 & 0 & 1 & 1 & 0 \\
1 & 1 & 1 & 0 & 0 & 1
\end{array}\right]
\end{array}
$$

$$
\text{Card 4} = \begin{array}{c} \\ \bar{8} \\ \bar{9} \end{array}
\begin{array}{cccc}
a_2 & 0_4 & p_3 & r_3 \\
\left[\begin{array}{cccc}
0 & 1 & 1 & 0 \\
1 & 0 & 1 & 1
\end{array}\right]
\end{array}
$$

Table II summarizes this partitioning, and illustrates how the pins required have been kept to a "minimum."

"Inherent" pins are those which carry input-output signals to the circuit being partitioned. "Induced" pins are defined as pins needed on each card, which are not inherent pins to the total circuit.

An alternate method allows $n/2$ vectors to select their optimal correspondents, where n is the total number of vectors. Then new vectors are written describing these pairs, and $n/4$ vectors select their optimal correspondents, etc., until every card is filled. This can be illustrated using Table I. The first $n/2 = 7$ pairs are:

$(\bar{1}, \bar{7})$; $(\bar{2}, \bar{4})$; $(\bar{3}, \overline{12})$; $(\bar{6}, \overline{11})$; $(\bar{5}, \overline{14})$; $(\bar{8}, \bar{9})$; and $(\overline{10}, \overline{13})$. These in turn select as cards: Card 1 $(\bar{8}, \bar{9}, \overline{10}, \overline{13})$; Card 2 $(\bar{2}, \bar{3}, \bar{4}, \overline{12})$; Card 3 $(\bar{5}, \bar{6}, \overline{11}, \overline{14})$; Card 4 $(\bar{1}, \bar{7})$.

The incidence matrices for these cards are:

$$
\text{Card 1} = \begin{array}{c} \\ \bar{8} \\ \bar{9} \\ \overline{10} \\ \overline{13} \end{array}
\begin{array}{ccccccc}
0_4 & p_3 & a_2 & r_3 & s_3 & f_3 & w_3 \\
\left[\begin{array}{ccccccc}
1 & 1 & 0 & 0 & 0 & 0 & 0 \\
0 & 1 & 1 & 1 & 0 & 0 & 0 \\
1 & 0 & 0 & 1 & 1 & 0 & 0 \\
1 & 0 & 0 & 1 & 0 & 1 & 1
\end{array}\right]
\end{array}
$$

$$
\text{Card 2} = \begin{array}{c} \\ \bar{2} \\ \bar{3} \\ \bar{4} \\ \overline{12} \end{array}
\begin{array}{cccccc}
b_4 & j_4 & l_3 & m_3 & u_4 & k_3 \\
\begin{bmatrix} 1 & 1 & 1 & 0 & 0 & 0 \\ 1 & 0 & 0 & 1 & 0 & 0 \\ 0 & 1 & 1 & 1 & 1 & 0 \\ 1 & 0 & 0 & 0 & 0 & 1 \end{bmatrix}
\end{array}
$$

$$
\text{Card 3} = \begin{array}{c} \\ \bar{5} \\ \bar{6} \\ \overline{11} \\ \overline{14} \end{array}
\begin{array}{cccccc}
u_4 & f_3 & s_3 & b_4 & k_3 & w_3 \\
\begin{bmatrix} 1 & 00 & 0 & 0 & 0 & 0 \\ 0 & 1 & 0 & 0 & 0 & 0 \\ 1 & 1 & 1 & 0 & 0 & 0 \\ 1 & 0 & 0 & 1 & 1 & 1 \end{bmatrix}
\end{array}
$$

$$
\text{Card 4} = \begin{array}{c} \\ \bar{1} \\ 7 \end{array}
\begin{array}{cccc}
a_2 & i_3 & i_4 & k_3 \\
\begin{bmatrix} 1 & 1 & 1 & 0 \\ 0 & 1 & 0 & 1 \end{bmatrix}
\end{array}
$$

Table III summarizes the results of this partitioning.

TABLE III

Induced pins	Card number	Number of blocks
4	1	4
4	2	4
6	3	4
3	4	2

POSITIONING THE CIRCUIT ELEMENTS OF A COMPUTER
ON A MATRIX OF DISCRETE POINTS

The Problem. Laying out high-density cards points up rather emphatically the necessity of generating an automated method of positioning miniaturized integrated logic circuits on a card so that they may be interconnected with a minimum of difficulty. It is a matter of utmost importance that a means be devised for positioning components on a matrix of discrete points in a manner designed to ease the wiring problems encountered. Thus, the implication is that an absolute minimum total wire length may not be essential. A local minimum of sorts will be satisfactory and perhaps preferred. Another way of stating the problem is that units which have connections between themselves should be placed in proximity so that wires need not extend over long distances. This problem and its refinements are of a nature so that no means of solution has even been found other than examination of all $n!$ arrangements of the blocks on the card where n is the number of units. For example, in a system of 200 blocks, the method of exhaustion is impractical, even for the fastest computers, as there are more than 10^{370} arrangements.

Past Work on the Problem. There have been several approaches for positioning components in the "geometric plane" in such a way that total wire length is minimized. In one case [2], the wire-length function

$$F(X_1, X_2, \ldots, X_n; Y_1, Y_2, \ldots, Y_n)$$

is defined as

$$
\begin{aligned}
F = {} & M_{12}\Big[(X_1 - X_2)^2 + (Y_1 - Y_2)^2\Big] + M_{13}\Big[(X_1 - X_3)^2 \\
& + (Y_1 - Y_3)^2\Big] + \cdots + M_{1n}\Big[(X_1 - X_n)^2 + (Y_1 - Y_n)^2\Big] \\
& + M_{23}\Big[(X_2 - X_3)^2 + (Y_2 - Y_3)^2\Big] + M_{24}\Big[(X_2 - X_4)^2 \\
& + (Y_2 - Y_4)^2\Big] + \cdots + M_{2n}\Big[(X_2 - X_n)^2 + (Y_2 - Y_n)^2\Big] \\
& + \cdots + M_{n-1,n}\Big[(X_{n-1} - X_n)^2 + (Y_{n-1} - Y_n)^2\Big]
\end{aligned}
$$

where M_{ij} is the weight (number of connections) associated with block i and block j. (X_i, Y_i) are the coordinates of block i. Then a classical extremum problem is solved. A system of simultaneous linear equations is obtained by taking the various $\partial F/\partial X_i = 0$ for variables X_i and $\partial F/\partial Y_i = 0$ for variables Y_i. Of course, some of the (X_i, Y_i) must be fixed (set equal to constants) or degenerate solutions will result. An alternate distance function [3] is obtained by replacing

$$
\Big[(X_i - X_j)^2 + (Y_i - Y_j)^2\Big]
$$

by

$$
\sqrt{(X_i - X_j)^2 + (Y_i - Y_j)^2}, \quad i \neq j
$$

in which case a solution to the extremum problem involves a system of nonlinear equations, which in general are not easily solved. Either procedure could be used to approximate a solution to the discrete positioning problem. Neither procedure, however, can be regarded as an exact answer to the minimum wire-length problem as it is defined on a matrix of discrete points. The method which will be described in this paper differs from most of the existing methods in that a decision-making process is used at the time of placement to produce an arrangement. As will become apparent, the decision-making process is extremely flexible and, though it does not necessarily provide a minimum wire-length array in specific examples, it gives layouts which tend to realize design objectives.

The Procedure. Once the contents of a card have been determined, each element is positioned on a matrix of discrete points. It can readily be seen that in the formation of a layout it is desirable to have components with the greatest number of potential points in common in nearby sockets. By this means, not only does a minimum length of hookup wire need to be used, but it also decreases the possibility of not being able to route the necessary wires, because short wires, in general, corrupt less space on the wiring board than long wires.

The algorithm used employs the Q-matrix for manipulation. First a permutation matrix P is sought such that

$$
\text{trace}\Big[(PQ)(PQ)' R\Big] = \text{minimum}
$$

where Q is an $m \times n$ Q-matrix representing a circuit, P is $m \times n$, and $R = ||r_{ij}||$ is an $m \times n$ distance matrix where

$$
r_{ij} = j - i \quad \text{if } j \geq i
$$

$$
r_{ij} = 0 \text{ otherwise}
$$

An approximate solution to this problem is obtained by utilizing the method discussed earlier.

First one chooses a row vector in the Q-matrix with a large number of incident wires and successively compares all other row vectors in the Q-matrix with it; then at least one of the other components will have a maximum conjunction and minimum disjunction (conjunction is always to take precedence over disjunction). This component will then be placed in a socket next to the first. A third vector can then be associated with the first two by taking each remaining component and placing it at all of the positions on the border of the first two. Again, maximum conjunction and minimum disjunction monitor the process. Now, however, there is a possibility of intersection with a unit two sockets away. A weighting factor less than 1 handles such a situation. In a similar manner successive components can be added to the group. At each step there will be no possible addition which will give a greater concentration of interconnection and lesser outputs from the group. Constraints on the problem can be handled by allowing the addition process to continue until the available card spaces are filled.

An example of the positioning scheme is given in Fig. 4.

Automatic Orthogonal Wiring. Human trial-and-error techniques of wiring are tedious, expensive, and subject to errors. Therefore, a wiring pattern which is amenable to automation is desirable. A wiring board has been designed which circumvents some of the more difficult wiring problems. It is proposed that the necessary interconnections be made in two planes. Imposing the constraint that the necessary crossings between planes be made only at socket pins unduly restricts the wiring and creates a formidable problem. By using a standard pattern of plated through holes in the wiring board, "orthogonal" wiring can be accomplished. An orthogonal lattice of wires is formed by superimposing an $"X"$ plane of wires on a $"Y"$ plane of wires. In the $"X"$ plane, wires travel only on lines parallel to a horizontal reference line. In the $"Y"$ plane, wires pass only on lines parallel to a vertical reference line.

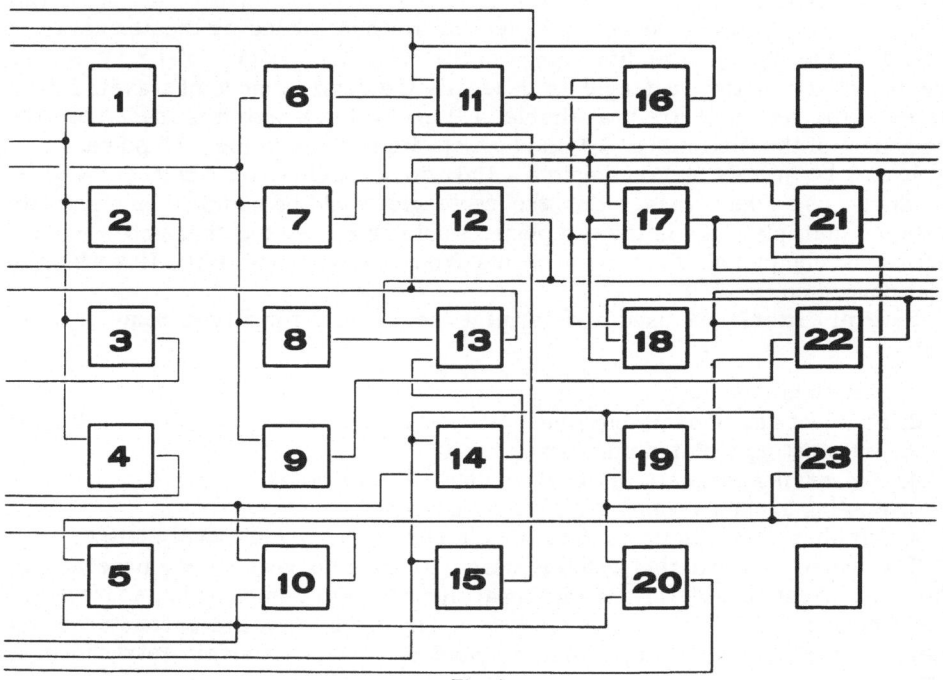

Fig. 4

The power of the orthogonal wiring pattern is that for a given array of
"blocks" a precise upper bound for the requisite number of plated-through holes
and wiring paths may be determined, which will guarantee that any circuit of the
same "size" may be interconnected. The "pattern" also obviates the need for an
exhaustive search for paths through a difficult maze. There are, however, critical
line limitations due to the physical restrictions on the size of the wiring board
and the land areas of the plated-through holes. Thus, the problem of automating
the interconnection of a circuit is not entirely trivial.

A wire list is made in terms of the name of the wire and the (X, Y) coordinates
of the signal pins in the net. A net is the single continuous wire which
interconnects a set of blocks. Each net is separated into links, where a link is
the binary connection between two signal pins. The links are then ordered on the
basis of a procedure given by A. Weinberger and H. Lobermann [4]. This procedure essentially entails describing the $n(n-1)/2$ paths which can connect n
terminals, making a monotonic sort of these paths in terms of increasing length,
and accepting or rejecting each path (in sequence) according to whether it is
redundant or not; i.e., the minimum number of paths which can connect n terminals is $(n-1)$. Then the minimum length D_i, through net i is computed. B_i is
the number of blocks in net i. The nets are then sorted monotonically in terms
of increasing

$$L_i = \frac{D_i}{B_i}$$

The images of two rectangular lattices (representing the vertical and horizontal planes) are then placed in the memory of a large-scale digital computer. The
actual path a link takes is indicated by making the appropriate nodes on the lattice "used" (i.e., by placing a "1" in their corresponding memory locations).
The first link in the first net, (X_i, Y_i) to (X_j, Y_j), is then inspected. Such a link
determines a rectangle (possibly a degenerate one, whenever the abscissas or
ordinates are equal). The dimensions of the rectangle are $|X_i - X_j| \times |Y_i - Y_j|$.
The periphery of the rectangle is generally the desired mode of travel. If both
halves of the rectangle are unavailable, a "line-by-line" search is made until the
component of the link in the "X" plane meets its partner in the "Y" plane.

It may be impossible to complete a link with a single line or half-rectangle—
in which case other lines within the rectangle would be sampled on a priority
basis, or multiple lines in the two planes would be pieced together until the connection is completed. Each path begins from the terminal of the link which is
most congested.

Several criteria are used to determine the best path for a link. Some of these
are:

1. Minimum wire length
2. Efficient use of available lines
3. Anticipating isolated pins
4. Preserving accessibility to the plated-through holes
5. Routing around areas of critical congestion
6. Avoiding travel in more than one dimension in the individual planes

The above criteria find general usage in our automatic wiring program. In
addition to these considerations additional algorithms are embodied in subroutines
which may be called up whenever particular difficulties are encountered. An
example of a particular algorithm, dictated by certain physical restrictions, is
given below.

At present we have an $l \times w$ wiring board, where $l >> w$. Thus, the ratio of possible vertical wires to possible horizontal wires is approximately $2 : 1$. We have found that if there are enough horizontal lines, the necessary interconnections can easily be completed. An algorithm has been designed which makes a "global analysis" of the horizontal components of the various links and assigns them to their "optimal" lines, "optimal" in the sense that under this assignment most efficient use could be made of the available horizontal channels. The algorithm may be applied to the entire board or to a critically congested portion of the board. The link connections are given as before:

$$(X_i, Y_i) \text{ to } (X'_i, Y'_i), \text{ and } (X_j, Y_j) \text{ to } (X'_j, Y'_j)$$

Therefore, two horizontal segments can be placed on the same horizontal line if and only if

$$\min (X_i, X_j) > \max (X'_i, X'_j)$$

or

$$\max (X_i, X_j) < \min (X'_i, X'_j)$$

By utilizing these inequalities the number of lines, n, which m u s t be placed on different horizontal lines may be determined. Let these segments be $I_1, I_2, \ldots, I_n$, where I_i is the length of the horizontal component of link i. These segments will be placed on n distinct horizontal lines, to be selected on the basis of their associated ordinates. That is, available horizontal lines which utilize the least length of vertical wires should be chosen. Let X_i, X'_i be the endpoints of I_i. Compare all of the abscissas of the segments not in $(I_1, \ldots, I_n)$ with the abscissas of each of the $(I_1, I_2, \ldots, I_n)$. The segments having abscissas closest to an abscissa of one of the $(I_1, I_2, \ldots, I_n)$ are selected. If there is more than one segment equally close to $(I_1, I_2, \ldots, I_n)$ the longest segment is assigned to the line extending the corresponding I_i. This process is continued until the horizontal segments under consideration have been rearranged optimally.

REFERENCES

[1] O. Veblen, "Analysis Situs," Am. Math. Soc., Colloquium Publications, Part II, Vol. V (1931).
[2] U. R. Kodres, "Geometrical Positioning of Circuit Elements in a Computer," IBM Technical Report TR 00.01.110.685, (July, 1959).
[3] W. Miehle, "Link-Length Minimization in Networks," The Journal of the Operations Research Society of America, Vol. 6, pp. 232-243 (1958).
[4] "Formal Procedures for Connecting Terminals with a Minimum Total Wire Length," Journal of the Association for Computing Machinery, Vol. 4, pp. 428-437 (1957).

DISCUSSION

Question: Leo Fiderer, RCA, Van Nuys. How many logic elements do you place in one pluggable unit? Is the interconnection between these logic blocks routed through the pluggable pins or is the interconnection between the blocks better performed in the module itself?

Answer: In order to keep the number of pluggable levels down to a minimum we forsee doing no logic on them but doing the logical interconnections down on a "mother" card or back panel. This also allows all the logic to be expressed in terms of the wiring done on the mother board. As to your first question; the optimum number of logic blocks or modules on the smallest pluggable unit is a complex function of block size, interconnection size, field servicing procedures, and repair and rework procedures, among others. A compromise size is usually selected. We have some prejudices and biases but so does everybody.

Question: (Mr. Fiderer) In high-speed circuits, routing the logic interconnections through these pluggable connectors creates a problem in delays, impedance mismatch, etc.

Answer: Yes, it does. For each circuit type, one must develop the specific rules which describe the fan out and interlevel communication allowed.

Question: Martin Camen, Bendix Corporation. I would be interested in knowing some of your "biases" or "prejudices." Consider the organization of the arithmetic unit of a general purpose computer: Do you find it better to organize on a bit basis, two bit basis, or in other words on what bit basis should it be organized?

Answer: As a matter of fact there has been considerable work along these lines. The answer depends very much on the machine type you are considering, unfortunately. It's different if you are considering a machine which operates on binary words, or a machine which operates on alphanumeric characters which are parallel by bit. On machines which operate serial by character and by parallel by bit one can certainly get some interesting functional packages. We tend to organize it, not by bit particularly, but more by "character." You can derive a "translator" module as it were, which enables you to translate a character from any given code, say BCD to biquinary, and with the addition of a few redundant gates can be used in several different places in a "CPU." You can obviously have a character register module with some form of checking on the output. I think each machine has to be considered on its own terms.

Question: What I had in mind was a binary machine parallel, on the order of 20 bits. This is a problem we ran up against.

Answer: The advantages of going in either word or bit directions in a parallel binary machine was not too clear in the applications I encountered. Another factor is that one would prefer the functional packages to be approximately the same "size."

ILLOGICAL PACKAGING DESIGN

Edward F. Uber and Kenneth L. Jones

Lockheed Missiles and Space Company

INTRODUCTION

Most of the papers presented at this symposium deal with advanced packaging designs and techniques. This paper, however, will attempt to describe some of the best regressive thinking of packaging design, which we choose to term "illogical packaging design." Illogical design is not a concept; rather, it may be termed a lack of concept. Occasionally, illogical design is accidental rather than premeditated. The illogic of design stems from many sources and takes many forms. It is our aim to bring to your attention some of the bad design habits existing in industry today. By describing them, the job of eliminating these bad design habits and features should be somewhat easier.

HISTORICAL BACKGROUND

It would be extremely difficult to find a beginning point for illogical packaging design. The problem is probably as old as the electronics industry itself. But as circuits become more and more complicated, and the demand is increased for smaller and smaller packages, the problems of good design become more acute. Each time the allotted package size has been halved, the problems of designing an acceptable package have more than doubled. The probability of design errors occurring are greatly increased with each size reduction. In order to be aware of these design errors, the illogical packaging techniques can be catalogued as follows.

ILLOGICAL DESIGN "TECHNIQUES"

The Latest Fad. Designing electronic packages by the latest fad is a common occurrence in the work of some designers. It is unfortunate that many of these people will "go off the deep end" when designing a unit simply because the trade magazines are all running articles on certain techniques which are supposed to be the hottest packaging techniques in twenty years. Occasionally the designer who designs by fad will consult with the production people to see if facilities and capabilities exist to properly fabricate the unit. More often than not, the designer assumes that since the technique has been described in the trade magazines, it is common practice and anybody can build units this way. Usually, it is only after the design hits the production facility that the designer learns there is more to the latest fad than the magazine articles he read had implied. By then, however, the completion date is usually too close at hand to complete the equipment by any fabrication technique. This necessitates alternatives, such as: (1) creating a production capacity for the fad; (2) having the design fabricated by the company which did the basic development of the technique; or (3) completely redesigning the package on a "crash" basis. Any of these alternatives can lead to trouble.

One of the pitfalls of "the latest fad" technique is that the fad is seldom completely documented or sufficiently tested at the time it is first described in maga-

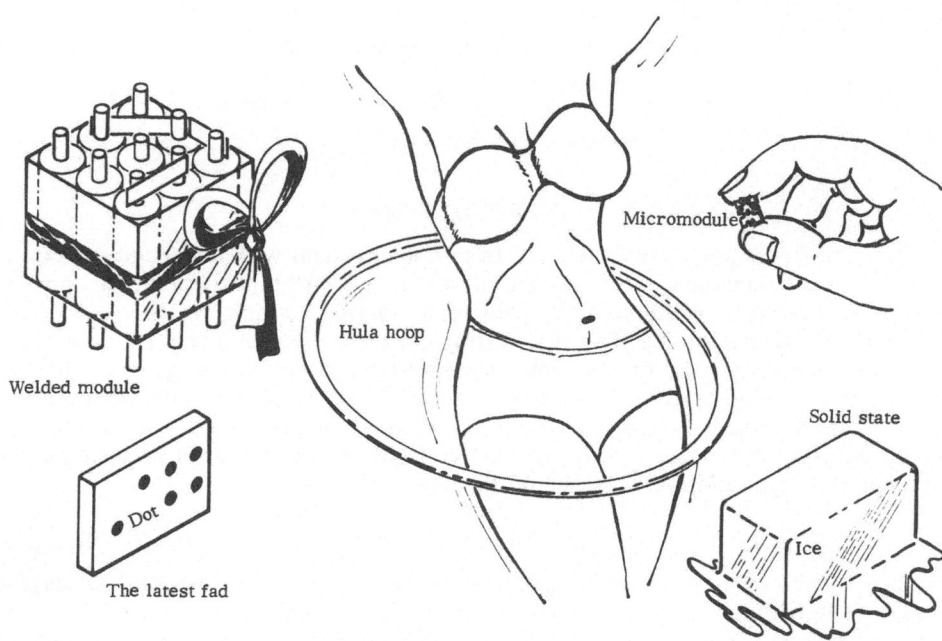

Welded module

Hula hoop

Micromodule

Solid state

Dot

The latest fad

Ice

zines. One or more physical parameters of the package (such as thermal considerations or encapsulation effects) is usually unexplored, so an uninitiated designer cannot arrive at an adequate design. Designing without adequate research data to substantiate theories is one of the most illogical design gambles in existence.

The intention here is not to say "never try new ideas," but rather to say "never try new ideas which have inadequate substantiating data," which do not fit the job, or are unnecessary to an adequate design. If data are lacking, they should be obtained prior to beginning a new design.

The Crash Design. One of the most wasteful illogical design techniques is the "crash" design. Designing by crash program is expensive to the producer and to the customer, and usually results in an inferior product. Crash programs usually result from the telescoping of the development time schedule from any of several causes, such as: (1) the overly optimistic estimation of the length of time necessary for completing the packaging design; (2) the overly long time for circuit development and research, resulting in missing the target date for that phase; or (3) the necessity for redesign to accomplish initial goals within the specified time after a previous design has proved unsuccessful or unproducible.

In crash programs, the possibility of having an error slipping through is greatly increased because of the rushed nature of the job and the hastiness of design checkout. Design parameters are sometimes purposely ignored with an airy, "We haven't time to look into this thoroughly right now. It will probably work all right and if it doesn't we'll catch it in redesign." And thus is born a piece of equipment which is plagued with problems during fabrication and all during its years of use in the field.

Crash programs can, however, turn out good designs. Some designers think best under pressure. In addition, much time consumed by some designers—

especially young, inexperienced designers—is decision time: trying to decide which of two equally good design concepts should be used. Crash programs force one to make decisions. Good or bad, the decision is made and is made quickly. Radical approaches are often used which would not have been otherwise used in the conservatism of a normal design program.

These designs are usually gambles on the designer's part. He figures he has nothing to lose and everything to gain if the design succeeds. If the radical approach works on the case in question, it is added to the designer's repertoire of favorite tricks. Further, he may go on a "kick" and design everything, regardless of applicability, in this manner for the following six months.

Crash program

The crash design is a paradox. It is the result of insufficient time to complete a design by regular methods, yet it is often the cause of large expenditures of time for redesign. Crash programs are expensive luxuries which few companies can afford, but most companies experience them year after year without attempting to eliminate the cause. Some companies operate on a constant "crash" basis.

The "Original" Design. Let us now consider the "original" design. Designers are artists whether they admit it or not. They are artists by the nature of their jobs, by temperament, and by the pride they have in their designs. It is this "pride of authorship" which occasionally leads to trouble in the fabrication of a design. This stems from the fact that designers may include a new "original" detail in the design which "nobody ever thought of before." When these designs arrive at the fabrication facility, trouble develops. The designer is called upon to aid the fabrication facility, and, more often than not, he blames the lack of skill of the production personnel, the materials procured, the density of the supervision, the weather, the humidity, and any other reason—real or imagined—which can be thought of—but never his design. For, after all, the design is a small part of him and, therefore, is perfect. It must be defended to the bitter end. For what could possibly be wrong with the packaging design?

Perhaps an illustration is appropriate here. Let us consider a packaging designer who invented a new method of applying terminals to a welded diode block.

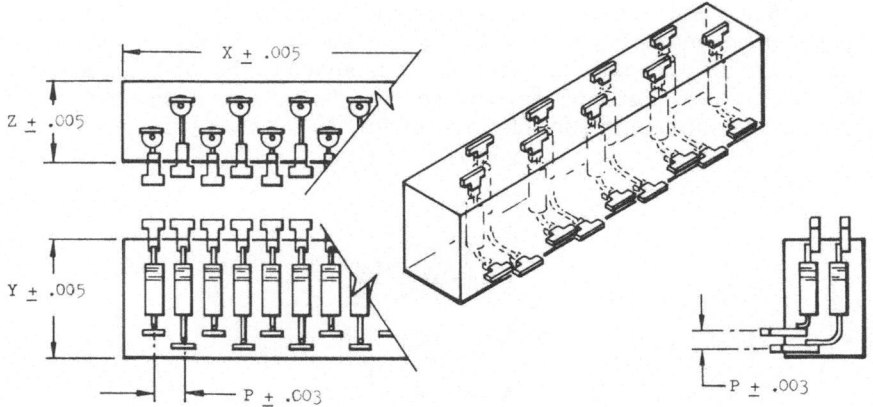

Diode block—an original design

It was very simple—all that had to be done was to weld 0.017-in.-diam diode leads to punched nickel T-shaped terminals which had a cross section 0.060 in.[2] One lead of the diode was bent to form an "L." The diodes were then arrayed in a fixture and potted. The terminals had to fall within ± 0.003 in. of the design position, and the potted block was to be cast within ± 0.005 in. of the design dimension. The entire block was connected into the system by soldering to the nickel "T's." Soon after production began working with the design the following problems presented themselves:

1. The diode leads were a different material on each end of the diodes, necessitating two sets of welding schedules;
2. Further, the size of the diode leads compared to the size of the terminals was so small that welding schedules produced could not meet specification requirements;
3. The potting fixture developed by the tool designers consistently produced products which had unacceptable tolerances in terminal placement and block size;
4. When parts were finally produced, it was found that it was difficult to solder to the heavy nickel lugs and damage often resulted to the diodes.

When the designer was called in for assistance he said, "I don't know what's wrong with you production people. I know there is nothing wrong with the package design—I did it myself."

The "Overdesigned" Package. We are probably all very familiar with the over-designed package. Packages are not overdesigned in any single way but in many different ways. Perhaps the most common type of overdesign is reducing the over-all package size where space is available for much larger packages. Where size reduction is done solely as an exercise to prove it can be done, rather than for some valid reason, such as limited space, the packaging concept is truly illogical. True, with modern components and techniques, the circuitry for a television set could be reduced to a few cubic inches of space. But it would be unreasonable to do so when the large display tubes required by the purchasing public leave unused space around them many times that amount. In addition, by accomplishing such a size reduction, fabrication and maintenance, and often signal difficulties, would be increased. It is true that large computers which occupy large rooms could be reduced to desk or console size—and will soon be— but what is the purpose of doing so if space is no problem and if the reliability is sufficient in a larger size? If reliability is the objective, size reduction produces the ability to put into the package redundant or standby circuits where they would have been prohibited before. Miniaturization does not, however, inherently produce improved reliability.

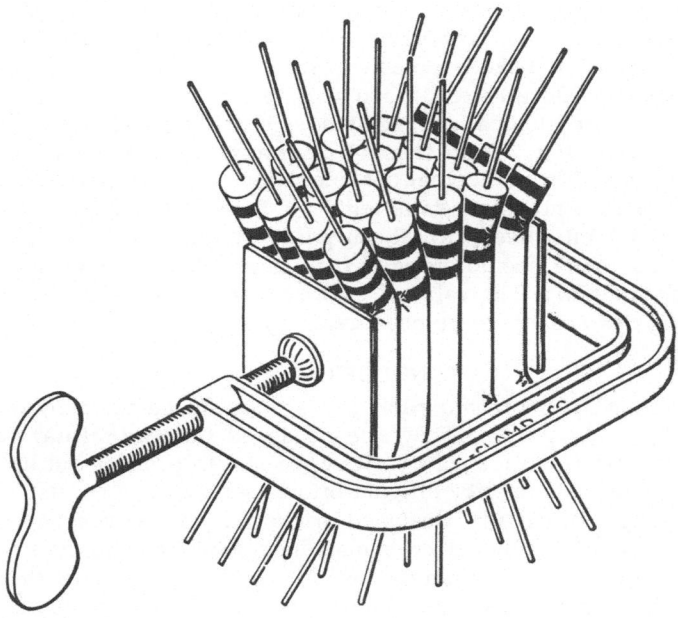

Overdesigned package

Overdesigning is also seen in the hardware and controls associated with the package. Mechanical tie-downs having sufficient strength to hold a package many times the weight of the actual package through the worst shock and vibration forces which can possibly be encountered are commonplace. In these conditions even the hardware to support individual components is usually overdesigned.

In general, the illogic of overdesigning manifests itself in inefficient use of allotted space, increased weight from overdesigned mechanical components, unnecessarily compact packages, unusual thermal problems, and increased difficulty of production and maintenance.

The "Underdesigned" Package. At the other end of the scale of illogical design lies the underdesigned package. Its attributes are "180° out of phase" with the overdesigned package. The designer often fails to take advantage of newer, more reliable, more easily producible techniques. He ignores these, occasionally, as unproven or unworthy of consideration in one of his designs. He also tends to leave space unused or at best badly uses it.

Other designers underdesign by using marginal parts, hardware, encapsulants, and connection techniques. Underdesigned packages of this type are a bane to the existence of many production people. They never know from one unit to the next if it is going to pass acceptance tests or not.

Remedies for Illogical Design. One remedy for illogical package design is to be found in the organization of the design team. The design team should ideally consist of electronic and mechanical personnel in addition to the designer. These people should be well versed in the production techniques which will be employed in the production of their design. If they are insufficiently versed in production know-how, they should receive orientation in the production techniques available to them, or a third type of personnel—the production specialist—should be added to the team. In addition, the design team members should keep abreast of the production technique improvements which take place in the fabrication facility. As capability to produce more complicated and involved designs increases, the package designs may become more complex. By maintaining a good knowledge of the production facility and the problems involved in fabrication, a good design team will be able to produce better, more workable designs.

Another remedy for illogical packaging design can be found in a well thought-out designer's check list. This check list should ask questions which should obviate a poor design: "Have thermal considerations been calculated?" or "Have specification requirements been satisfied?" or "Are tolerances too tight or too loose for the materials used or the purpose for which the unit is intended?" Questions such as these should bring to the designer's consciousness any previously overlooked details. The value of such a check list must not be overlooked if good packaging designs are to be produced.

CONCLUSION

Illogically designed electronic packages are produced in many ways, from many causes, by many people. They are produced for both military and commercial products with equal frequency. They are easy to produce but hard to discover and eliminate. It is only by diligent effort that they can be discovered and eliminated. Perhaps illogically designed packages will never be completely eliminated for they are designed by human beings who are somewhat illogical themselves. In fact, when you examine human beings, it is evident that they are rather illogically designed packages themselves. So let us conclude by asking a question, "If electronic packages are to be designed by a machine (a human being) which is somewhat illogical in design, how can any electronic package escape being somewhat illogical in design?"

RADIATION-EFFECTS CONSIDERATIONS IN THE DESIGN OF ELECTRONIC CIRCUIT PACKAGING FOR NUCLEAR-POWERED VEHICLES

J. H. Levine

General Dynamics/Fort Worth

INTRODUCTION

The combination of induced and natural environments presents the designer of airborne electronic packages with an imposing range of conditions to consider in his design. The major induced and natural environments of importance to the electronic-package designer are as follows:

Induced Environment
  High temperature
  Thermal shock
  Noise and vibration
  Explosive atmospheres
  Acceleration
  Mechanical shock
  Corrosion-fuels, rocket gas, etc.
  Nuclear radiation

Natural Environment
  Altitude
  Solar, primary, cosmic, and van Allen
    radiation
  Rain, sand, and dust erosion
  Humidity, salt spray, ozone
  Biological-bacterial and fungal growth
  Low temperature

The purpose of this paper is to focus attention on the induced environment of nuclear radiation in combination with other environments, and its influence in the design of electronic packages.

Transient, permanent, and special radiation effects are described and examples given. Four alternatives available to the electronic-package designer are discussed, namely, (1) critical material and part replacement, (2) relocation, (3) shielding of the radiation source, and (4) local shielding. The degree to which these alternatives are applied is one of the major problems requiring attention in the design of equipment for use in vehicles operating in the presence of a nuclear environment.

The radiation resistance of various packaging materials and parts is discussed, such as derivation of functional thresholds and response relationships, and their application in electronic-package design. Results obtained from irradiation tests on dynamic assemblies, including a gyro and accelerometer and magnetic-flux-valve assemblies, are discussed.

The Convair Radiation Effects Data Analysis and Retrieval (CREDAR) and Radiation Effects Reference systems are discussed. Emphasis is placed on the increased importance attached to the role of information retrieval.

The major sources of nuclear radiation of concern to the electronic-package designer are primarily attributable to nuclear reactors, radioisotopes, primary cosmic rays, Van Allen belts, solar flares, and nuclear-weapons radiation. The relative importance of these sources of nuclear radiation is, of course, a function of the design application and is a topic beyond the scope of this paper.

RADIATION EFFECTS IN ELECTRONIC PACKAGING
MATERIALS AND PARTS

The radiation effects which have been noted in electronic materials and parts may be categorized into transient, permanent, and special effects:

1. Transient effects are changes in the properties of a material or part which occur during irradiation and disappear at some time after irradiation. This term is sometimes restricted to a radiation effect which disappears immediately upon removal of the irradiation environment.
2. Permanent effects are changes in electrical and mechanical properties which occur during irradiation and are irreversible in nature. These permanent effects may depend only on the total amount of radiation received by the material or part (dose effect) or on the intensity of the radiation (dose-rate effect).
3. Special effects, for purposes of this paper, include such radiation-induced phenomena as liberation of gas from organic materials and fluids, nuclear heating, air ionization, and electromagnetic-wave attenuation.

Transient Effects. The irradiation of light-sensitive parts provides an excellent illustration of a transient effect. Results from the irradiation of three types of phototubes are illustrated in Fig. 1, where dark-current dependence on the intensity of the radiation is clearly shown. Postirradiation tests revealed no significant change relative to preirradiation tests [1].

Another transient-effect example is provided by the conductivity dependence on radiation intensity commonly displayed by dielectric materials. The increase of conductivity of several dielectric materials as a function of radiation intensity is shown in Fig. 2. It can be shown that the conductivity in a dielectric is related to the radiation intensity by

$$\sigma \propto I^k$$

where σ is the conductivity (mho-cm), I is the gamma-ray intensity, and k is a constant depending on the specific dielectric (average: 0.5).

The recovery time of the dielectric materials after irradiation varies with the dielectric, but, in general, the time required for the conductivity to return to its preirradiation level is approximately 10-12 hr at room temperature [2].

It must be emphasized that even in parts and materials exhibiting significant transient effects, permanent damage may predominate after sufficiently high radiation exposures.

Permanent Effects. Permanent radiation effects have been observed in electrical insulations, electron tubes, semiconductors, potting compounds, capacitors, resistors, radomes, frequency control crystals, lubricating greases, and many other materials and components. It is generally accepted that inorganic materials are more radiation-resistant than organic materials; a notable exception to this generalization is the semiconductor class of materials.

Nuclear radiation produces changes in the various properties of organic materials by ionization and excitation processes, which cause chain scission, cross-linking, free radical formation, and polymerization of the molecules. Gamma rays are the ionizing radiation of primary concern in organic materials. Inorganic-material damage, such as lattice structure displacements in ionic- and metallic-bonded materials, is primarily caused by heavy, energetic particles, such as neutrons.

Figure 3 illustrates the relative radiation resistance of several electronic materials and components. The left scale on the chart is based on units of energy absorbed in carbon by gamma rays [3] and is used for organic materials and

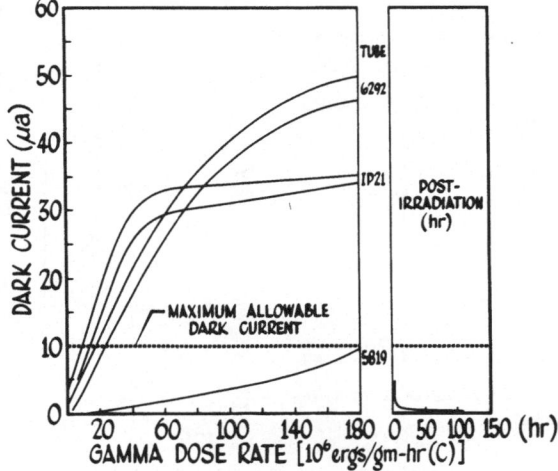

Fig. 1. Effect of gamma radiation on photomultipliers.

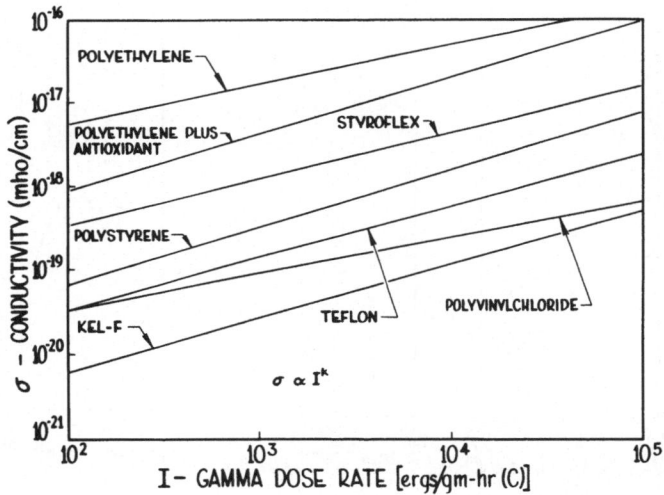

Fig. 2. Transient effect on dielectrics.

components. The right scale on the chart is based on units of fast neutrons per square centimeter and is used for inorganic materials, electron tubes, and semiconductors. The large ranges for particular classes of materials and material applications serve to point out that it is literally impossible to obtain simple "rule of thumb" numbers that are adequate for design in a nuclear environment.

There are instances where combination of transient and permanent radiation effects may be observed. An illustration of this effect is provided by the irradiation of 2-mil polyvinylfluoride film (Fig. 4; [4]). The left-hand portion of Fig. 4 illustrates the change in breakdown voltage as a function of radiation intensity. The middle portion of Fig. 4 shows the change in breakdown voltage as a function of the integrated radiation intensity (or dose). The right-hand portion of Fig. 4 illustrates the effect of annealing on the breakdown voltage after irradia-

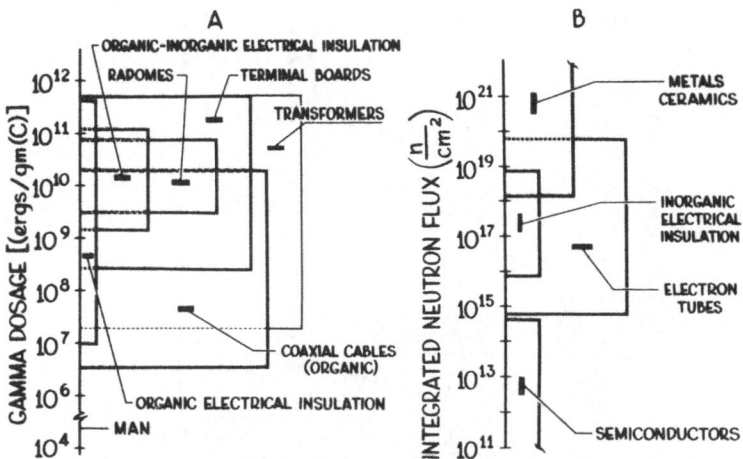

Fig. 3. Relative radiation resistance of various electronic materials and components.

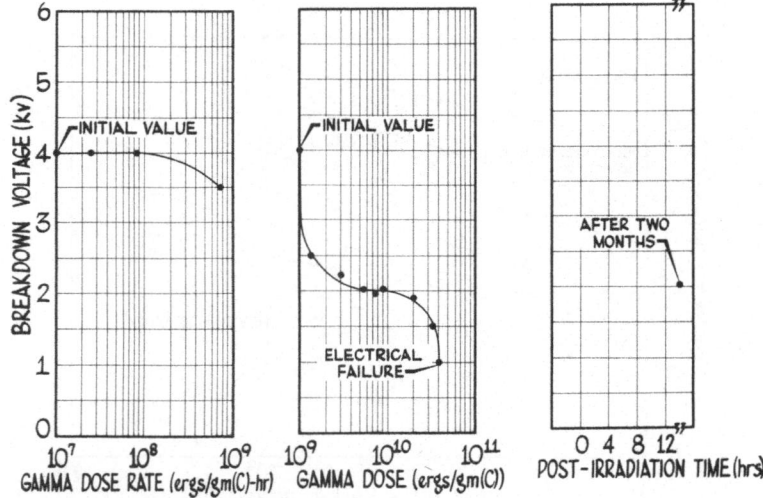

Fig. 4. Effect of radiation on dielectric strength, 2-mil polyvinylfluoride film.

tion. It is observed that even after two months of postirradiation time, complete restoration of the breakdown voltage to original levels was not accomplished.

The effect of sample size must also be considered in the materials and parts evaluation. This relates to the surface area exposed to a particular environment and the radiation attenuation afforded by the material under consideration. For example, very thin specimens will allow secondary particle capture and multiple scattering to be neglected when calculating energy absorbed within the specimens. Since radiation damage is directly related to the energy absorbed, life for the very thin specimens in a nuclear radiation environment is thereby lengthened. The maximum amount of energy absorption per unit of weight occurs in specimens varying from a few thousandths of an inch up to approximately one inch [5]. Self-shielding reduces the incident radiation, thus tending to protect the material at greater depths from damage. It should be emphasized that for most elec-

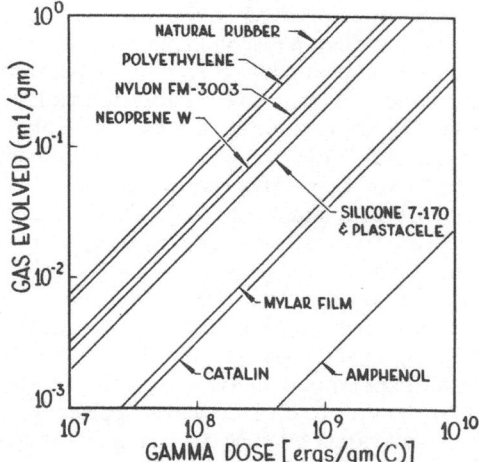

Fig. 5. Radiation-induced gas evolution from plastic
and elastomer electrical insulation.

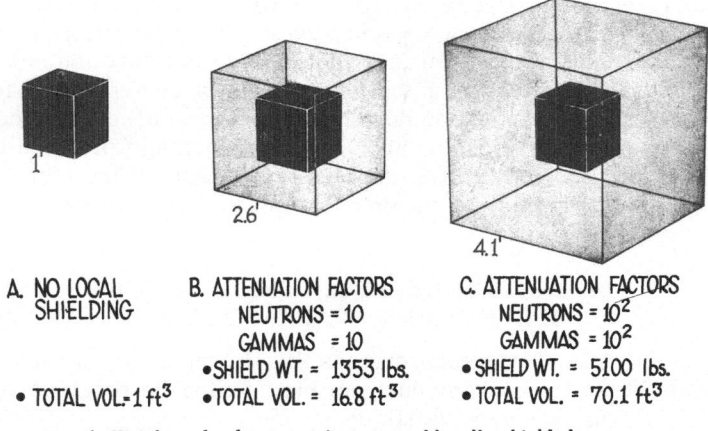

Fig. 6. Weight and volume requirements of locally shielded system.

tronic materials and parts, the self-shielding effect is small and is generally not a large factor in reducing radiation damage.

Special Effects.

Gas Evolution. The evolution of gas is a typical radiation effect noted in all organic materials, and results from bond cleavage. An example of gas evolution is illustrated in Fig. 5, where gas yield for a number of insulating materials is plotted as a function of radiation dose [1].

Consideration of gas evolution is extremely important in closed systems, such as hermetically sealed electronic parts. For example, in the design of a hermetically sealed relay, a knowledge of the amount and types of gases evolved is important if case rupture due to pressure buildup or the formation of explosive mixtures is to be prevented. In this example, the installation of a pressure-release mechanism would alleviate the problem. Damage caused by fluid "off-gassing" will be further illustrated in a subsequent section relating to dynamic assemblies irradiation.

Nuclear Heating. The energy deposited in materials and components by the scattering and absorption of incident neutrons and gamma rays may, in some cases, cause a significant rise in temperature. This is particularly true if the materials and components are located in a nuclear radiation field of high intensity. A temperature rise of about 15°F above ambient air has been observed at General Dynamics/Fort Worth in standard elastomeric compression-set buttons in a gamma radiation field of $3.2 \cdot 10^9$ ergs/g(C)-hr and a neutron radiation field of $2.2 \cdot 10^{11}$ n/cm²-sec (energy greater than 2.9 Mev). This effect must, therefore, be considered during the radiation-effects analysis and allowance made for establishing cooling requirements.

Air Ionization. Under certain conditions, the ionization of air produced by nuclear radiation can be of sufficient magnitude to create problems in part performance. An example of this ionization effect was observed by General Dynamics/Fort Worth [5] during an experiment in air-insulation resistance at the Los Alamos-Godiva Facility. Six hundred volts-dc were applied across two parallel metal plates, $3/4$ by $3/4$ in., separated by a $1/4$-in. air gap. At the peak dose rate during the burst [$\sim 5 \cdot 10^8$ ergs/g(C)-hr], the equivalent resistivity of the air was about 2.6 megohm-cm. The ratio, then, of the un-ionized air-gap resistivity to the ionized air-gap resistivity was about $1 \cdot 10^{15}$.

The free electrons produced through ionization of air by nuclear radiation can cause attenuation of the electromagnetic waves. This attenuation is, of course, a function of the free-electron density which in turn is a function of the radiation intensity and air density. In applications presently contemplated for nuclear-powered vehicles, this attenuation does not appear to be of importance. However, in public releases by the Department of Defense regarding Operation HARDTACK and, in particular, Project ARGUS, the observed effects of interference in radar frequencies and disruption of communications focus attention on this special radiation effect.

RADIATION-EFFECTS CONSIDERATIONS
IN ELECTRONIC-PACKAGE DESIGN

At this point, it is of interest to discuss the alternatives and radiation-effects methods available to the package designer in achieving the required design goals. At least four alternatives are available:

1. Relocate the system to a position of lower nuclear-radiation intensity. In some cases, relocation may be impossible from the standpoint of functional use. For example, it is desirable that an aft-looking airborne radar be located in proximity to the tail of the vehicle.

2. Shield the nuclear radiation source to permissible radiation levels. This entails vehicle weight, balance, and performance considerations. In certain instances, the weight penalties to shield the radiation sources to permissible levels would be prohibitive. Typical nuclear sources are nuclear reactors and radioisotopes.

3. Locally shield the system to permissible radiation levels. Local shielding is generally impractical because of the weight penalties involved. For example, assume that a "black box" as illustrated in Fig. 6, occupies a volume of 1 ft³ and that the neutron and gamma nuclear-radiation levels are to be attenuated equally. The weight and volume penalties imposed by attenuation factors of 10 and 100 are shown in Fig. 6.

4. Replace the critical components in a given system with components of greater radiation resistance. Attention should be given to a

system design which allows for component degradation. This alternative is further discussed in subsequent paragraphs.

The design "tradeoffs" required usually suggest a combination of the above alternatives. The degree to which a given alternative is applied is a function of the particular design and performance penalties that the designer is willing to assume. However, a thorough understanding of the limitations of each of the above design alternatives is mandatory if efficient vehicle design is to be obtained.

Materials and Parts Evaluation. Ideally, a materials and parts evaluation would be accomplished through the use of theoretical relationships between nucelar radiation and changes in physical and electrical parameters. Unfortunately, theory has not been developed to the degree that the macroscopic behavior of a material or component can be established on the basis of microscopic changes alone. For this reason, the field of radiation effects is dependent on experimental data and empirically derived relationships.

System materials and parts may be evaluated by the use of "functional thresholds" or "response relationships," provided sufficient radiation-effects information is available. A functional threshold is defined as the minimum amount of nuclear radiation required to change the properties of a material or functioning component to values outside the specification limits. Response relationships consider the total environment and the relative effect of each environment to the property changes of materials and parts.

A natural question that may be posed is, "How reliable is a functional threshold derived by the use of specifications limits as applied to an actual service test?" This is, of course, a problem which exists without nuclear radiation. The answer to this question depends on several considerations. For example, if general specifications are written that cover a large number of design problems, one would expect a rather large amount of conservatism to be "built in" when compared to actual service conditions. On the other hand, individual specifications covering a particular design application would have very little "built in" conservatism. The functional thresholds currently in use are usually derived from general specifications and are, therefore, generally conservative. It should be pointed out that functional thresholds should include, in addition to the functional-

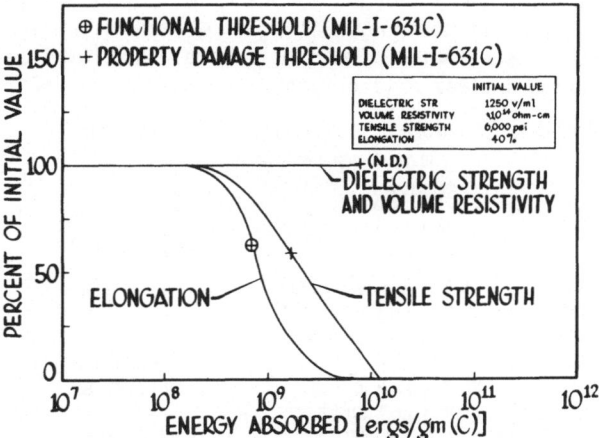

Fig. 7. Determination of functional threshold for ethylcellulose "ethocel R-2" electrical insulation.

threshold number, a statistical "confidence" which takes into consideration all possible applicable data.

Figure 7 illustrates the functional-threshold determination for ethylcellulose electrical insulation based on MIL-I-631C. Note that some four properties are plotted as a function of irradiation with the functional threshold based on the property of elongation. It must be emphasized that data were not available on additional properties, such as softening temperature (penetration), lengthwise shrinkage, dissipation factor, and dielectric constant. Both dielectric strength and volume-resistivity thresholds are based on "no damage" (N.D.) thresholds. A "no damage" threshold exists when property data have not exceeded specification limits at the conclusion of irradiation.

Unfortunately, in the derivation of functional thresholds, insufficient attention has been given to other environmental parameters and their contribution to the total response of a material or part. A more general approach to evaluation of radiation-effects problems lies in the solution of the combined environmental-effects response equation:

$$R = b_0 + b_1 x_1 + b_2 x_2 + b_3 x_3 + b_4 x_4 + b_5 x_5 + b_6 x_6 + b_7 x_7$$
$$+ b_8 x_1 x_2 + b_9 x_1 x_3 + b_{10} x_1 x_4 + b_{11} x_2 x_3 + \cdots$$

$$\underbrace{\qquad\qquad}\;\underbrace{\qquad\qquad}\;\underbrace{\qquad\qquad}\;\underbrace{\qquad\qquad}$$

Interaction Terms

where the b's are constants for a given material and environmental envelope and

x_1 = temperature
x_2 = vibration
x_3 = gamma dose
x_4 = neutron integrated flux
x_5 = gamma dose rate
x_6 = neutron flux
x_7 = gamma-to-neutron ratio

The use of this response relationship involves irradiation testing within the desired environmental envelope and subsequently performing a least-squares fit of the equation. The response relationship serves to emphasize the significance of the various environmental conditions to the total response of properties of a particular material, part, or component.

Nuclear Environmental Specifications. Several general specifications have been prepared which include nuclear environmental considerations. WADD specifica-

TABLE I. Nuclear Environmental Specification Requirements

ENVIRONMENTAL CLASSIFICATION	REACTOR NEUTRONS ($\frac{\text{Fast neut}}{\text{cm}^2\text{-sec}}$)	REACTOR GAMMAS ($\frac{\text{ergs}}{\text{gram(C)-hr}}$)	TIME (hr)	PULSE NEUTRONS ($\frac{\text{Fast neut}}{\text{cm}^2\text{-sec}}$)	PULSE GAMMAS ($\frac{\text{ergs}}{\text{gram(C)-hr}}$)	TIME (μ sec)
GRADE E *	1×10^9	1.6×10^7	1000	—	—	—
GRADE F *	1×10^{11}	1.6×10^9	1000	—	—	—
GRADE G *	1×10^{13}	1.6×10^{11}	1000	—	—	—
GROUP IV **	—	—	—	1×10^{17}	3.2×10^{13}	80
GROUP VI **	1×10^{10}	1.6×10^7	1000	1×10^{17}	3.2×10^{13}	80
GROUP VII **	1×10^{10}	1.6×10^7	1000	—	—	—

* WADD SPEC. R & D EXHIBIT WCRE 56-18 (6 MARCH 1956)
** TECHNICAL SERVICES, DEPT. OF COMMERCE ECP-2 (1959)
(NOW DOCUMENTED AS MIL-STD-446A)

tion WCRE 56-1B, "Environmental Design Requirements and Test Methods for Electronic Component Parts for Use in Airborne Equipment" [6], lists three nuclear environmental classifications (Grades E-G) of increasing severity as shown in Table I. No requirement is listed in this specification for pulse radiation testing.

ECP-2, "Environmental Requirements Guide for Electronic Component Parts," was prepared by the Advisory Groups on Electronic Parts and on Electron Tubes through the Office of Director of Defense, Research and Engineering [7]. This specification groups the nuclear environmental requirements in terms of both reactor and pulse radiations (Table I).

The greatest deficiency in available nuclear environmental specifications appears to be in test methods and procedures.

DYNAMIC-ASSEMBLY IRRADIATION TESTS

Although the knowledge gained from irradiation tests on individual parts and materials is necessary to establish functional thresholds or response relationships, the knowledge gained from dynamic-assembly testing is invaluable in checking predictive methods and data, and in obtaining results normally unattainable on the material and part level. The introduction of dynamic forces simulating actual operation of the parts and materials in a nuclear-powered vehicle may affect the test results in a way that cannot be theoretically predicted. For this reason, perhaps the most important aspect of these tests is that they provide data on the performance of materials and parts under combined environments and typical operational conditions.

Gyro and Accelerometer Dynamic Assembly. During the latter part of 1959, a U. S. Time Company Rate Gyro and Dynamics Measurements Accelerometer dynamic assembly was modified for radiation resistance and subsequently irradiated. The unmodified gyro and accelerometer dynamic assembly as it existed prior to modification is presently used in conjunction with the General Dynamics/Fort Worth B-58 autopilot. Figure 8 shows the gyro and accelerometer dynamic assembly following irradiation to a maximum gamma dose of $1.0 \cdot 10^{10}$ ergs/g(C) and an integrated neutron flux of $1.0 \cdot 10^{15}$ n/cm^2 ($E > 2.9$ Mev) [8, 9].

The results of the postirradiation tests revealed that the normal axis of the accelerometer failed after receiving a gamma dose of $2.0 \cdot 10^9$ ergs/g(C) and accompanying neutrons, and that the lateral accelerometer axis failed after receiving a gamma dose of $7.0 \cdot 10^9$ ergs/g(C) and accompanying neutrons. These accelerometer failures were attributed to coagulation of the damping fluid, a mixture of DC-550 and DC-710. The gyros lost most of their DC-710 damping fluid during irradiation, because of fluid off-gassing, which resulted in rupture of the temperature-compensating bellows. Unfortunately, the exact dose when the gyro-case rupture occurred could not be determined from the data. Preirradiation analysis of the gyro and accelerometer dynamic assembly predicted the limitations and "off-gassing" noted for the damping fluids in the above-described applications.

Sperry Magnetic-Flux-Valve Dynamic Assembly. During the early part of 1961, a Sperry flux-valve dynamic assembly was irradiated. Figure 9 shows the flux valve prior to irradiation. A flux valve is a device which transposes the earth's magnetic field into an interpretable directional heading.

The Sperry flux valve was irradiated to a gamma dose of approximately $2.6 \cdot 10^{10}$ ergs/g(C) and an integrated neutron flux of $2.4 \cdot 10^{15}$ n/cm^2 ($E > 2.9$ Mev) [9]. Preirradiation modifications to increase the radiation resistance of the flux

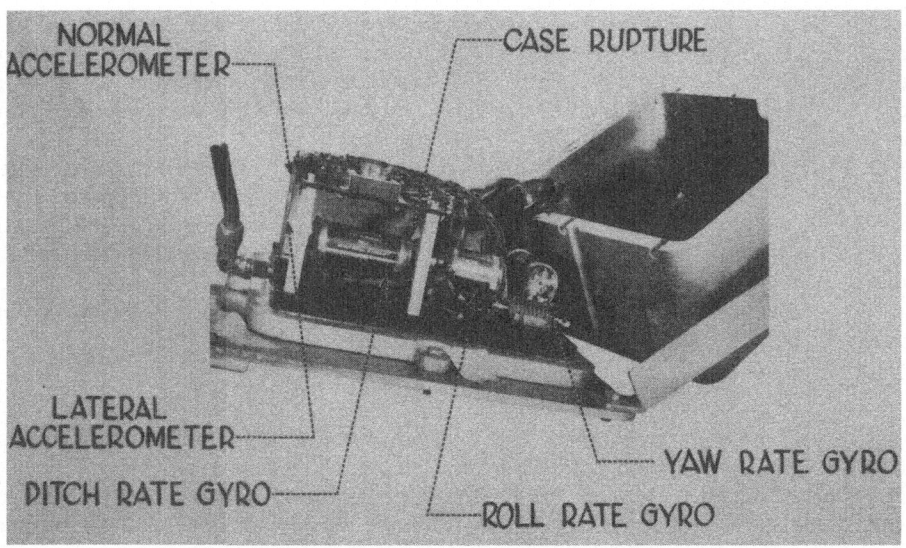

Fig. 8. Gyro and accelerometer irradiated dynamic assembly.

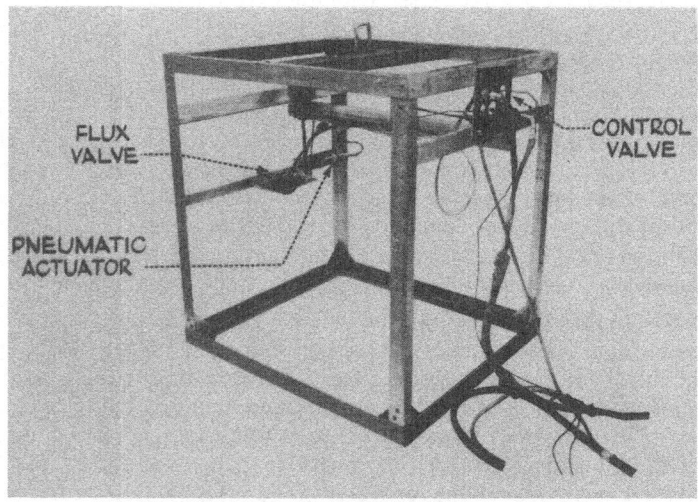

Fig. 9. Sperry magnetic-flux-valve dynamic assembly.

valve included change-out of magnet coil wire and replacement of all Teflon electrical standoffs. Glass-coated silver coil wire was substituted for the previously used lacquered copper wire. A melamine formaldehyde plastic was used for the spool modification; other modifications included replacement of all Teflon with ceramic materials. The damping fluid (DCF-60) did not require replacement, because of the relatively high radiation resistance of this fluid compared to other candidate damping fluids.

Gelation of the DCF-60 damping fluid occurred as expected at a gamma dose of approximately $2.0 \cdot 10^{10}$ ergs/g(C) and accompanying neutrons. As was predicted, a linear pressure build-up occurred in the valve case because of radia-

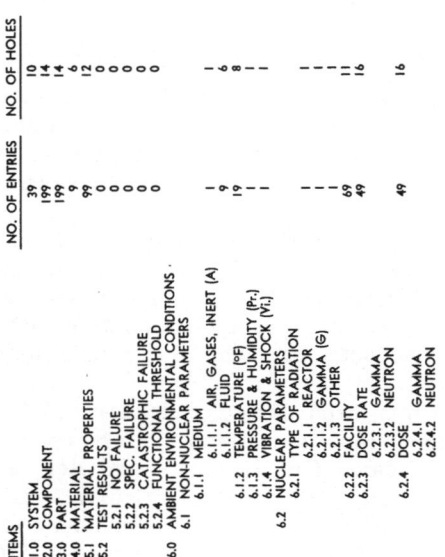

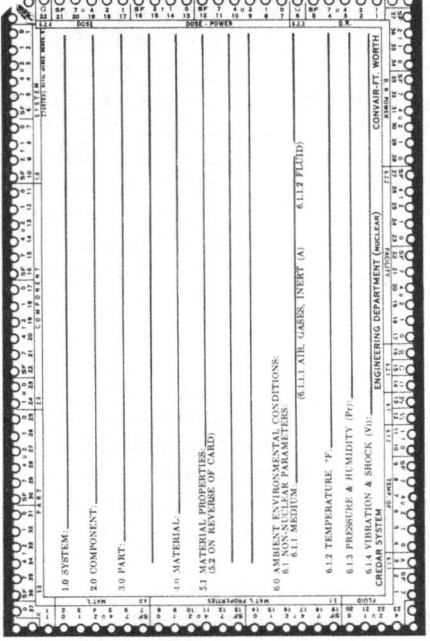

Fig. 10. Convair radiation-effects data analysis and retrieving system (CREDAR).

tion-induced off-gassing of the damping fluid. Final postirradiation results of the flux-valve dynamic assembly are not available.

RADIATION-EFFECTS DATA MANAGEMENT

The increased importance attached to the role of information retrieval in the world's technical organizations is evidenced by the number and frequency of articles being published on this subject. It has been estimated that the world's technical journals carried about 60 million pages of scientific reports during 1960 [10].

Past experience has indicated that progress in technical fields has been seriously impeded by the lack of data handling methods. This difficulty has arisen from the fact that the amount of pertinent data has become so staggering that efficient indexing now becomes mandatory. Data accumulation without efficient indexing and retrieval methods accounts for much of the duplication in the research field at the present time.

At General Dynamics/Fort Worth, hand-sorted, punched-card filing systems have been developed (1) to store radiation-effects reference data and (2) to record analyzed radiation-effects data. The Punched-Card Reference System [11] contains such information as author, title of the particular item, publication date, agency, and a brief abstract of the reference. The periphery of the card is coded to permit rapid retrieval of cards on the basis of publication date, material topic, author, material properties, agency, and radiation. This reference system is of particular value for maintaining cognizance of the radiation-effects field.

The Convair Radiation Effects Data Analysis and Retrieval System (CREDAR) was designed to bridge the gap between the published raw radiation-effects data and analyzed processed radiation-effects data. The basic analyzed results of the irradiation test are typed on the face of the card and include system application, component, part, material, property changes, nuclear and nonnuclear data, test results, empirical response relationships or functional threshold and basis, facility, and references. Coded data around the periphery of the card include the system application, component, part, material or part properties, facility, and environmental test conditions (nuclear and nonnuclear). The time to retrieve information using these systems is in the order of minutes compared to the weeks often required to obtain specific information. Figure 10 illustrates the format for the CREDAR system keypunch cards.

CONCLUDING REMARKS

The induced environment of nuclear radiation is one additional environment of concern to the electronic-package design engineer. Although considerable data are available on radiation effects for materials and parts, there still exists large data gaps requiring additional irradiation testing. This data deficiency is particularly evident with regard to the irradiation of materials and parts under combined environmental conditions.

Where equipment is required to operate in a radiation environment, specifications should be prepared that include nuclear environmental conditions, pertinent nuclear test methods, and candidate materials and components. The nuclear test methods must encompass combined environmental conditions.

A data-management system is mandatory for maintaining cognizance and avoiding duplication in the radiation-effects field. The investment made in data management often prevents the possibility of literally redesigning the "wheel."

REFERENCES

[1] "Radiation Effects — Methods and Data," Convair-Fort Worth Report FZK-9-134, NARF-58-43T (Oct., 1958).
[2] P. M. Johnson, Private Communication.
[3] W. R. Burrus, "Standard Instrumentation Techniques for Nuclear Environmental Testing," Wright Air Development Center Technical Note 57-207 (Dec., 1957).
[4] A. B. Spears, "Transient and Dose Effects of Reactor Radiation on the Dielectric Strength of Several Insulating Materials," Convair-Fort Worth Report FZM-2050 (Jan., 1961).
[5] C. G. Collins and V. P. Calkins, "Radiation Damage to Elastomers, Plastic and Organic Liquids," General Electric Company, Atomic Power Division, Aircraft Nuclear Propulsion Dept. Report APEX 261 (Sept., 1956).
[6] "Environmental Design Requirements and Test Methods for Electronic Components Parts for Use in Airborne Equipment," Wright Air Development Center, Directorate of Research, Electronic Components Laboratory R and D Exhibit WCRE 56-1B (March, 1956).
[7] "Environmental Requirements Guide for Electronic Component Parts," Advisory Groups on Electronic Parts and on Electron Tubes of the Office of Director of Defense Research and Engineering; available through Office of Technical Services, Dept. of Commerce, Washington 25, D.C., ECP-2 (1959).
[8] "Radiation Effects," NARF Progress Report (Aug. 1, 1959-Jan. 31, 1960), Convair-Fort Worth Report NARF-60-3P.
[9] "Radiation Effects," NARF Progress Report (Aug. 1, 1960-Jan. 31, 1961), Convair-Fort Worth Report NARF-61-3P.
[10] "Research Trendletter," Industrial Research, (Scientific Research Publishing Company, Chicago, Apr.-May, 1961) p. 29.
[11] J. W. Gordon and P. M. Johnson, "A Punched-Card Reference System of Radiation-Effects Data," Convair-Fort Worth Report FZM-1983 (Apr., 1961).

DISCUSSION

Question: Jim Milligan, Dynamic Astronautics. I would like to ask Mr. Levine about this gassing effect of liquids which was apparent in this dynamic assembly experiment. Were the various containers designed to take this gas pressure? Is this a transient condition where they would again return to a liquid form?

Answer: In response to your first question relative to the gyro and accelerometer dynamic assembly case structure, there were no special structural provisions made in the case. I should emphasize that the modifications were only those that could be performed easily without recourse to extensive redesign. Relative to your second question, as to the off-gassing of the gyro fluid being a transient condition, this is not a transient condition since recombination of the gas with the liquid would be considered negligible. An important point to be made is that with recognition of the gas problem, special provision for removal of the gas is possible. I should emphasize that off-gassing in closed containers can cause significant damage as evidenced in irradiated-oil-filled capacitors where case ruptures have occurred.

Question: Anton Oswald, Kearfott, Clifton, New Jersey. What is the material that you are using in the nuclear radiation shielding?

Answer: I should first emphasize that I am not a shielding specialist; however, I can answer you in a general way. The materials used in shielding are dependent upon the type of radiation you are attempting to shield. For example, gamma radiation will be most effectively attenuated by high-density materials such as lead and steel. Neutron radiation is most effectively attenuated by hydrogenous materials or materials of low atomic number, such as water and most organic compounds. Where water or liquids cannot be used easily, plastics such as polyethylene may be used. Other considerations in selecting a shield material would be its fire resistance, radiation resistance, degree of toxicity, resistance to aging, resistance to oil or solvents, ease of application, insulating properties, etc.

Question: George Pettibone, General Electric, Johnston City. I have two questions. The first concerns test equipment. Do you have any recommendations as to how a firm can provide test equipment, or are facilities available in the country? Second, is there any correlation between radiation effects and temperature effects? Could you get preliminary tests by providing a temperature test that would give an indication of what you might expect in radiation?

Answer: In response to your first question, as to how a firm can provide test equipment, I am not sure I understand what you are referring to when you say "test equipment." There are a number of companies that are manufacturing test equipment for use in nuclear facilities. Some of this equipment is fairly radiation resistant. With regard to your second question as to the correlation of temperature and radiation effects, I would say in some materials the damage mechanisms are similar—particularly in elastomers—but it would be extremely difficult to predict performance of the material in a radiation environment by preliminary temperature tests alone.

Comment: (Mr. Pettibone). I really implied what kind of test equipment would be required to test devices that were being made; in other words, sources of nuclear energy. I presume this is a very difficult type of test equipment to provide and very few people have it.

Answer: Yes, I would say that the number of test facilities available for conducting extensive irradiation tests are rather limited in number. However, there exists an adequate capacity to take care of future needs as they are anticipated today. At General Dynamics/Fort Worth, we have an elaborate nuclear facility with a 3-Mw radiation-effects testing reactor of the swimming pool type. Equipment can be lowered to three sides of the reactor face on the dry side of the facility pool. Pallet capacities of several thousand pounds and volumes per pallet in the order of 4 by 4 by 5 feet are available for test specimens. I fully realize the need by industry for specialized nuclear facilities of this type and feel that an industrial participation program could successfully answer your nuclear environmental requirements. I would appreciate any suggestions you might have on a participation program of this type.

Question: Harry Nash, General Instrument Corporation. Could you tell me what data you have to show the maximum radiation level for semiconductors?

Answer: There has been a considerable amount of data reported in the literature. I would suggest a report prepared by GD/FW NARF (Nuclear Aircraft Research Facility) titled, "Radiation Effects—Methods and Data," FZK-9-134. I have a copy in my possession, which I would be happy to show you. GD/FW is continuing to study the radiation-effects problem with semiconductors as are other companies. I should emphasize that the transient-effect problem with semiconductors might be somewhat different that the steady-state problem. I am referring to the effects you might see under steady-state reactor radiation conditions versus nuclear detonations.

Question: Bill Lyall, General Electric. I would like to extend Mr. Pettibone's question a step further. Could you comment or tell us anything you know about a good controllable measurable source of pulse radiation available for testing or evaluating larger electronic assemblies?

Answer: There are available several pulse sources which are used to study transient effects. These are the General Atomic Triga and the Los Alomos Godiva Facilities. The test volumes are somewhat limited. With the lifting of the nuclear test moratorium by the Russians, I would expect opportunities to be available for exposing large assemblies during nuclear weapon tests.

Question: (Mr. Lyall). Is Godiva II reasonably measurable and controllable?

Answer: Yes, it is. I believe the earlier Godiva was severely damaged due to an intense radiation pulse. This damage was due to the amount of moderating material present in the earlier Godiva installation. There are limitations, at present, as to the amount of test material that may be placed near the reactor safety screen.

Question: Charles Vannamen, Lear Inc., Grand Rapids. I would like to know what you based your size and weight increase on for the gamma and neutron shielding.

Answer: I assume you are referring to the example used in Fig. 7 in the paper. A combination of lead and plastic materials were assumed in the shielding calculations. However, I should emphasize that by using more sophisticated shielding methods and materials than were used in this illustration, some weight and volume savings might be realized. The purpose of this example was merely to emphasize the penalities involved when resorting to the shielding alternative alone. If we are going to be concerned with nuclear powered mobile vehicles or spacecraft, weight and volume certainly will be important.

Question: (Mr. Vannamen). Does this use solid metals or have you considered powdered metals also?

Answer: Are you talking about shielding?

Question: Yes.

Answer: The shielding assumed in the example in my paper was considered to be solid; however, I would assume powdered metals could be used, provided you knew the nuclear properties of these materials.

Question: Fred Cohen, RCA, Summerville. In your chart of relative radiation resistance, you show that metals and ceramics are more resistant to neutron radiation than, say, semiconductors. Does this indicate that the use of metals and ceramics for packaging semiconductors would be more effective against radiation than glasses or epoxies?

Answer: The point to make is that the semiconductor will probably be more sensitive to radiation than the materials used for case construction. For this reason, using metals and ceramics in packaging would not in itself protect the semiconductors, with the exception of the slight amount of shielding the metal would provide.

Question: Art Shafran, Philco Corporation. In Fig. 6 you indicate that the gases evolve from the different materials. Have you been able to detect what gases these different materials evolve?

Answer: Yes, hydrogen is rather dominant in the light hydrocarbon evolved gases. For this reason, explosive mixtures are possible, particularly in closed containers such as relays.

Question: Frank Kottwitz, Collins Radio, Cedar Rapids. In relation to the equation on combined response, do I take it to mean that would be an over-all effect on the chemical deterioration or something similar?

Answer: The response relationship referred to in the paper would be used to fit empirical data obtained in performing environmental testing within the desired envelope. Once the parameters are evaluated in the response relationship, the response of the property of a material may be evaluated at any point within the environmental envelope without recourse to an infinite amount of testing. This permits a real saving in test time.

Comment: Then the response we are talking about might be one of several things. It might be tensile strength.

Answer: Yes, that is correct. You could fit the response equation to empirical data of any property.

Question: Having established that, then I would like to ask, where did you dig this equation up? Is this strictly an empirical thing?

Answer: Yes, the response relationship is used to fit empirical data of the type previously described.

Comment: The thing that actually prompted my question was the nature of the temperature term in terms of the chemical reactions. I would observe that this is certainly more unusual than usual in that respect. And would, in fact, question the validity of the temperature term used in an equation in that fashion. In those instances when we are talking about a chemical reaction I would think that it could be some other kind of term.

Answer: The response relationship illustrated in the paper was a linear equation; whereas, it is entirely possible that the relationship required to fit certain empirical data would be a relationship of higher order.

ELECTRONIC PACKAGING FOR OIL-WELL LOGGING

Lyman M. Edwards

Pan Geo Atlas Corporation, Houston, Texas

INTRODUCTION

Energy is a most important factor in the development of the modern world. Although energy can be obtained from many different sources, the greater share of industrial energy comes from coal, natural gas, and petroleum. In 1958 the United States required 41,000 trillion Btu, and it is estimated that by 1975 at least a 70% increase in the 1958 requirement will be needed for the United States alone. Thermal and mechanical energy will be most important but electrical energy consumption will expand most rapidly. Petroleum will continue to provide the largest share of the growing need for energy sources. Known domestic reserves are not adequate to support future demand. By 1975 the United States will need to import nearly one-third of its petroleum needs and one-fourth of its natural gas requirements. The world supply, known and estimated reserves, is sufficient for the immediate future, but improved methods of exploration and production must continually be sought to improve the economics of the oil industry, and to conserve and locate further oil-bearing areas to keep abreast of world demands.

Special mechanical and electronic instruments, packaged to withstand the terrific pressures and high temperatures present in deep-drilled oil wells of today, have played an important role in the location and recovery of much of the world's oil and gas. The petroleum industry is quite familiar with the multitude of various special services provided by these ingenious instruments, and has long accepted them as necessary and essential to their continued success. To the vast field of experts in the modern day rocket and missile field and to those occupied by advancements in the fast moving space age, the science of packaging delicate electronic circuitry containing vacuum tubes and semiconductors to be lowered several miles below the earth's surface in fluids whose hydrostatic heads often exceed 10,000 lb/in.^{2} and whose temperatures are often 400°F or greater, is relatively obscure.

In order to understand the problems encountered in this field more clearly, a very brief review of the procedure used in drilling an oil or gas well is in order.

These wells are drilled to great depths through many formations in the attempt to locate and bring the oil or gas to the surface so that it can be made useful to industry. This drilling is accomplished by a "bit" placed at the lower end of a drill string and rotated so as to cut through the formations to a planned depth. The cuttings thus obtained are brought to the surface by fluid or "drilling mud" forced down inside the drill pipe by high pressure pumps and returning to the surface carrying the cuttings.

When a depth of interest has been reached, or when the total depth intended to be drilled has been reached, the bit and drill string are withdrawn to the surface to permit entry of a logging instrument so that a "survey" may be made. A survey is made by lowering the desired instrument into the hole on the end of a cable containing insulated conductors which permit power to be transmitted from equipment at the surface to the instrument and enable signals, proportional to the parameters being measured, to be transmitted from the instrument in the hole to the surface equipment, where they are recorded.

The practical and economic limits of cable design generally result in a very limited number of individual conductors or, as is common in many circumstances, may even be reduced to a single conductor over which it is necessary to transmit power in one direction and one or more signals in the other direction—sometimes on each individual conductor. To accomplish this many various and ingenious signal-transmission systems are used.

Since the holes drilled are more or less of cylindrical shape, the maximum use of space necessitates the design of a protective case, cylindrical in configuration. To withstand the high pressures due to several thousand feet of high density fluid with its resultant hydrostatic head of rather large proportions, it is necessary to carefully select the type of material used for the cylindrical housing, maintain proper diameter-to-length ratios, include a sufficient safety factor in case thickness, provide proper attachment means from instrument housing to the cable itself and, at the same time, permit easy access to the electronics circuitry inside for inspection and maintenance.

The parameters of the earth's formations usually measured and sent to the surface for permanent recording consist basically of the following: lithological identification of the formations traversed by the drill; determination of characteristics of the formations favorable for the accumulation of oil or gas (such as computation of porosity, permeability, and saturation); identification of fluids or gases present; reservoir characteristics permitting quantitative evaluation; and location of mineral deposits in or near the borehole.

The most common types of logs used to obtain this information are an Electrical log (a measurement of formation resistivity); a Radiation or Nuclear log (a measurement of natural gamma radiation and measurement of induced gamma effects); an Induction log (a measurement of formation conductivity); an Acoustic or Velocity log (a measurement of the acoustic properties of the earth's formation, i.e., velocity, amplitude, and attenuation); a Nuclear Magnetism log (a proton-precession log, measuring the free-fluid index and the thermal-relaxation time); and a multitude of additional logs such as Temperature, Caliper, Dipmeter, Photoclinometer, and the Microfocused logs. There are, also, additional services performed in the borehole not directly classified as logs, but just as essential. These are casing perforation (Bullet and Jet), Sidewall Coring or Sampling, Fluid Sampling, Collar Location, and Depth Control. All of the equipment and services are basically of electronic design and require many special electronic components.

Boreholes drilled in the ground are cylindrical in shape; therefore, to make maximum use of the geometry of the borehole, most oil-well logging tools are encased in protective cylindrical pressure vessels. The diameter of these vessels or housings generally varies from slightly over 1 in. in diameter to 4 or 5 in. in diameter, and from 3 or 4 ft in length to 20 or 30 ft in length, depending on the geometrical spacing required for coils or electrodes, and upon the cubic capacity necessary to contain the electronics. These tools are secured to the end of a special single- or multiple-conductor cable designed to satisfy the rigorous requirements of downhole conditions. This cable is spooled on a drum on a truck and the drum is rotated by power from the truck in order to lower the tool or instrument into the hole, and to retrieve the instrument during or after the logging operation.

Let us consider the conditions that are expected to be encountered in the borehole.

TABLE I. Hydrostatic-Well Pressures

Type of fluid	Weight, lb/ft^3	Density	Depth of hole, ft	psi
Fresh water	62.43	1.0	20,000	8,660
9-lb drilling mud	67.3	1.08	20,000	9,352
Salt water	72.0	1.154	20,000	9,994
15-lb drilling mud	112.2	1.80	20,000	14,588
18-lb drilling mud	134.6	2.16	20,000	18,706

SUBSURFACE TEMPERATURES

The temperatures encountered in a borehole in the earth's surface generally increase with depth, although the rate of increase varies considerably in different areas due to many factors. First, it is generally acknowledged that the core of the earth itself has an extremely high temperature, that this heat energy is radiated outward to the surface, and so produces a temperature gradient related to the depth, but not consistent in all areas in the slopes of the gradients. Normally, we think of heat as being transferred from one point to another by conduction, convection, and radiation; but conduction is the largest factor in the temperature of the earth's subsurface. Due to the many types of geological formations below the surface, the factor of conduction is altered in accordance with these geological conduction properties of the beds. To reach a more practical understanding of the temperatures about which we are talking, the average temperatures in wells along the Gulf Coast are generally equal to the average surface temperature, plus the depth of the point in question divided by 64. Due to the compaction of certain highly conductive geological formations in many of the areas now being drilled, temperatures of 400°F are sometimes encountered in wells of from 10,000 to 20,000 ft in depth. This means, of course, that instruments placed in wells for the purpose of making earth measurements must have components capable of operation when well-fluid temperatures of 400°F or over are encountered. Some of the subsurface tools used must, of necessity, employ the use of many heat-producing components such as vacuum tubes, transformers, and power-dissipating resistors. When the temperature in the downhole instrument is likely to exceed the borehole fluid temperature, means are found to employ the exterior housing of the instrument case as a heat sink for the heat-producing components in order to pass the heat generated internally to the borehole fluid. In some cases, certain measuring devices or components are used within the instrument that are not heat generating in themselves and, furthermore, cannot operate successfully at temperatures normally encountered in the borehole. In these cases, many different means are used to reduce the temperatures of the container encompassing this particular component. In many cases, the method thus employed is an electrochemical refrigeration process and, more recently, thermoelectric cooling devices have been employed.

It can be seen from the table of well pressures related to depth of the well, and well temperatures also related to depth, that any instrument placed in a deep borehole must be capable of withstanding and operating successfully in pressures of 15,000 to 20,000 psi and at temperatures of 400°F or higher. Furthermore, the electrical measurements thus obtained under these conditions must remain un-affected by both temperature and pressure changes during the measurement time which, in some cases, may be several hours except, of course, for instruments employed solely for the measurement of pressure and temperature.

Many instruments employed in making borehole measurements have all of their components completely enclosed in pressure cylinders or housings, and are thus relieved of the necessity of withstanding these pressures themselves, since the outer case eliminates the pressure effect on components within. There are, however, many instruments and tools which have external moving parts which are actuated by signals or controls from the interior instrument housing. In these cases, means must be employed to allow mechanical motion to be imparted from the interior of the tool to the exterior part. For example, many devices employ prime movers inside the case, such as stepping switches or electric motors, which drive shafts extending through the housing to some external part. This means, of course, that there must be an adequate seal around the shaft which protects the internal components from external pressures while, at the same time, allows the shaft to rotate without undue loss of driving power, due to reduction of friction on the pressure seal. The number of cir-cumstances in which this principle is employed is rather small and, most generally, the same effect is accomplished by eliminating the pressure differ-ential and making the internal pressure equal to the exterior pressure. An example of this is the use of an electrically nonconducting fluid, such as silicone fluid, sealing the internal chamber and its components, and providing a suitable pressure compensating device, such as a bellows or a moving piston. This allows motion of a driving shaft without the friction loss normally encountered where pressure differential is maintained. This, however, presents another problem, in that the components which are immersed in the insulating fluid must all be capable of withstanding the same pressures that are encountered in the borehole fluid. Many electronic components will withstand high pressures pro-viding no space voids exist within them. It is also necessary, for observation of instrumentation and for routine maintenance of the instrumentation, that a rather rapid and simple means be employed to remove and replace the pressure hous-ing on the downhole instrument. One of the most popular sealing methods used, both for this purpose and also for the purpose of sealing around a revolving shaft, is the use of an O-ring. Figures 1, 2, and 3 illustrate the principle and effective use of O-ring and quad ring seals described above. It is also necessary to have a rapid method of connecting and disconnecting each of the various downhole instruments from the conductors and armor of the cable. This is ac-complished by an insulated pressure-proof head. Figures 4, 5, and 6 show the various connectors generally employed in this head. It is also desirable, in some cases, to isolate the various parts of downhole instrumentation from each other, and to be able to substitute other units in their places without discon-necting the entire tool from the cable. This is accomplished within the tool by the use of rotary-plug and slip-ring assemblies. Figure 7 shows a typical slip-ring assembly and Fig. 8 shows some downhole instrumentation. The use of transistorized circuitry for downhole use is becoming more popular, both for the purpose of miniaturization and because most signal-receiving semiconductor circuitry does not present the problem of additional heat generation which is encountered by the use of vacuum tubes. Extreme care must be exercised in the design of suitable stabilization and compensation networks to minimize the change in functions of semiconductor circuitry operating over so wide a temperature range.

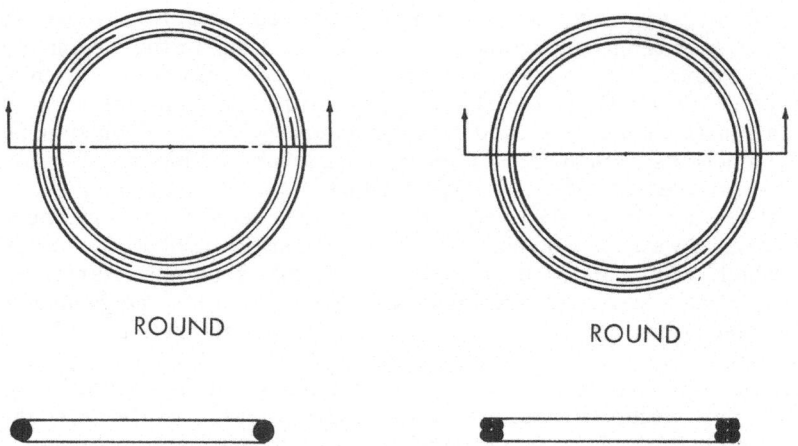

ROUND   ROUND

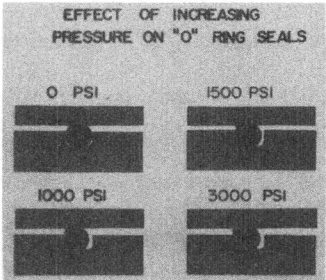

Fig. 2

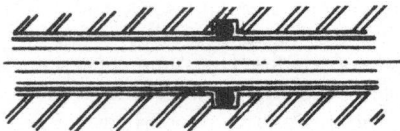

Fig. 3

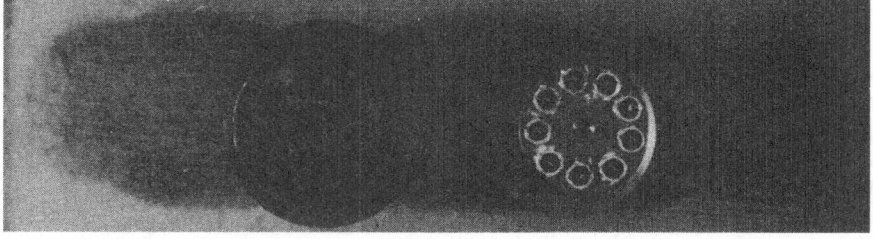

Fig. 4

Fig. 5

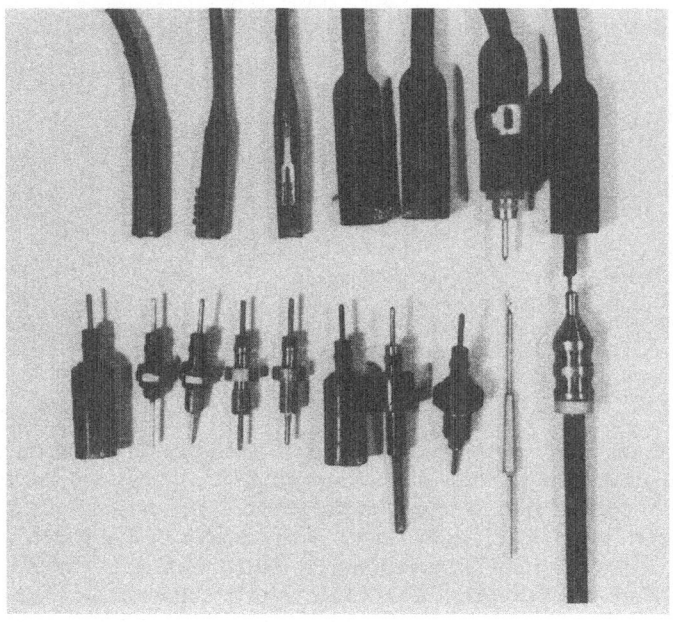

Fig. 6

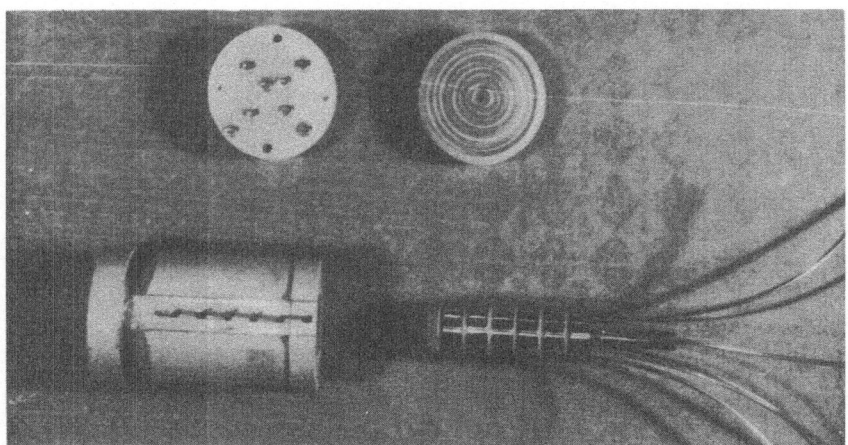

Fig. 7

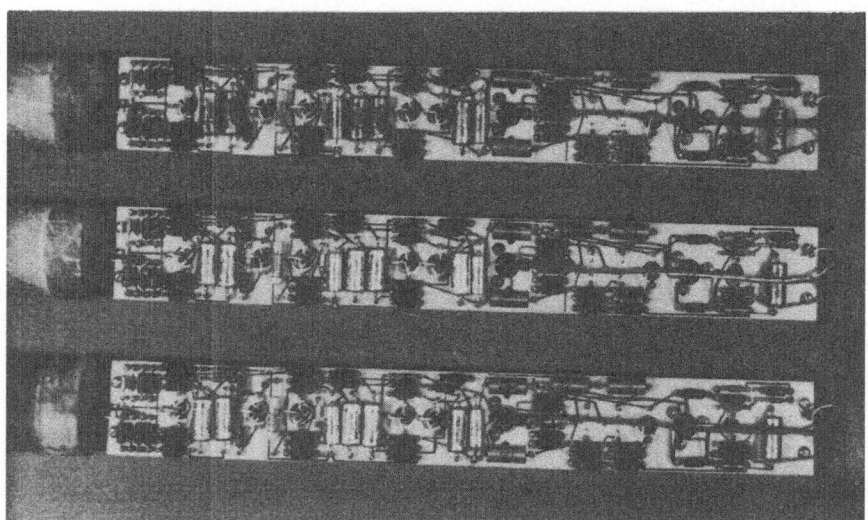

Fig. 8

Much has been said of the packaging and operation of instrumentation employed in the downhole portion of well logging. This phase of the operation is an extremely important one, because of the physical limitations of the boreholes, which are becoming smaller in diameter with greater depth of drilling, and because of the many difficult limitations of instrument components due to high pressure and high temperature. The cable which connects all of this downhole instrumentation to the control and recording panels at the surface is necessarily a very important part of successful field operations.

SURFACE EQUIPMENT FOR DOWNHOLE INSTRUMENTS

The remainder of this paper will be devoted to the description of equipment employed on the surface to power the downhole instruments and to convert the

signals received from them for suitable presentation in record form. The scope
of oil-well logging started with extremely simple devices and consisted of only
one or two services actually performed in the well. Within the last 25 years, however, the number of completely different services performed in a drilled hole by
most of the larger companies in this field has increased from the one or two
originally employed to 20 or 30 now considered desirable in various combinations. Large trucks are used to transport the tools, surface panels, cable, and
power supplies to the oil well, and to provide motive power for lowering and
raising the several thousand feet of cable and instruments for the purpose of
making the survey. Most of these services which provide a record of the results
obtained are recorded on film by the use of a multigalvanometer oscillographic
camera.

The necessary surface equipment generally consists of the camera, surface
control panels for the various instruments being run, and power which is supplied
to the downhole instrument. On instruments in which several different parameters
are measured simultaneously with the same tool, it can easily be seen that because of physical limitations of the borehole, all of these measuring transducers
will be situated at the same depth reference point in the borehole. It is therefore
necessary to provide, generally at the surface, a delay system which accepts signals from transducers in the upper portion of the tool, stores them in the memory
circuits of the delay system, and reproduces them at a time coinciding with the
location of the lower transducer's arrival at the same depth point. This applies
when the tool is being moved up the hole during the logging process. This allows
all of the signals to be placed on film with a common depth reference point, even
though the instruments producing them may be separated physically by several
feet. The trend in surface instrumentation is such that all of the equipment
generally considered necessary for the performance of each one of the special
services previously described can be constructed in small compact units, and

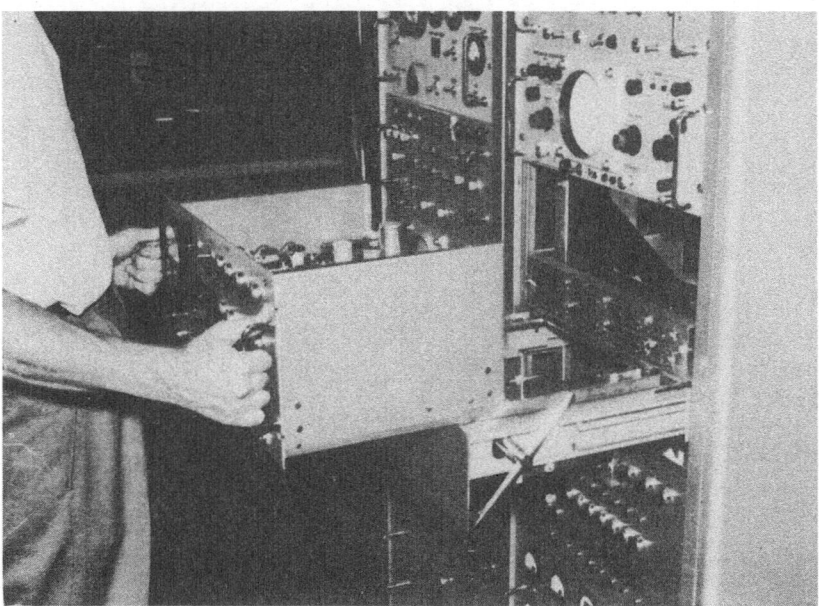

Fig. 9

Fig. 10

that the control panels pertinent only to the performance of a specific service be limited in size and circuitry to the functions necessary for that service alone. This allows much greater flexibility when new instruments are conceived and integrated into the already established general services. A minimum of circuitry changes is necessary due to the application and use of programing plugs and panel plug-in devices. Figures 9 and 10 show some of the typical panels and the method of programing and interchanging devices employed in their use.

It is believed that the experience gained and the solution to problems obtained as the result of many years of experience in the field of subsurface oil-well logging may well benefit those in the space and missile field and, more particularly, those interested in the solution of pressure-seal and packaging problems in connection with the underwater acoustic antisubmarine.

DISCUSSION

Question: Martin Camen, Bendix Corporation. For these long runs of cable that you must have, I am curious to know what kind of effects you have due to losses in the cable when you get to your circuitry.

Answer: This is a very major problem in well logging and all circuits have to be designed so that they will tolerate these large changes in resistance, due to the change in the temperature of the cable that is lowered in and out of the hole; it is also important to minimize the effects due to a change of capacitance between conductors, because there is a change of capacitance. Most generally these signals are converted to rather high-level signals in the downhole tool, so that these effects are negligible. They have to be that way because these are measurements, absolute measurements, such as resistivity and so on, and have to be repeatable because they are of great value.

Question: (Mr. Camen). Are these voltage levels ac or dc?

Answer: These voltage levels are everything from dc to ac, 60 cycle, 400 cycle up to 200,000 kc and also at low-frequency power. It would take me some time to explain why, but many services can be done only by one or more of these methods, generally tied in with sending a great number of signals over the same conductor.

Question: Beck, Bendix Research. Who makes your high-pressure cable?

Answer: The high-pressure cable is made by several companies. The principle maker is American Steel and Wire, a division of U.S. Steel; another company doing quite well in the business is the Vector Manufacturing Company of Houston, Texas.

Question: On your chassis, I presume you're working up to 400°. Do you have any trouble with your components, and do you have to use any special components?

Answer: Oh yes, that's one of our biggest troubles. We have had over the years to develop a great number of special components ourselves because the field of requirement of some of the components is very small. And therefore we have developed our own and had them accepted. All capacitors have to be of the "Teflon" type and if you have had very many capacitors made of high capacitance of "Teflon" you will find that it's pretty expensive.

Question: How about transistors?

Answer: They are getting better and better, temperaturewise, and I might, if you want me to, mention a couple of makes that we have had very good success with. I am not sure of the type but there are some types that you can test and some methods of compensation you can use that will do the job. In fact you just saw one tool up here which is entirely transistorized. This is one of the tools that provides the pulse and measures the time in microseconds of travel from the pulse to the receiving transducers. Both G.E. and T.I. have excellent transistors in this field now.

Question: Bob Condon, Boeing. In one of your early pictures, showing the end cap used on the instrument case, you showed two O-ring seals and then a small seal over the end of the cap itself. Last time I messed around in hydraulics, which was some years ago, we had difficulty using multiple seal on the interface between high and low pressure. Has there been some change in the state of the art in these seals?

Answer: Yes, there has. Quite often there are many O-ring seals used in one seal. Generally the first seal is just a safety seal and designed, if possible, to withstand the pressure, also to keep mud from getting into the thread of the barrel. And then sometimes a second and third one are used more for safety. One O-ring, properly seated, will withstand the pressure. There has also been quite an improvement in the materials in O-rings too. Of course, we are using "Teflon" and many other materials which were not available a few years ago.

INSTRUMENTATION EQUIPMENT FOR POLARIS-FIRING SUBMARINES INSTALLED IN A UNIQUE MODULAR DESIGN

Aubrey H. Jones

Lockheed Electronics Company, Avionics and Industrial Products Division*

INTRODUCTION

This paper discusses the parameters and solutions used in establishing a unique packaging concept for electronic instrumentation equipment aboard the latest Polaris-firing nuclear submarines. Equipment of this type is normally installed in rack or console cabinets, or in a smaller bench-mounted housing, but does not fulfill the requirements of installation within a submarine.

The rack or console type of mounting limits the portability of equipment making up a system while the bench-mounted type of housing doesn't lend itself to the integration of many pieces of equipment into a well arranged system (see Fig. 2). The system described herein is one in which the desired features of both the rack or console type of mounting and the bench type of mounting are used to obtain a portable and well integrated system using the standard and conventional chassis-mounted electronic equipment. Other required and desirable features are also introduced into the design concept as can be seen from the technical discussion.

Basically, this design provides accommodations for the installation of 19-in.-wide chassis-and-panel type of electronic equipment into individual module shells such that when stacked they will provide the structural integrity of a rack installation (see Fig. 3). The module shells, through variation of height, are easily adaptable to many sizes of "chassis and panel" equipment and yet are essentially no larger than their normal front panel dimensions. Assembly (and disassembly) of the modules is relatively easy and can essentially be achieved with front access only. It will also be noted that the module shells provide for protection against handling abuse of individual chassis units during transportation, installation, and system operation.

It is stressed that the design discussed was engineered to fulfill a particular need with a specific set of design criteria. However, the design concept need not be limited to this particular application. It is believed that after the following discussion of the design, it will be evident that the technique can be used in many and varied applications. Also, it is hoped that the various design criteria used and discussed can be expanded upon and be used to improve the stereotyped packaging techniques of "chassis and panel" type of equipment.

PROGRAM HISTORY

This design was a result of the necessity to install electronic instrumentation equipment in an area presenting more than the normal installation problems. In fact, the installation had many of the requirements of tactical equipment as well as those of instrumentation equipment.

The installation of instrumentation equipment is usually required in those areas remaining after location assignments have been made for the permanent

*Formerly, Director of Mechanical Engineering, Interstate Electronics Corporation.

Fig. 1. Modular packaging of instrumentation equipment aboard nuclear-firing submarine.

BENCH-MOUNTED EQUIPMENT

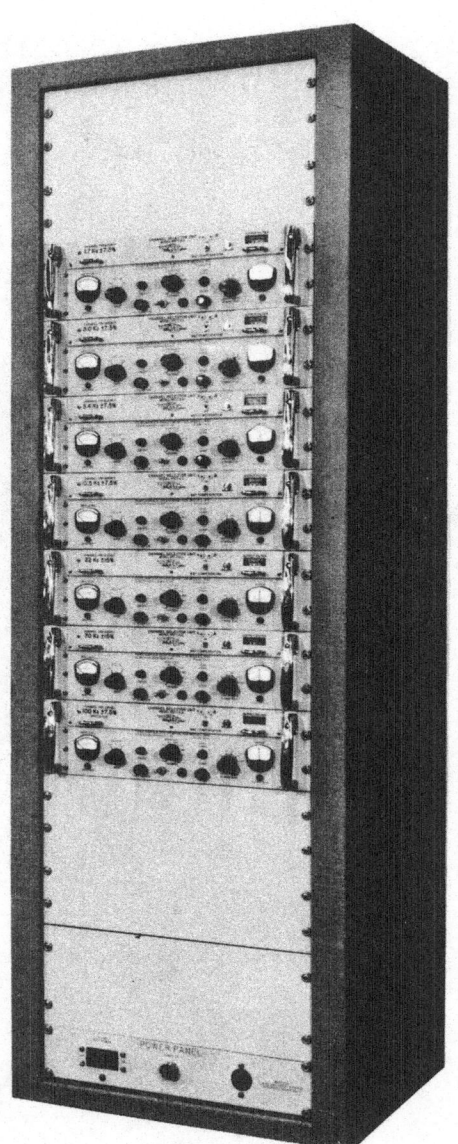

RACK-MOUNTED EQUIPMENT

Fig. 2. Typical conventional installations.

Fig. 3. Stack installation.

tactical equipment. This is usually complicated further by the fact that the instrumentation equipment needed may not yet be designed, and in some cases, the functions to be performed are not fully known. It then becomes evident that a flexible system packaging concept is needed for the mounting of equipment.

During the system development phase, the selected design must be capable of accepting the various equipment sizes as well as changes in sizes and locations of equipment. All of this must be accomplished without causing a major change in the system packaging concept.

The equipment to be mounted is usually purchased as a standard or modified unit, or is specifically designed for the needed application. These units are assembled on a chassis with front panels attached, and are standardized in size and mounting arrangement. Normally, the mounting of this type of equipment into a rack cabinet allows all the necessary flexibility for a well integrated system. However, when various additional system installation requirements are presented, the standard rack-housing type of design becomes obviously unacceptable.

One major requirement for this type of installation is that it be a system that can be installed aboard a completed submarine and later removed and used on another submarine. This means that the equipment must be passed through the 25-in.-diameter hatches as well as the 20 in. by 38 in. rectangular doors. This one requirement tended to force the design into a modular concept where units of the system could be taken into the submarine individually. These units must be small enough to pass through the openings and yet provide protection for the unit from damage due to handling.

Now the problem arose as to how the units could be structurally held together at the allocated system area. The first thing that came to mind was an arrangement using a breakdown-rack type of structure. This method primarily involves taking parts necessary to assemble a rack structure into the system area of the submarine. The individual electronic units are then brought in for installation into the rack structure. However, the normal amount of adjusting and fitting required in installing this type of equipment makes this technique undesirable. This is particularly true when the system is assembled and disassembled in a limited area.

Another undesirable feature of the breakdown rack includes the problem of having electronic units carried through a maze of hatches and doorways unprotected against handling damage. If cartons are provided to prevent this, the problem of removing or repacking the cartons is cumbersome and slows the assembly or disassembly operations.

It became obvious that what was ideally needed was a housing that provided protection for an electronic unit or units and just required being stacked, one upon the other, in a predetermined sequence to complete the system installation. This type of design allows for all the mechanical and electrical alignments to be performed prior to installation in the submarine and under "factory-type" conditions. It further eliminates the necessity for uncrating the equipment in the operational area of limited space since the housing will provide the protection as well as the structure.

With the system packaging design concept agreed upon, the real task of engineering was started. This was divided into two areas for consideration. One was the design of the "module shell." This is the individual housing that encloses the electronic unit or units. The assembled "module shell" with the electronic unit or units installed constitutes a "module" (see Fig. 4). The second design

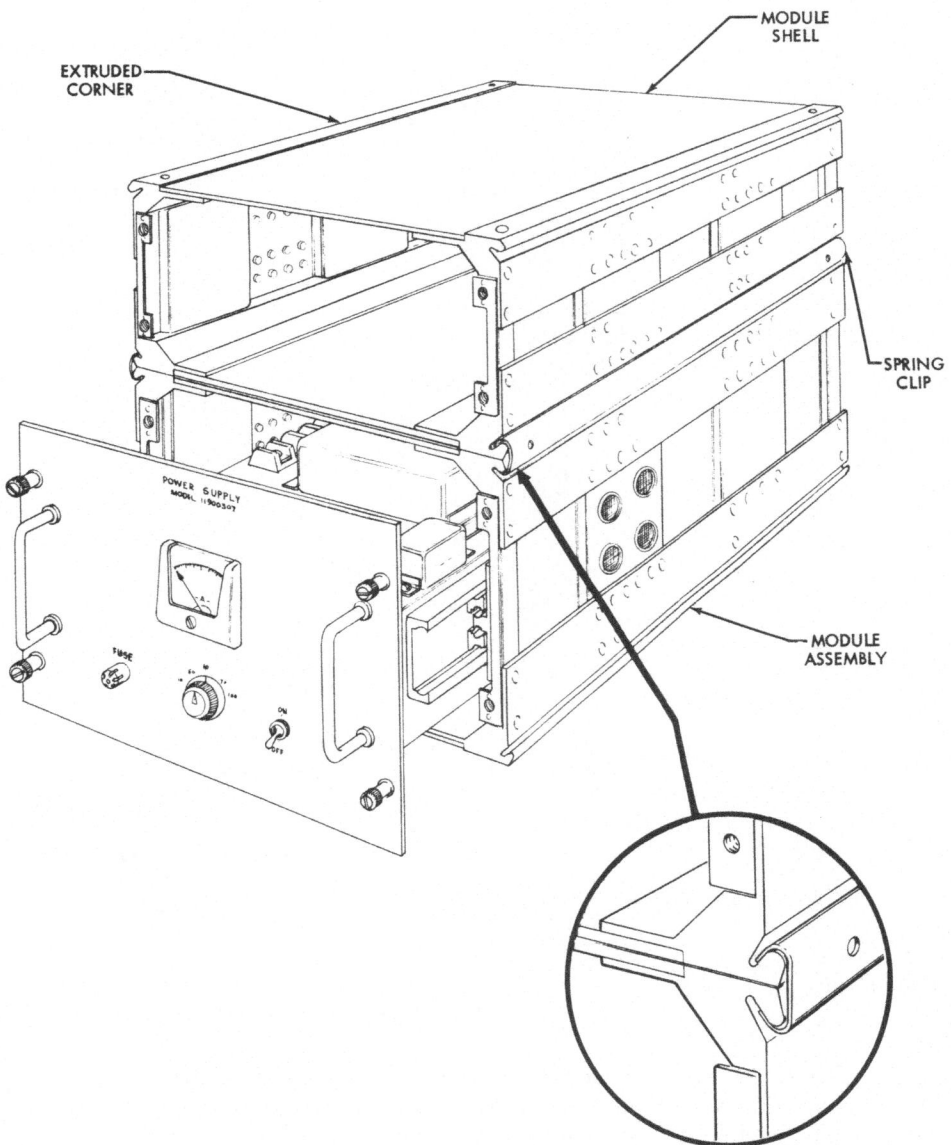

Fig. 4. Module and module shell.

consideration was the "stack" itself. This included the final assembly of a number of "modules" one upon the other, making the desired height (see Fig. 5).

Three specific requirements arose during the design of the "module shell" and the "stack." The first was the limitation on the size of the "module shell." Every effort had to be made to keep the width and height of the "module shell" as near as possible to the same size as the electronic unit's front panel. Second, the "module shell" and the "stack" must be able to provide a housing and structure adequate in strength to withstand the operating environment. Third, the

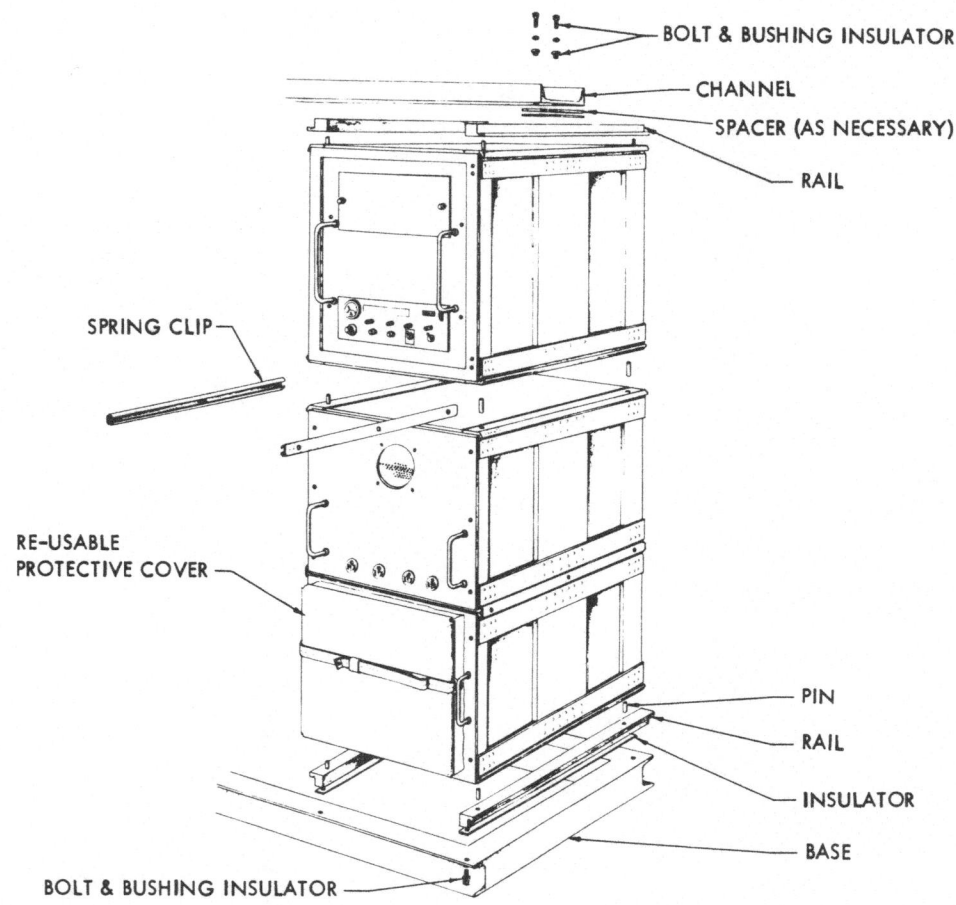

Fig. 5. Stack assembly.

method of attaching one "module" to another had to be solved so that the attachment could be accomplished easily, quickly, and with only limited access to the units.

In addition to the specific design problems, the general considerations for cooling, serviceability, human engineering, etc., as encountered in electronic instrumentation system design, had to be evaluated and solved.

The structural design was determined to be particularly critical due to the emphasis for maintaining an optimum structure with sections and thicknesses no larger than necessary. This meant that a careful mathematical analysis had to be performed in order not to "overdesign" for strength and rigidity but yet meet the operational environmental requirements. Of course, the true analysis came after static and dynamic testing to simulated operational conditions of the system.

TECHNICAL REQUIREMENTS AND SOLUTIONS

Every system must be designed using the requirements determined from an analysis of the system's end use. As in this case, the design of the "module shells," when integrated as "stacks" into the completed system, must meet the applicable environmental conditions. These can be summarized as:

1. Vibration: sinusoidal from 0 to 20 cps with a maximum double amplitude of 0.06 in.
2. Shock: a minimum of 3 g's for 0.050 sec.
3. Attitude: The equipment shall function with a maximum rotation (roll) of 10° about any horizontal axis, and shall not be damaged by rotation of 25°.

The various other design requirements demanded by the system in addition to the environmental considerations were also evaluated. These were considered very carefully and were incorporated into two sets of criteria—one set up for the "module shells" and the other for the "stacks" making up the system.

The criteria used in establishing the final design of the "module shell" are as follow:

1. The enclosure will be in modular form capable of being stacked up to a height of 72 in. and able to endure the environmental conditions during operation.
2. The enclosure must not greatly exceed the normal front-panel size of the equipment. (In other approaches there were many inches wasted in modules with bezels 3 or 4 in. larger than the front panels.)
3. Each module must be capable of being passed through two sizes of submarine hatches:
    a. A circular opening, 25 in. in diameter.
    b. A rectangular opening, 20 in. wide by 38 in. high with 10 in. corner radii.
4. Each module, standing alone, must provide protection to the equipment that it houses.
5. Each equipment-loaded module should weigh less than 180 lb.
6. Mounting within the module must accommodate electronic equipment in similar fashion as the mounting within a rack, accounting for such things as extension slides, cooling, grounding, etc.

The following criteria were used in establishing the system design consisting of the module "stacks."

1. System installation (consisting of 18 stacks of modules) must be accomplished within 24 hours.
2. Due to space restrictions, access will primarily be limited to the front of the modules in their stacking for system installation.
3. No equipment must be removed or withdrawn on slides from any modules during the installation.
4. The modules, when stacked, must provide rigid stack construction equivalent to "rack type" of construction.

The solutions to the criteria for both the "module shell" and the "stack" had to be evaluated concurrently in order to obtain the optimum system design during the development phase.

The unique design technique is primarily illustrated by Fig. 4 showing the construction of the module shell. As can be seen, each module shell has aluminum extruded corners of a special design as the main structural and functional elements of the enclosure. The extruded corners serve to rigidify the module corner and tie the side panels to the top and bottom panels. They also provide the structural material needed for adding dowel pins in order to carry the shear loads and bending moments between modules when placed into the stack. The extruded corner functionally provides, as part of its design, a groove running its length upon which are thrust heavy spring clips to tie the modules together in the stack.

The dowel pins and the spring clips are the key to rapid system installation since they are the only pieces of hardware used to hold the modules to each other. In addition, the same hardware is used to mount the stack to a floor base rail and any overhead rail tie-in that may be used. Pictorially, Fig. 5 shows how a system installation consists simply of placing one module on a base rail and then continuing the stacking process of one module upon the other in a predetermined sequence. The dowel pins are merely hand inserted during the stacking process to prevent shear action between the modules and to assist in limiting the bending. As a final installation step, the spring clips are inserted into adjacent extruded module corners in order to prevent the vertical forces tending to separate the modules (see Figs. 4 and 5).

The spring clips are unique in that they are toleranced to be readily installed and removed, yet have an extremely high spring rate to minimize separation of the module under load. (See sketch of spring clip in the Calculations Section.) The module corners are hard anodized and "dri-lubed" in the area of spring contact in order to reduce friction and galling during spring insertion.

A comment at this time seems to be in order concerning the manufacturing tolerances related to the spring clip and the related surfaces on the extrusion. Tolerances can be best evaluated if it is assumed that once the extrusion die is made all extrusions will fall within a fairly close and defined dimensional tolerance. This is exactly what was experienced when inspecting the extrusions. This was true even after the hard anodizing process. Based upon the relatively constant dimensions of the extrusions, the spring-clip tolerances can then be determined. Of course, good tooling is necessary in the making of the spring clip in order to obtain as much consistency in the dimensions as possible. The whole material and process control is critical in order to establish a tolerancing probability acceptable over the entire manufacturing operations. However, it was found that once the technique was determined and a careful quality control was maintained that no problems were encountered in maintaining consistency.

As can be seen from the Calculations Section, the spring clip has about 600 lb of retaining force for every 0.01 in. of spring deflection. This is based upon a spring that is 24 in. long and has its nominal "gap" dimensions "snugly" fitted on the extrusions.

The basic design of the module shell allows for a minimum of space to be used by the structural elements but does create a critical design with regard to its strength and rigidity. Basically, it is not the normal way to build a "box" when the loads are primarily vertical. However, calculations, proven out by an environmental testing program, assured the structural integrity of the design.

The side panels of the "module shell," which become the complete side of the "stack" after installation, were determined to be the most critical in determining the strength of the system. The sketch on the next page shows the typical side panel cross section determined for minimum size in relation to strength. This corrugated side construction also provided indented areas for recessing mounting hardware. The top and bottom panels of the "module shell" were determined to be flat sheets only and 0.160 in. thick.

Other evaluations made indicated that bracing in the rear corners of the "module shell" would be desirable to completely assure a rigid design with a relatively high vibration damping. The back connector and outlet panels with their mounting angles normally provide this required stiffness to the installation.

CALCULATIONS AND TESTING

The theoretical analysis of the design proved to be very involved when consideration had to be given to both the structural and vibration problems. The major difficulty is the determination of a correct set of assumptions that can be used in the calculations in order to provide for an analysis compatible with the end use of such a complex structure. The following have been determined to be the most appropriate assumptions for use in calculation.

1. A stack of modules is considered to be fixed at both bottom and top since the submarine installation provides anchoring at the base and "stack-to-stack" anchoring at the top.

2. The stack, when turned on its side, will behave under load as an elastic beam. The properties of an elastic beam are developed by the dowel pins and spring clips which transmit shear and bending moments from module to module.

3. All stresses will be carried by the side panels with the top and bottom panels serving to transmit the load between the side panels. Front and rear panels will slightly stiffen the stack, but because they are attached with screws and clearance holes, it will be presumed that they carry no load. The maximum bending moment will be resisted by a couple imposing tensile stresses on the bottom side panels and compression stresses on the top side panel with the assembled stack turned on its side. The assumption that the stack is turned on its side is conservative, since it implies 90° rotation of the submarine. However, since this restriction simplifies the analysis and is conservative, it is deemed valid.

4. A 3 g loading of a maximum 720-lb stack will be uniformly distributed from top to bottom of stack. This gives a distributed load of 30 lb/in.

Care must be taken to insure strict conformance to the design concept during the development and manufacturing phases. Any additional requirements involving cutting or the deleting of metal must be carefully analyzed as to their effect on the over-all structural integrity. The structure has not been designed on the theory "when in doubt of the strength, double the thickness." In order to conserve space, every precaution has been given to establish an adequate but not overdesigned structure.

The following calculations have been limited to only those concerning the stress as applied to the module side plate. This analysis proved to be the most critical when determining the actual success or failure of the module's structural strength.

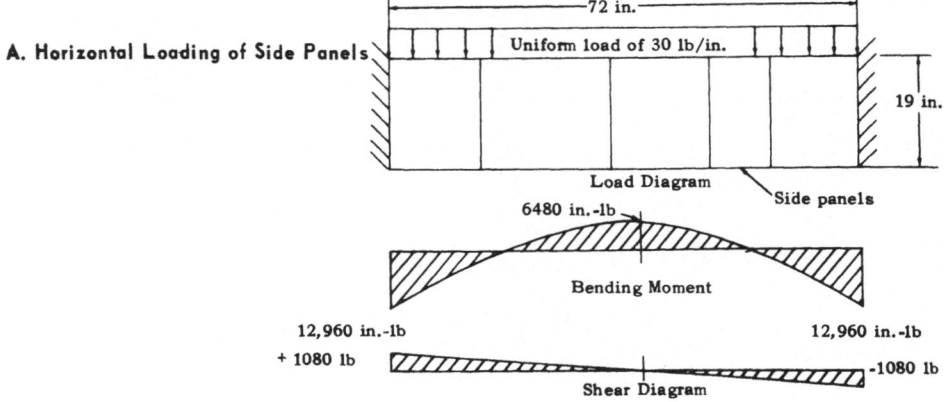

A. Horizontal Loading of Side Panels

Maximum bending moment:

$$M_{max} = \frac{WL}{12} = \frac{2160 \cdot 72}{12} = 12,960 \text{ in.-lb}$$

Coupling force:

$$F = \frac{M}{D} = \frac{12,960}{19} = 682 \text{ lb/side}$$

Shear load:

$$F = \frac{WL}{2} = \frac{30 \cdot 72}{2} = 1080 \text{ lb}$$

B. Vertical Loading of Side Panel

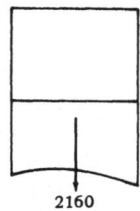

2160

$$F = \frac{2160}{2} = 1080 \text{ lb/side}$$

C. Stress Analysis

Area $= 0.09 \cdot 24 = 2.16 \text{ in.}^2$

$I_{x-x} = \frac{B}{12}(H^3 - h^3)$ $B = \frac{24}{2} = 12$

$\quad = \frac{12}{12}\left[(0.375)^3 - (0.195)^3\right]$

$\quad = 0.0453$

Equivalent flat plate thickness:

$$t = \sqrt[3]{\frac{12\,I}{b}} = \sqrt[3]{\frac{12 \cdot 0.0453}{24}}$$

$$= 0.283$$

Material: 5052-H34 aluminum

Compression stress:

$$S_c = \frac{P}{A} = \frac{1080}{2.16} = 500 \text{ psi}$$

Critical compression stress: (Roark, Formulas for Stress and Strain, Case A-6)

$$s' = K\frac{E}{1-u^2}\left(\frac{t}{b}\right)^2 \qquad \frac{a}{b} = \frac{30}{24} = 1.25$$

$$\therefore \ K = 4.72$$

$$E = 10.2 \cdot 10^6 \qquad u = 0.33 \qquad t = 0.283 \qquad b = 24$$

$$s' = 4.72\,\frac{10.2 \cdot 10^6}{1-(0.33)^2}\left(\frac{0.283}{24}\right)^2 = 7500 \text{ psi}$$

0.090 in. thick

X   X

24 in.   0.375 in.

1080 lb

30 in.

Side Panel Configuration

Margin of safety:

$$M.S. = \frac{7500}{500} - 1 = +14$$

D. Spring Design

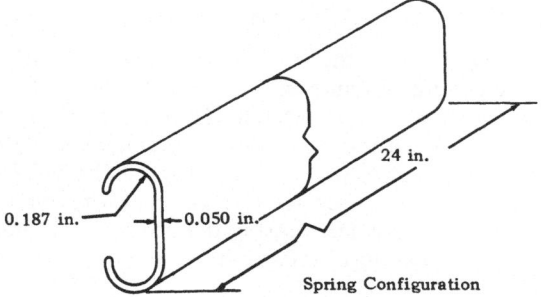

Spring Configuration

Force required to open spring 0.01 in.:

$$E = 30 \cdot 10^6 \text{ (high-carbon steel)}$$

$$F = \text{deflection}, \quad b = \text{length}, \quad h = \text{thickness}$$

$$P = \frac{EFbh^3}{14.15D^3} = \frac{30 \cdot 10^6 (0.01)(24)(0.050)^3}{14.15(0.475)^3} = 593 \text{ lb/0.01 in.}$$

Stress encountered in opening spring 0.01 in.:

$$S = \frac{FEK}{3h} \quad \frac{D}{h} = \frac{0.475}{0.050} = 9.5 \quad \therefore K = 0.\overline{013}$$

$$= \frac{(0.01)(30 \cdot 10^6)(0.013)}{3(0.050)}$$

$$= 26,000 \text{ psi/0.01 in. of deflection}$$

Therefore, a deflection up to 0.05 in. is within the yield of 1095 heat-treated high-carbon steel.

The testing program was conducted based on the design assumptions; however, the procedure was in accordance with the Military Specification for shipboard equipment.

Testing was conducted using a stack made up of modules loaded with weights to simulate the maximum conditions that could be encountered with the equipment. The tests consisted of pitch and roll, vibration, and shock. These tests confirmed that the design principle used in the spring clip was very satisfactory and had essentially no "opening up" between modules even under the most extreme conditions of pitch and roll or resonance in vibration.

The vibration test was conducted with 3 g's on the equipment up to a frequency of approximately 25 cps. At this point, the resonance of the fixture and the higher frequency on the stack caused an amplified excursion. However, this frequency range was determined to be adequate in view of the equipment's end usage and the consideration that the test was extreme.

The stack withstood repeated shock of 13 g's for 0.024 sec duration without any noticeable damage.

It can be concluded that the design is adequate when based upon the results of calculations, the test results, and mature judgment used the application of the design's concept.

MANUFACTURING AND COST CONSIDERATIONS

The fabrication and assembly of the module shell can be easily accomplished with standard equipment normally found in sheet metal shops. In higher production runs, the side panels can best be produced for lower costs with the use of sheet metal forming rollers rather than with a metal brake. Required panels for sides can then be merely cut to the desired length. The 5052-H34 aluminum material has good structural properties as well as good forming ability.

The most economical method of fabricating the corner element of the module is by extrusion. A milling or shaping process can be used if a prototype or model is needed.

The most complicated problem is involved in making the spring clip. The close tolerances requisite for dimensions with high-carbon steel and the requirements for heat treating necessitate extensive tooling in order to maintain quality and consistency.

Basically, it can be stated that the cost of production is highly dependent upon the quality of tooling and the quantities of production. When compared to the standard rack or cabinet construction, the cost will be slightly higher. This is mainly due to the need for the more specialized tooling and care needed to assure the maintainance of dimensions. For example, the tolerances necessary in the module shell are approximately as follows:

1. Perpendicularity based upon weight of module—approximately 0.020 in. per 10 in.
2. Parallelism between top and bottom of module based on number of modules in a stack — approximately 0.015 in.

This tolerance is necessary to assure maintenance of a perpendicularity of the stack after assembly of all modules.

The spring clip gap opening tolerance was from 0.545 to 0.570 with the nominal and desired dimension of 0.555. Since trouble was encountered in maintaining a dimension as close as possible to the 0.555 in., it was necessary to establish various tolerances for gap dimensions based on a percentage of the total length of the spring. This was required in order to establish a good quality while limiting excessive rejection due to close tolerances in the specifications.

The extruded module corner was made of 6061-T6 aluminum providing the structural and bearing strength as well as being a desirable extrusion material. The tolerances were established as:

1. Angularity: ±1° (maximum)
2. Twist: $\frac{1}{4}$° per foot
3. Flatness: 0.004 in./in. of width

The dowel pins used to carry shear loads and bending moments are inserted into four holes drilled with a fixture with a hole diameter tolerance of +0.003 and −0.000.

The cost of the spring clip and the extruded corner as established after tooling cost and based on orders of 900 ft are approximately as shown below:

1. Spring Clip – $3.30 for 14 in. to $8.50 for 24 in.
2. Extruded Corner – $1.05 per foot.

It can be concluded that the relative cost differences between the standard type of construction and the module construction of similar quality are minor.

This is especially true when the quantity and the special features of the module warrant its use.

OTHER USES AND POSSIBILITIES

The use of this design concept for application other than that for which it was originally designed needs but a wandering of one's imagination. The design has proved itself in its original use on the present Polaris-firing nuclear submarines and is being expanded for use on many more of the advanced and new submarines. Admiral Raborn, Director of the Navy's Special Projects, has commended the instrumentation as reflecting "a portable, modularized and hence unconventional concept which has required considerable developmental effort and individual component testing."

Other uses are presently being contemplated using the design concept for field equipment requiring extreme portability.

The design of laboratory equipment is also being planned utilizing some of the concepts incorporated in submarine instrumentation packaging.

It is hoped that the techniques discussed within this paper have stimulated interest in this type of packaging and that they will be incorporated and expanded into other uses.

ACKNOWLEDGMENT

To all those individuals who cooperated in the preparation of this paper from Interstate Electronics Corporation, prime contractor for the Polaris Instrumentation, and Lockheed Aircraft Corporation, prime contractor for the Polaris Missile.

DISCUSSION

Question: Louis Polaski, General Electric, Philadelphia. It was brought up that there was to be no access in the rear, only front access. Would you go into how you made your electrical innerconnection?

Answer: Yes, but let me clarify that it was mentioned that there was limited access to the rear with the primary access from the front only. The electrical connections were made by utilizing a sequenced assembly procedure. This allowed access to many areas before the next assemblies were installed and also allowed individual connections to be made as the modules were stacked.

Question: One other question on the cabling. How was the cabling tied into the modules? Or was it tied onto the submarine structure?

Answer: Since this particular installation is temporary equipment, the cabling was not tied into the submarine structure. It was merely routed from module to module and from stack to stack. The cabling in many cases did have to be supported to the stack structure to relieve strain due to the weight of the cables.

Question: Harold Ferris, Bell Helicopter. I would like to comment on the excellent package design; I am particularly interested in the extrusions. Were you able to get the mills to make the run economically for the small amount that you needed?

Answer: Yes, however, we did have a requirement for over 1000 feet of extrusion for this particular installation. The cost of this extrusion ran $1.02 a foot. In addition, the cost of the die must be amortized over the required amount. I am not sure, but I believe the die costs $300 or $400. I believe that you will find that special extrusion costs can be very well within economical reason and are certainly much more reasonable than they were a few years ago.

Question: Julius Jodele, Jet Propulsion Laboratory. This unique approach of yours—I am having a hard time following in the area of its advantages. You are actually in a permanent installation using an approach of stacking the so-called modules in their shipping containers; then once they are installed all you have left are just inconveniences. One is that of cabling; you don't have anymore the flexibility you mentioned, and I kind of doubt that you can cable from one unit to the other. A system is normally much more complex and we have intercabling probably between these boxes. Secondly, it's very doubtful that you can have any system-cooling approach if you have any of these boxes in a container. What do you do if you have more power in that box, and how do you handle the cooling problem? Thirdly, once they are stacked together you cannot service them. I don't believe you have the equipment there that would be permanently installed. Then, once in the field and in operation, you have to take the unit out anyway without the box because if you take a lower box out you have to collapse the whole structure. So you are still handling in the field for operation. Also, you said that you did not have time to install in racks; therefore, you built the box, but you still have to handle the boxes and equipment more than if in a rack.

Answer: Let me answer these before you go on, because I believe we are getting a little afar at this point. Number one, there has been no claim in this paper or presentation that you have any advantages over a rack installation once the equipment is installed. This system provides the same problems that a rack would once you are assembled. Now, as far as getting it through a small opening and ending up with the structure integrity of a rack after it is stacked, it does do this. In fact, it is now being operated aboard the mentioned submarines and has worked. The cabling has its problems and must be coped with. You did, however, mention a good point and that is the problem of cooling. In a rack installation, you do have a chimney effect and can therefore mount units one upon the other and yet utilize a central cooling system. In this particular design, one of the drawbacks is that an individual cooling system must be used. This can be seen from the unit here on display, which happens to be one of the hottest units encountered. In this case, small high-velocity fans were used in the front panel to direct air through the desirable path as determined by smoke tests and other normal means. As I see it, the heat problem must be solved just as it must be solved in a rack installation but maybe in more places than in the rack.

Question: Leo Fiderer, RCA, Van Nuys. About the cooling problem: one suggestion that I could make is that providing the horizontal piece with some openings would provide a chimney effect from one box to the other. That could not impair the rigidity very much, but it could provide some openings to transmit cooling air from one box to another. That was just a comment. The question I have is about the spring clip. Have you thought about any provision for actually securing these spring clips in a way other than just the force of the spring itself holding it? Have you tried clamping it with a screw to something to make sure the vibrations of a depth charge would not shake the spring loose?

Answer: In answer to your first suggestion, I can briefly say this certainly has merit and was thought of. We did, however, decide that we would be too dependent upon the placing of units one upon the other, thus limiting our flexibility. In answer to your question concerning the spring clip, I think that we thought about this problem, but we did prove through tests that no "locking feature" was required. In fact, during our development we actually tried and throught of many approaches but had to settle on one approach fairly quickly. This was our solution. I would like to admit to a mistake we encountered at first, but the solution turned out to be fairly simple. Our first approach was to have the spring clip have high clamping ability in its nominal position. We soon found that our friction calculations were wrong and the spring could not be inserted. However, by changing the principle to a high-rate spring that clamps upon small deflections, we solved the problem and allowed a "slip-fit" of the spring on the extrusions.

Question: John Lewko, ITT Federal Laboratories, Ft. Wayne. I wonder if you could comment on the reliability of this type of a system in view of the fact that it's composed of a series of individual boxes that constantly require your connectors on the back to be assembled and disassembled. And what type of a logistic problem did your initial contract call for in supporting a system like this when you had malfunctions that occurred prior to the shelf test that your system probably went through?

Answer: Let me see if I can get the meaning here. To comment on reliability on this type equipment, I don't know if it's pertinent because I would like to limit our discussion more or less to the packaging job. Generally, the type of equipment you mount in here, whether it's especially designed or purchased equipment, I think would be the same as you would run into in any shipboard installation. For instance, you do have on a ship the rack type of structure which has drawers and equipment. You have the same problem here. We also have shock pins when these are fully installed. Each type of system has to be analyzed, particularly when you are buying a lot of commercial equipment. This field is a little bit unique in some respects but is common to the instrumentation field. But in this particular case the requirements of our contract required specifications adequate to fulfill the required instrumentation use of the equipment. It did, however, have to go through essentially the same requirements as the tactical gear but without the arbitrary specifications.

Question: Now generally on a cabinet you take quite a lot of pains to suspend the harness assembly with the cabinet to guarantee that you ensure good reliability and good assembly techniques. Your cabinet cabling consists of large cables environmentally protected so there is a considerable weight factor there. How did you suspend these cables on the back to guarantee that the cable wouldn't damage itself by its own weight after long-term use?

Answer: There were a number of ways used, but each system will dictate a particular consideration. Basically, two methods were used in this installation. If the cables weren't too large, it could suspend itself. In other cases, we actually did tie to a side panel structure, or to a portion of the ship in some cases. But you did have to use judgment in this aspect.

Question: It looks like a very nice design. There were two questions I had here. One was, have you tried shock mounting on this type of assembly? And the second is, what is the strength of the assembly without the chassis in it? Can it take some side load?

Answer: Yes, we have tried shock mounting. It was not incorporated but it was certainly considered and where necessary in some units was individually considered. Now can we pull the chassis out of the shell? Yes, and without failure. I, however, would mention this one thing: in this particular design we did not overdesign. In other words, this design has been carefully analyzed and tested but has no large safety factors over the required specifications.

Comment: Although not a part of the discussed design, I believe it would be of interest to some to know how we solved the problem of protecting the front panels. It will be noticed that the front panel is the only part not protected by the module shell. We used, as can be seen here, a foam sheet material on an aluminum plate which in turn is compressed against the delicate instruments on the unit's front panel. The material used is a polyether Sta-foam. This cover is clamped onto the unit's handle thus not increasing the over-all module dimensions.

MINIATURIZED POSITIONERS FOR MODULAR PACKAGING

Harold A. Brill

Radio Corporation of America, Camden, New Jersey

The most radical mechanical change in present-day airborne electronic equipment, especially in military systems, is modular packaging. A radio package designed for today's aircraft is composed of stacked, functional modules, each plugged into a common base which supplies both electrical connections and cooling air. These modules, each a complete chassis with its own functions, performance requirements, and electronic capabilities, are major subdivisions of the electronic system. The sizes of these modules are standardized to permit their use as building blocks in the equipment construction.

The advantages of modular packaging are: (1) ease of maintenance and test, (2) interchangeability; (3) simplified apportioning of cooling air; (4) low-cost repackaging for different configurations to meet the needs of specific aircraft; (5) multiple use of the module or use of the same module in several equipments; and (6) facilitation of future modification by easy replacement of obsolete modules.

These modules frequently contain variable condensers, potentiometers, multiple circuit switches, magnetic tapes, coded discs, and other components which must be positioned or tuned several times during flight. Since the operating controls are located some distance from the equipment, a remotely controlled positioner is incorporated. Earlier modular sets used a central positioner module which drove the other modules through gears, couplings, and/or sprocket chains, and occupied a volume of some 50 in^3.

The advanced design UHF Communications Set AN/ARC-62, designed for the U. S. Air Force by RCA, uses individual precision positioners for each module requiring positioning. To fulfill the requirements of this program to design an improved UHF Communications Set, and to meet the real intent of the modular concept, it was decided that individual modules must be interchangeable on a plug-in basis. This prohibited mechanical coupling between modules, and suggested the use of individual positioners. To prove out the modular concept, it was necessary to reduce the size of each positioner sufficiently for incorporation as part of the module, without sacrificing accuracy. Once the decision was made to completely modularize the equipment, an intensive design effort was initiated, resulting in the successful development of the AN/ARC-62. This modularized equipment offers many advantages: (1) modules can be replaced without tracking or adjusting for the proper interface between module and positioner; (2) the need for a precision base filled with gears and shafts is eliminated; (3) repackaging is facilitated, since the relative position of the modules is unimportant (they need not even be in the same package). Figure 1 shows one of the AN/ARC-62 modules, complete with positioner.

Although the individual positioners must be extremely small and accurate, they are simpler in operation than a large central positioner with its many output shafts, each positioned differently.

The requirements of the many movable devices in these modules are diverse.

These requirements determine the proper choice of positioning technique. The pertinent factors to be resolved are: (1) the number of positions or stations; (2) the space allocated; (3) accuracy, backlash, and torque at output; (4) number of outputs; (5) input information available; (6) allowable time for positioning or tuning; (7) duty cycle and life requirements; and (8) operating and storage environment (temperature, vibration, humidity, etc.). The simplest device which will perform adequately should be chosen.

Fig. 1. Typical module with miniaturized positioner.

Devices which are useful as positioners include: (1) rotary selectors, (2) Geneva and similar indexing drives, (3) multiple input positioners, (4) magnetic and mechanical clutches, (5) servomotors, (6) stepping motors and stepping relays, etc. A number of these mechanisms have been used for miniaturized precision positioners in the AN/ARC-62 modules.

ROTARY SELECTORS

A very simple and useful positioner is the rotary selector, which is an assembly of a rotary solenoid, an interrupter switch, and a ratchet mechanism. These units are very often used to position one or more rotary wafer switches. Figure 2 shows the second injection module of the AN/ARC-62, employing two $1\frac{1}{8}$-in.-diameter rotary selectors as positioners. Each positioner drives a gang of 12-position wafer switches. Ratchet drives for 12- and 18-position switches are available as standard; but drives have been produced for other indexing requirements. As a positioner, the accuracy of the rotary selector is limited by the precision of the ratchet mechanism. This selector can operate at speeds of up to 30 or 40 steps per second. Use is limited to low-inertia loads unless sufficient space is provided for the larger solenoids required, and enough dc current is available to operate them.

The rotary solenoid without a ratchet makes an excellent device for operating single- or double-throw switches, small valves, etc. Strokes of 25, 35, 45, $67\frac{1}{2}$, and 95° are available.

Fig. 2. Second injection module (two rotary selectors used for positioning).

GENEVA AND SIMILAR DRIVES

When a limited number of stations is required of a positioner, a Geneva mechanism or intermittent gear drive may be used advantageously. With such a mechanism, the index wheel is stationary during an appreciable portion of the rotation of the drive wheel. This gives adequate opportunity to bring the motor to rest, either by coasting or (as is commonly done) by braking. Braking may be either mechanical (friction) or electrical (shorting motor leads or reversing polarity). The rest position of the drive wheel can vary over a considerable angle without influencing the positioner output. Figure 4 shows the two 10-position Geneva drives in the AN/ARC-62 first IF module. This figure is discussed in more detail below, in the description of multiple input positioners.

In a practical Geneva drive, the number of stations is limited. As the number of stations is increased, the locking arc becomes smaller, and it becomes increasingly difficult to constrain the index wheel. With more than nine or ten stations, accurate positioning is difficult to achieve, the mechanism tends to jam, and operation becomes marginal.

A NOVEL INTERMITTENT DRIVE

An interesting positioner was developed for use on the first injection module of the AN/ARC-62, shown in Fig. 3. It consists mainly of an intermittent-motion mechanism with a unique locking device. The index wheel is advanced intermittently by a pin mounted eccentrically on a rotating drive shaft. A link connects this drive pin to another pin on the opposite side of the index wheel. This second, or locating, pin is constrained to move in a straight line in and out of the index wheel slots. The length of the link is such that, as the drive pin exits from a slot in the index plate, the locating pin enters an opposite slot. Then, during the portion of the revolution of the drive shaft when the output is at a standstill, the index plate is kept from rotating by the locating pin, sliding first inward and then outward in the slot. At the point in the cycle where the drive pin enters the next slot, the locating pin exits from the slot in the index plate in which it has been riding. It remains away from the index plate during the period of engagement of the drive pin.

A drive such as this eliminates the problems encountered when the number of positions is larger than can be accommodated by a Geneva wheel. In the module shown in Fig. 3, one output shaft has 18 positions in 360°. The second output shaft requires 18 positions in 180°. A two-to-one gear reduction is used together with a false-position cam and switch to prevent tune-up in the unused 180°. Less than 7 in.^{3} are required for the positioner and motor.

Fig. 3. First injection module (positioner is intermittent motion mechanism).

MULTIPLE INPUT POSITIONERS

As the number of positions required becomes greater, the positioner must become more complex. A positioning device with a moderate number of stations geared down to increase the number of output stations can be used if the input information can separate this duplicity. Or, stages may be added. Several separate inputs may be used with an adding mechanism. In the AN/ARC-62 first IF module (Figs. 4 and 5), an output is required to provide 100 positions in

Fig. 4. First IF module (separate Geneva drives provide input).

Fig. 5. First IF module (two inputs combined in a differential).

180°. Two inputs are used, each a 10-position Geneva driven by a separate motor. The adder is a spur gear differential. The input from Geneva No. 1 (10 positions in 360°) becomes an output of 10 positions in 180° (two-to-one reduction through the differential). A similar input from Geneva No. 2, through the expedience of a cam, subdivides the 10 output positions by 10, thus providing the 100 needed stations. A false-position switch operated by a cam prevents tune-up in the unused 180° of the variable condensers.

The entire precision positioner, including two motors and electrical control system, requires a space of only 2 by 2 by 3 in.

ELECTRICAL CONTROL

The AN/ARC-62 positioners described above are controlled by a bridge balance circuit, one-half of which is in the remote control box and the other half in the module. The remote control box contains a set of wafer switches, with each switch contact at a different voltage. In the positioner, each input shaft, in addition to driving the positioner, turns a matching wafer switch. The contacts of this switch are also at various potentials. A relay, which controls motor-driving current, is connected between these switches. It is kept closed by the difference of potential across this bridge.

When a control box switch is set on any position, the motor on its circuit is driven until it rotates its switch to the contact which is at the same potential as the control switch. At this point, with no difference of potential across it, the relay opens. The motor stops, and stays at rest until its control switch is moved to a new position, upsetting the voltage balance.

CLUTCH SYSTEMS

As the number of positions a component must assume increases, additional subdivisions of the positioner output are obtainable by introducing additional inputs. A point may be reached where it is advantageous to drive all inputs with a single motor. Figure 6 is a diagram of such a system (in this case, not used in the AN/ARC-62). Magnetic clutches engage each input with the motor drive until the desired position is reached. The clutch, when released, determines true shaft position by engaging a pin in one of the precisely located radial slots in a stationary plate. The motor is made to run until the last clutch has released.

A trade-off analysis of motors versus clutches, listing size, weight, tune-up time, reliability, etc., is necessary for complex systems.

SERVO POSITIONERS

Another device useful for obtaining a great many output divisions is the servo positioner. In the AN/ARC-62, such a system is used to tune assemblies in the RF module to 1800 discrete positions in 180°. (Actually 1750 positions within 175°; the remaining 5° are not used.) Figures 7 and 8 show the module and its positioner. Positioning information is received from the remote control box by way of the first injection module and the second injection module. The positioned shafts of these modules provide the 1800 voltage steps which control the servomotor. Wafer switches are installed on these shafts: two on the 18-position first injection module shaft, two on one of the 10-position second injection module shafts, and one on the other 10-position shaft. Figure 9 illustrates the method used to obtain a constant impedance voltage divider.

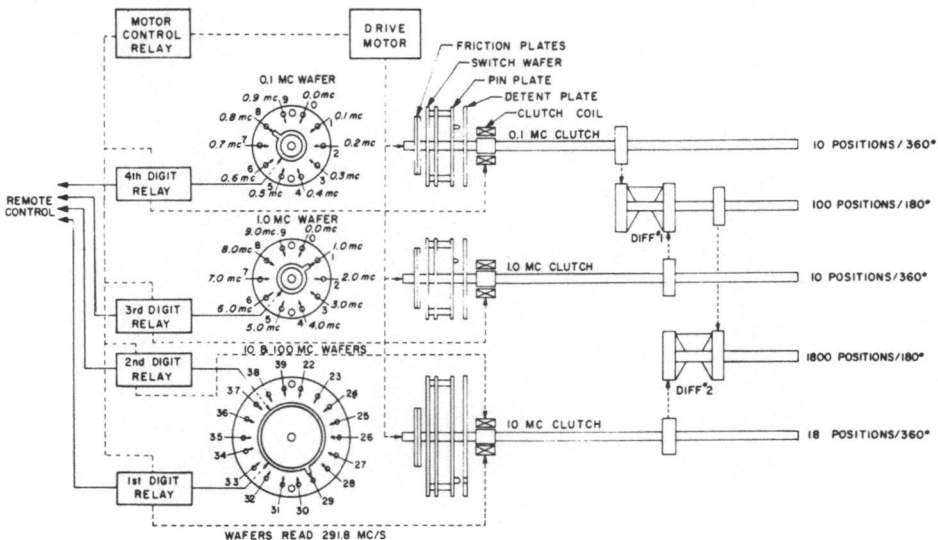

Fig. 6. Multiple input positioner (magnetic clutches control each input).

Fig. 7. RF module (servo system provides 1800 discrete positions).

Fig. 8. RF module positioner (consists of servomotor, follow-up potentiometer, and gear train).

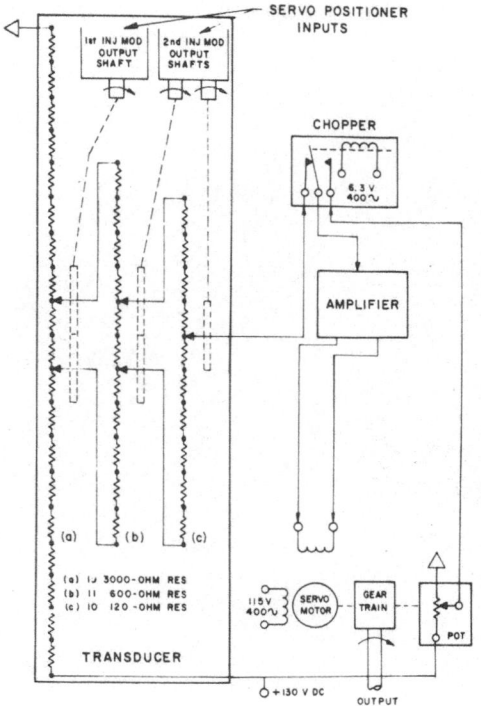

Fig. 9. Servo positioner system (voltage steps determine output shaft position).

The transducer bridge is energized at the top of the first bank of bridge resistors. The same voltage is connected to the top of the servo potentiometer. Any variation in the supply will affect the transducer and the potentiometer in the same manner, and therefore will not cause errors in the operation of the system.

The transducer output voltage and the potentiometer voltage are fed to opposite poles of a chopper. The chopper has a vibrating reed which continually switches between these two voltages; the voltage difference produces an ac error signal which is amplified and connected to the servomotor control winding.

STEPPING MOTORS AND STEPPING RELAYS

Incremental steppers take several forms. The most common are phase-pulsed synchronous motors and solenoid-operated pawl-and-ratchet devices. The former are detented magnetically and the latter mechanically. Adequate off-the-shelf hardware is available for many tasks. However, neither of these types was used on the AN/ARC-62 design because of their limited load-handling capability. Use was precluded, in cases where they were otherwise satisfactory, because inertia of the loads was greater than the miniaturized versions could handle. Pawl-driven mechanisms produce large accelerations and, with heavier loads, cause chattering and even faulty indexing. Magnetically operated steppers require that momentum of the component be damped for effective detenting. Though no illustrations are given here of incremental steppers as positioners, a series of articles [1] is recommended for further information.

ACCURACY

Reliable, accurate hardware is, of course, the result of good machine design practice. Recalling the specifics of precision gearing or kinematics is not the purpose of this paper. Full information on these techniques is available from other sources. A few words on positional accuracy, however, are warranted.

Wafer switches in the AN/ARC-62 second injection module were driven with a ratcheted rotary solenoid. Off-the-shelf selectors provided all the accuracy needed in this case.

The positioner in the first IF module tunes variable condensers. Angular accuracy of the mechanism was held within 15 minutes. Repeatability was even more critical, and was held to a maximum of 6 minutes.

The positional accuracy of the servo in the RF module depends not only on the mechanical accuracy of the mechanism and the stability of the servo amplifier, but also upon the linearity of the input transducer and therefore upon the precision of the resistors in the voltage divider circuit. Over-all output accuracy is held within 0.4°.

CONCLUSION

This discussion identifies the major requirements peculiar to modular positioners. Several designs are described, illustrative of the many techniques available to satisfy each set of criteria. Each of the positioners illustrated provides the required module tuning as a function of electrical information fed to it.

In all of the designs described here, particular care was taken to provide an easily accessible electrical system. This allows wiring and testing to be done after mechanical assembly. It also permits electrical troubleshooting to be performed on a mechanically operable package. In several modules, the control system was laid open by merely unscrewing the bottom plate.

The individual positioners designed for the AN/ARC-62 provide a new freedom in packaging, resulting in a very versatile equipment. The set is designed for UHF communications in high-performance aircraft. It has proved out the modular concept, and is setting a pattern for future designs of airborne electronic equipment.

ACKNOWLEDGMENTS

The author thanks the Radio Corporation of America and the Wright Air Development Division of the United States Air Force for granting permission to use photographs and information included in this paper. The AN/ARC-62 UHF Communication Equipment was developed by the Radio Corporation of America, Camden, New Jersey, under Air Force Contract AF33(600)34622.

REFERENCES

1. S. J. Bailey, "Incremental Servos," Control Engineering (Nov., 1960; Dec., 1960; Jan., 1961; Mar., 1961; May, 1961).

DISCUSSION

Question: Frank Keefe, Collins Radio, Dallas. I noticed that in a nimber of your modules, between the lower drive plate and the module itself, you had a modified Oldham coupling for a drive. I did not notice any means of being able to properly align these couplings for putting them back together again. I wonder if you could describe that. I have had difficulty with this same problem.

Answer: Two of the modules I have described use modified Oldham couplings. In the first injection module, Fig. 3, the drive is connected directly to the positioned component and the coupling is between the drive and the wafer switch. Since a wafer switch is all that is driven through the coupling, alignment is not critical. To couple the drive shaft to the wafer, the wafer is oriented to approximately the angular position of the coupling, and then placed in position. The switch rotates rather easily and is brought to true position by the force of engagement against the coupling's rounded engagement bar. It isn't difficult to view the procedure from the side, and if the positioning is not close enough, the wafer can be rotated with a probe until the coupling is able to cam itself in. It is possible for the wafer switch to be assembled 180° out of phase with the drive, but this is immediately apparent and the change can be made. In the RF module, Fig. 7, the coupling is between the drive and the tuned components. To eliminate the problem of alignment for reassembly, the grips of the coupling are made of spring stock. These grips are stiff enough to drive without angular distortion, but they can accept a variation in angular alignment during assembly of the drive to the module. Since the grips are stiff enough to overcome the torque of the system, they bring the driven shafts into accurate alignment as soon as connection is made.

Question: Jim Kuller, Bendix Radio, Baltimore. Is this equipment now in production?

Answer: The requirements of the contract were fulfilled with the design and manufacture of a limited number of prototypes. I am now on a new project and not up to date on the status of the equipment. I understand it is still undergoing Air Force tests.

Question: Pete Dreesen, Electro Instruments, San Diego. What plating are you using on your printed circuit switches and are you having any problem with them?

Answer: The printed circuit switches are seen clearly in Figs. 3, 4, and 5. After etching, we plate the copper surface with nickel and then rhodium flash. This produces a very hard, corrosion-resistant surface. For the wiper on the rotor we have used a coin silver spherical button. The silver button has high electrical conductivity and is reasonably low in cost. Its relative softness is used here to advantage. Since wear must occur somewhere during operation, we ensure that it occurs on the thicker button rather than on the thin switch surface. We have found this combination very satisfactory.

Question: Jim Siebert, Magnavox Co. Fort Wayne. Have you run into any reliability problems using the Ledex rotary solenoid?

Answer: Reliability problems that we have encountered with rotary selectors on earlier equipments seem to have been cleared up by the selector manufacturers. We are not experiencing problems with them on the AN/ARC-62. Also, the control system we use is not dependent on the number of pulses to which the selector responds. It is a null-seeking system. In other words, it continues to operate until a position is found with no difference of potential across the bridge. If the solenoid misses a pulse, well, then the next pulse will operate it. If it has overtraveled, it will operate through another revolution until it reaches the null. It does not depend on pulses, but upon true position. I would like to repeat a point made in the paper. The inertia of the load must be within the capability of the rotary selector. These devices produce high accelerations. Added damping can be used to improve the action, but this will increase the size of the solenoid and the current requirement.

DEVELOPMENT OF MINIATURE AND MICROMINIATURIZED ELECTRONIC PACKAGES

Walter Prise

Lockheed Missile and Space Company, Sunnyvale, California

INTRODUCTION

The packaging of electronic circuitry today is going through an evolutionary change in which old methods are being altered and new concepts are coming into existence. The field today could be divided into three categories:

1. High-density packaging.
2. Integrated circuitry.
3. Solid state and molecular electronics.

Every new concept is associated with a "numbers game" where a higher number of parts per cubic inch or cubic foot is being claimed. The higher the component density, the better the package. The big questions are why miniaturization is necessary and whether or not miniaturization is the main aspect and/or aid of the new packaging evolution. It is the intention of this paper to outline the major characteristics of miniaturized packages and to compare them with the "ideal" package. An evaluating survey of various existing techniques was made, bearing in mind their relationship to the "ideal" package. The "ideal" miniaturized package should have:

1. Simplicity in design procedures.
2. High reliability—long life.
3. Electrical, mechanical, and thermal stability.
4. Compactness, i.e., small size and light weight.
5. High producibility rate.
6. Low cost.
7. Simplicity and ease of connections.
8. Small number of parts.
9. Resistance to shock and vibration.
10. Satisfactory behavior in a dry-circuit condition.
11. Easy mounting.
12. Easy cooling within the operational range.
13. Suitability for automation.
14. Resistance to nuclear radiation.

Most of the new techniques have claimed desirable characteristics. This evaluation process provides a list of positive and negative aspects of each packaging method; however, negative characteristics were emphasized. In this manner exaggerated claims of each method could be properly balanced. The following concepts were analyzed and evaluated:

1. Soldered sandwich } High-density packaging.
2. Welded construction }
3. Modules with disciplined geometry } Integrated circuitry.
4. Thin films }
5. Solid state

No attempt was made to classify any of the evaluated packages as an "ideal" concept because this goal is unattainable. Effort was made to establish the working characteristics of various concepts as a guide for prospective applications.

DESIGN DEVELOPMENT

The first step in design development is the idea, followed in electrical work by the schematic diagram. From a schematic diagram a list of component parts and types of equipment needed can be established. Power limits, voltage, current, and frequency requirements are evident from schematic diagrams and the performance specifications. Thermal analysis follows, showing maximum dissipation of power and temperature limits. Based on the findings of thermal analysis, classification of component materials and parts can be made. Environmental conditions and the design limitations of weight and space will provide a clue to the selection of a packaging concept.

Desired life expectancy plays an important part in the selection of packaging methods. The life of a package depends on inherent characteristics of processes involved. The effect of a process on the behavior of a finished package should be considered. This evaluation outlines steps in each process and lists deficiencies associated with each step. Potential problems associated with each packaging method may provide guidance for selection of a packaging method to meet the desired operational, functional, and environmental conditions.

Circuit-design difficulties in miniaturized electronic assemblies are intensified by the proximity of parts to one another. Most of the new packaging assemblies must be bound together by some encapsulating material. Thousands of encapsulants are available on the market but only a few of those should be considered for electronic usage. A design engineer must have a guide-line to assist him in the selection of a proper encapsulant. To obtain a suitable encapsulant for a specific application in the selected circuit, the requirements for thermal, mechanical, and electrical characteristics must be balanced.

Most of the miniaturized packages are using substrates for deposition of passive networks and installation of active elements. Demands on the substrate are of a conflicting nature and must be properly balanced for a good design.

It is a common practice that an original design goes through a process of transformation where modifications and alterations may be introduced, thus flexibility in the packaging method is of great importance.

In miniaturized concepts, detailed planning of the project is necessary. Use of new and unusual components can seriously jeopardize the over-all progress by delays in their procurement. A lack of suitable specifications also may slow the programs. New methods and processes require attention to minute details; therefore, they must be properly planned and controlled. Testing specifications and procedures are to be written in advance. Drafting and design practices present many unique problems in the miniaturized field. Assembly and subassembly drawings must be drawn on the enlarged scale to meet the demands of photographic processes commonly employed in the art of electronic miniaturization. Production control measures are necessary to keep the process from developing undesirable characteristics. Assembly and fabrication of new packaging methods require ingenuity, imagination, and good judgment. Inspection procedures and techniques adaptable to miniaturized assemblies have to be developed. Producibility aspects must be considered for each phase in fabrication and assembly. All difficulties experienced in design layout and manufacturing processes of prototypes should be recorded for future guidance in production areas. One of the main advantages of miniaturization is the possibility of eventual introduction of automation.

EVALUATION OF PACKAGING TECHNIQUES

Soldered Sandwich. Module construction using soldering processes has certain advantages which should not be overlooked in the selection of a packaging method. It is a simple, well-known process. With introduction of controllable soldering, much higher reliability could be achieved than previously experienced in this area. Part density comparable with welded packages can be obtained by soldering. The main deficiency of soldering is the extended application of a high temperature, and there is still lack of good inspection techniques. Soldering cannot be used on heat-sensitive devices. Due consideration should be given to the use of soldered-welded construction.

Welded Packaging. The introduction of welding techniques into electronic packaging was based on the promise of higher reliability. The main advantages of welding procedures are:

1. They are low heat processes which do not affect behavior of heat-sensitive components.
2. High density.
3. Lighter weight.
4. Smaller size.
5. Uniformity of joints.
6. Faster production procedures.

The main disadvantages of this process are:

1. Difficulty in welding of certain component lead materials to interconnecting ribbon.
2. Lack of a definition of a good weld and the absence of "dud" detecting devices.
3. Difficulty in maintaining a consistent process.
4. Need for encapsulation to assure increase in mechanical strength.
5. Effect of aging process on behavior of encapsulated package.
6. Some of the components are not available with weldable leads.
7. The electrodes must be continuously dressed and cleaned to obtain uniformity in pressure and energy requirements.
8. Some of the lead materials require special preparation.
9. The alignment of parts into a desirable configuration requires additional fixtures.
10. Encapsulating materials and processes cause undesirable movement and stressing of the parts (see Figs. 2 and 3).
11. The weld schedules must be either developed by the designing agency or obtained from some reliable source.
12. There must be some assurance of the consistency of lead material when it reaches the production floor. Unexpected substitutes in lead materials will throw the schedule out of order.

Transistors installed

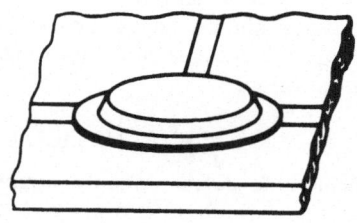

Fig. 1

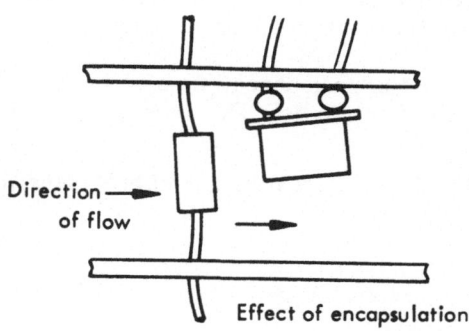

Direction of flow

Effect of encapsulation

Fig. 2

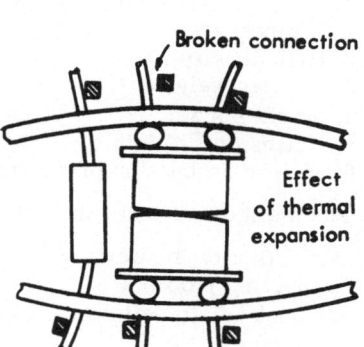

Broken connection

Effect of thermal expansion

Fig. 3

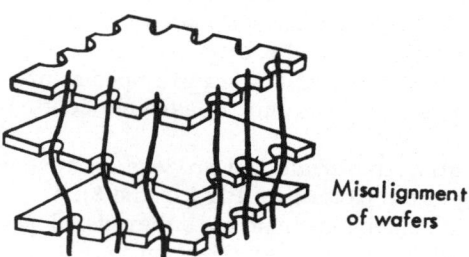

Misalignment of wafers

Fig. 4

Modules with Disciplined Geometry. Miniaturized modules can be used as a part of high-density techniques or as integrated packages. They offer compactness, certain flexibility, improved order of reliability, and reduction in weight and volume. They do, however, have some inherent negative characteristics:

1. Misalignment of the wafer can take place (see Fig. 4).
2. A limited number of connections per wafer (12 to be exact) (see Fig. 5).
3. The grooved area at the end of the wafers must be properly prepared for making connections (see Fig. 7).
4. Mounting of components requires specific concentrated efforts (Figs. 1, 8, 9).
5. Use of bare wires requires caution.
6. Material for the substrate used in capacitors is different from the ordinary substrate and does not possess the desired mechanical strength.
7. Encapsulating materials are necessary for structural support of the wafer assembly.
8. Work must be performed with a microscope.
9. Procurement of parts may be slow due to the limited supply.
10. Behavior under shock and vibration conditions is questionable (Fig. 6).

Perforated Wafers (also part of modules with disciplined geometry). In this method of assembly a special component part is inserted into perforations in the substrate.

1. The components must fit the openings. A mismatch may occur because the components are out-of-round, egg-shaped, undersized, or oversized (see Fig. 11).
2. The making of components into this shape may be difficult and sometimes contrary to the nature of a component.
3. The drilling of holes is difficult as the side walls in the perforation may be damaged (see Fig. 12).
4. The irregularities in the shapes of the openings will make the installation procedure difficult.
5. Insertion of the components into openings may damage the walls of the opening as well as the inserted components (see Figs. 10, 13).
6. The adhesive used in cementing the components into holes may overflow and coat conductive surfaces (see Fig. 14). This condition will lead to an open circuit in the conductive trace (see Fig. 15).
7. Connection of the component with conductive trace is a possible point of trouble; blisters, cracks, or voids may develop and cause difficulties.
8. The heat transfer from embedded parts may be difficult.
9. The installation of a component in a tight opening in the substrate is undesirable from the standpoint of heat transfer and circulation of air.
10. Differences in the thermal coefficient of expansion between ceramics of the perforated boards and the metallic surfaces of the components may cause a break in the seal between component and trace (see Figs. 16, 17).
11. After the pattern of the component layout is made and the holes are drilled, changes in the configuration are impossible (see Fig. 18).
12. The use of contour cables or flat cables for interconnections between individual boards creates many difficulties commonly associated with flat cables.

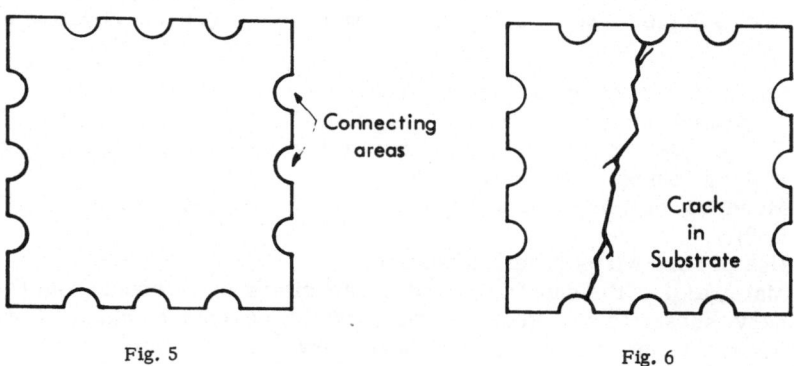

Fig. 5   Fig. 6

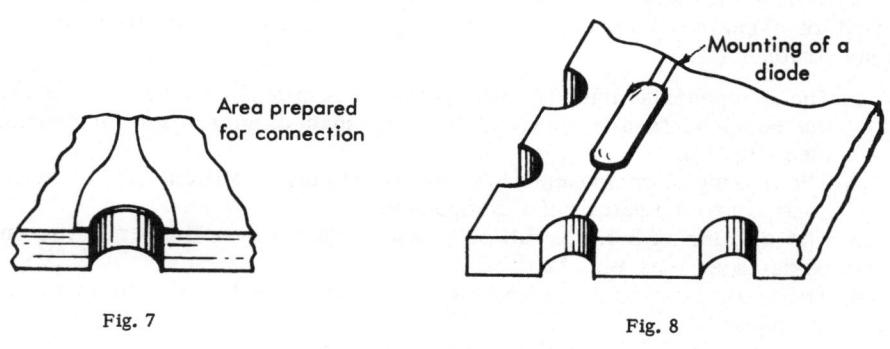

Fig. 7   Fig. 8

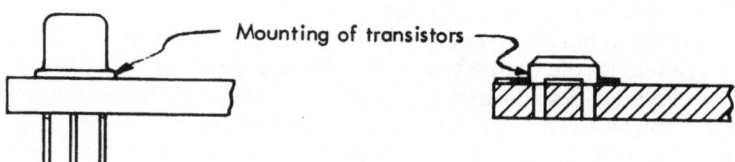

Fig. 9

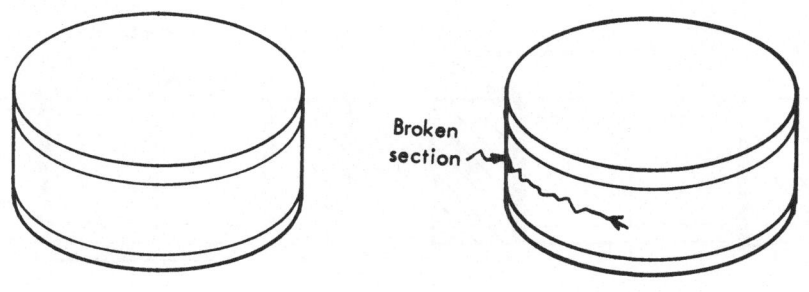

Fig. 10

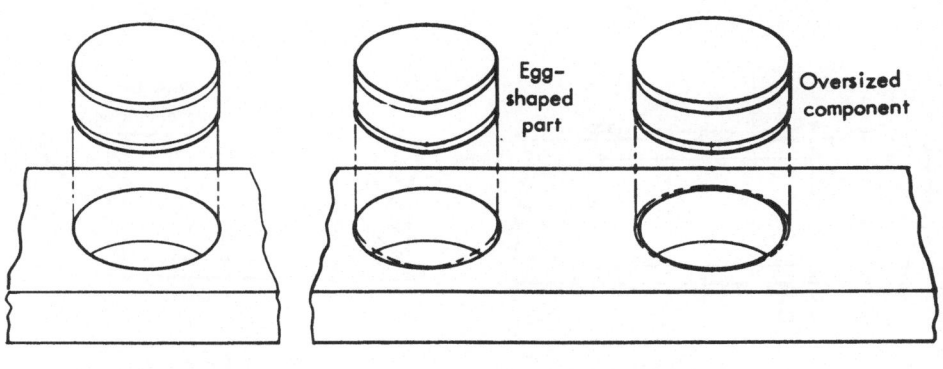

ACCEPTABLE   NOT ACCEPTABLE

Fig. 11

Fig. 12

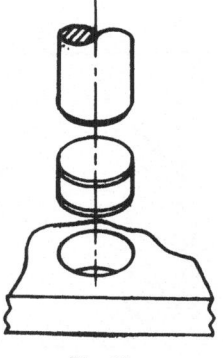

Fig. 13

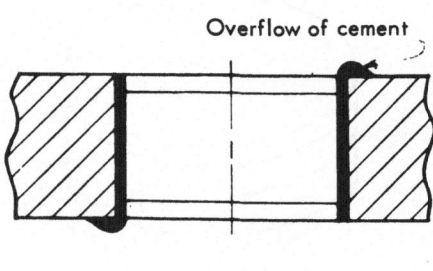

Overflow of cement

Fig. 14

Opening in trace
caused by overflow
of cement

Fig. 15

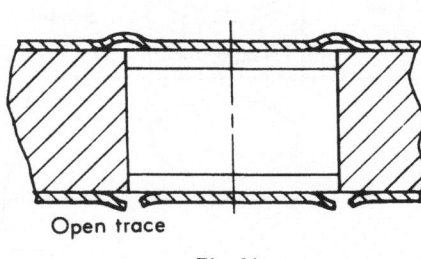

Open trace

Fig. 16

Broken top
of component

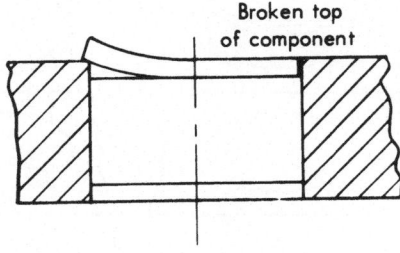

Fig. 17

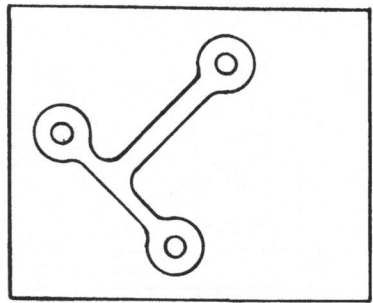

Original pattern

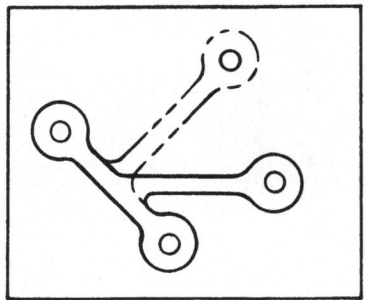

Desired change

Fig. 18

Many of the difficulties outlined above are not noticeable under laboratory conditions when and where people are driven by the glamour of a new idea, but they are quite possible in a production line where the level of interest is much less and the working man does not possess the high level of skill needed for a highly reliable package.

Thin Films

1. Offer very reliable and stable trace. .
2. Passive elements and traces are stable and do not change their characteristics with fluctuation of temperature and pressure.
3. In thin films built by chemical means the adhesive bond between substrate and metallic films is very strong.
4. Thin films could be developed and changed by an etching process.
5. Anodic processes offer possibility of converting a thin film into alternate layers of conductive and dielectric layers, making possible manufacturing of capacitors.
6. Surfaces of holes and cavities could be coated with homogeneous coating of the thin films offering means for interphase connections.
7. Films can be made both by the vacuum deposition and chemical means depending on circuit requirements.
8. Control of circuit parameters, resistances, and capacitance requires effort and sensitive instruments.
9. Surface condition (finish) of the substrate has effect on resistance and capacitive values. It is desirable to obtain smooth finishes on substrates.
10. Connections between components and traces require special effort.
11. Number of interconnections is reduced because parts of the same films are serving as components through modification by chemical means.
12. Both vacuum deposition and chemical means require special processes which must be developed and handled properly.
13. Size of the module can be reduced and still retain original resistance value, but at a reduced wattage.

Solid State Devices

1. Practical in low-power applications.
2. Must be developed for each circuit, unless available as a standard device.
3. Changes are impossible after the solid state devices are made.
4. A building process for these devices is very complex.
5. Very tight controls in production are necessary to assure a reliable product.
6. Connections between solid state devices and exterior wires are critical; the device could be damaged in this process.
7. At the present time devices are expensive.
8. Provide a compact and light-weight assembly.
9. Have potentially higher reliability.
10. This method may not be usable with some circuits.

CONCLUSIONS

Additional investigations are necessary to obtain electrical characteristics of the selected circuit built by various packaging methods. This program is currently under way. Comparison between available packaging methods was based on available information and samples. Improvement and change in design of packaged assemblies would change described characteristics.

The emphasis on negative aspects of the evaluated concept was made because, as a rule, advocates of packaging methods neglect to mention them.

Progress in the field of miniaturization of electronic circuits can be expedited if each method is evaluated at the beginning. When only desirable aspects are emphasized and accepted by prospective users, unexpected occurrence of disadvantages leads to disappointment. Interest of both the seller and the buyer suffers when exaggerated claims for new concepts are made.

DISCUSSION

Question: Martin Camen, Bendix Corporation. In your last slide there was a term called bionics. Would you care to elaborate a little more on this?

Answer: Bionics is a term that is applied right now to use of elements similar to that of cells of a human brain. It's a new field which should play an important role in the future.

Question: Joe Ritter, Electronic Modules. In your paper you discussed soldered modules as well as welded modules; however, in your talk you made the assumption in a flat statement that welded modules are the coming thing, but in the paper there are about 13 or 14 points of difficulty in welding, and only one point in soldering. I would like to know on what basis you make this flat statement.

Answer: I feel that the deficiencies in solder are well-known facts. Welded electronics is a new field which should be approached with caution. We all know the deficiencies in soldered connections. I felt that right now the soldered type of construction is at a great advantage, because of the emphasis given to welded construction. I know from our investigations that under certain environmental conditions the soldered sandwiches can perform satisfactorily. However, this does not mean that welded sandwich construction does not have a proper place. As I explained in my paper, I am trying to approach all the packaging devices from a negative standpoint of view, because I feel there are too many advocates already telling about the positive characteristic of those concepts.

Comment: (Mr. Ritter). I agree with your approach and I feel it is the right one, however, I think that probably we should apply the same objective review to all the processes and evaluate them on the same basis. I don't believe this is being done. It's quite true we know the problems in soldering and in knowing them we know how to get around them, but we don't know the problems in welding.

Answer: We intend to do exactly what you are saying; to be very impartial.

THE HIDDEN VALUE—PACKAGING FOR APPEARANCE

J. R. Milligan

General Dynamics/Astronautics, San Diego, California

Discussions involving appearance design are often poorly received by engineering groups because in discussing appearance we are forced to consider the aesthetic quality of a product. This does not mean engineers are not aesthetically inclined. However, they take professional pride in an analytical and unemotional approach to their work while appearance design is dictated by sensory impressions of objects and not the objects themselves. Yet, appearance and engineering have much in common, for the ability to design a meaningful appearance into a product is based on intuitive power, insight, and knowledge. These same prerequisites are necessary in all creative work, not only art, poetry, and philosophy, but science and engineering as well. The basic creativity in appearance design eliminates graphs, scales, and charts which can be consulted by designers, and from these evolve an acceptable appearance. Each design has its own unique criteria from which the final appearance is developed. Nevertheless, it is possible to present standard procedures and methods using these as guidelines to develop specific designs.

If sensory appeal were the only attribute of appearance, it would receive little attention in everyday product design. It was discovered long ago that appearance not only has basic appeal—it sells! The saleability of appearance has given rise to an entirely new area of product design. This area is sometimes referred to as industrial design, appearance design, or styling. The word "styling" carries with it the unfortunate connotation of artiness or lack of consideration for the product's function. This connotation results from the use of the word "styling" by the clothing industry which has never been known for its functional designs. Also, during the past twenty-five years, advertisements of consumer goods have placed increased emphasis on the prestige factor of "the latest design." This has helped sell everything from can openers to cars. At the same time it has created an artificial obsolescence.

Because appearance design has occasionally been used only as a sales device, its use has been avoided by some manufacturers whose products are primarily of an engineering nature and not necessarily directed toward a mass market. The ability of these manufacturers to sell their products is based on functional use, reliability, and maintainability. By not placing the proper emphasis on appearance, their products not only lack aesthetic appeal, but their functional qualities are not enhanced by the abilities of industrial designers.

In American industry there are three primary approaches to the use of appearance. The first is most often found in mass-produced consumer goods where the designer or stylist is given primary control of the product. This leaves the engineer to struggle with the problem of fitting his mechanism or circuitry into a predetermined package. The fact that these products work well is a monument to the skill and ingenuity of the engineers.

The other extreme is a product which is completely engineered and packaged without any consideration for the appearance. After the product has been checked out functionally, an appearance consultant is summoned. In spite of an anodized aluminum miracle, the results are often disappointing and neither enhance nor express the function of the equipment.

The third approach requires engineers and appearance designers to work together. They place appearance in proper perspective and at the same time increase the quality of the product. This last approach has several requirements which must be met. The first requisite is a designer with an intimate knowledge of appearance and the necessary background and training to use this knowledge. Colleges and universities are presently training people in this area. Design students receive a basic engineering background including mathematics, industrial processes, materials, and product design. They are also given intensive courses in two- and three-dimensional arts. Upon completion of college the student is awarded a degree in Industrial Design. The results of this training are individuals who, while specializing in appearance design, can communicate and coordinate their work with any of the engineering or technical disciplines. Many companies employ one or more industrial designers who work with various engineering departments and assist in maintaining a high quality of appearance in all products. Large and small firms alike have available the services of highly competent industrial design organizations which will work with them to improve the appearance and quality of their products.

Second, the appearance must be considered in the early stages of design development. A satisfactory appearance is not merely a matter of decorating a package but evolving a form which will best express the function. This can be done only by designing the appearance from the inside out.

Finally, it is necessary to recognize the freedoms and restrictions in each design and to be sure the final product fits within these limitations. Consideration must be given to the market toward which the product is directed, the type of manufacturing equipment and materials available, portions of the product to be bought or subcontracted, and lead time on delivery. Individual companies have further limitations such as military specifications. These limitations have their ultimate impact on the cost of the products produced and ability of manufacturers to compete with one another for a specific market.

Civilian products manufactured for the consumer market follow definite trends in styling. Some manufacturers wait for trends to develop and then point their design in this direction. There are difficulties, though, in following specific trends. Once a trend is recognizable, it has already begun to diminish and is of little benefit. Often following the trend only shows a lack of imagination on the part of those who continue to rely on it. An example of a trend is the use of an extended bezel often referred to as an "eyelid" around the panel face of various items of electronic equipment. The eyelid was first used to cut down glare on the faces of instruments and displays. Later it was transferred to the upper portion of the instrument panel where it shielded all controls and displays. Perhaps this can be justified by a combination of glare reduction and a certain dust-catching function. During the past year literally dozens of companies used this eyelid motif primarily as a sales device to show that their product was "up-to-date." The use of an eyelid may have been justified in some cases, but using it as decoration rather than a functional part of the design merely reflects the inability of an organization to design its equipment with an original appearance.

Original design is the result of a step-by-step analysis of the product and an evolutionary development. Often, when design time is limited and other factors, such as cost, prevent a complete analysis of each individual piece of equipment, it is possible to analyze the requirements of a line of equipment and from this analysis to develop individual items. Grouping together similar types of equipment enables a designer to evolve an appearance which can be adapted to many different items. Companies designing for the military find that a large portion of their electronic packaging is devoted to R&D checkout and test equipment. Often there is little consideration given to appearance design due to severe limitations. These limitations include: (1) tight schedules which allow little time for redesign; (2) proofing changes resulting in major circuit rework which must be incorporated into the design; (3) short production runs often less than three or four sets of a given type of equipment; and (4) limited budgets to develop the equipment. There are also restrictions imposed by military specifications which designate not only wire types but also a complete range of items from environmental testing to which the equipment must be submitted, down to the type color, and thickness of paint to be used. Designers often feel surrounded by these limitations, but it is possible to work within them and still develop a satisfactory appearance. When necessary, military specifications can be changed if the improvement justifies the change—it is difficult, but possible.

An analysis of various R&D packaging jobs has shown that the majority of this equipment can be adapted to the standard "19-in. relay rack" enclosure. There are many well-designed enclosures of this type presently on the market. They come in standard sizes and configurations. Some are of modular construction which permits an infinite variety of configurations. The appearance varies, but the manufacturers of these enclosures are quite willing to have their designers work with an electronic firm to develop a specific cabinet style. There is presently one enclosure manufacturer who has marketed a standard cabinet with a removable bezel which forms the front of the enclosure. The design of the bezel can easily be altered, and original designs become the proprietary property of the firm which orders them, giving a custom look to standard "off-the-shelf" cabinetry. Because production runs are small and development time limited, standard enclosures are sufficient to meet the needs of most R&D packaging jobs.

Enclosures have a decided effect on the final appearance, but over-all appearance is dependent not only on the relationship between the enclosures and instrument panels, but also the effect of the panels upon each other. The appearance of the panels is in turn determined by the chassis design because the manner in which the chassis is fastened to the panel and its position limits the panel area available for displays and controls. Further analysis has shown that for R&D packaging the ideal chassis must have the following features: flexible usage, simplicity of construction, and minimum engineering time required for packaging. There are many standard chassis on the market, but none of these fitted the particular needs of this equipment. The frequency of changes in circuitry resulted in major packaging changes, often requiring that whole new chassis be made. It was necessary, therefore, to develop a chassis which could be changed in a minimum amount of time. The result of this development was a standard chassis of three-piece construction affixed to the panel with simple gussets and a small L bracket (Fig. 1). This construction permitted rapid replacement of the top or rear of the chassis in event of component or circuit changes. By using the same hole pattern for the handles on the front of the panel to secure the chassis, protruding screws were eliminated and a maximum of the panel face was available for controls and instrumentation. Standard drawings of this chassis were made and packaging engineers now simply call out hole patterns for component mounting. This reduces the amount of time consumed by layout and draft-

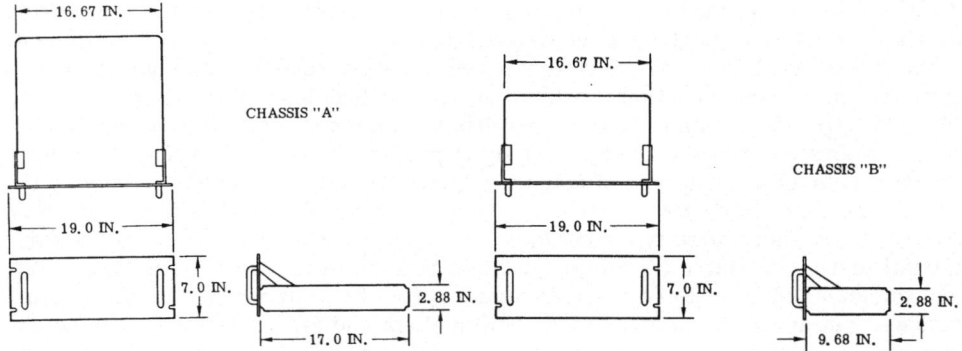

CHASSIS "A"

CHASSIS "B"

BOTH CHASSIS SHOWN WITH STANDARD 7 INCH FRONT PANEL.

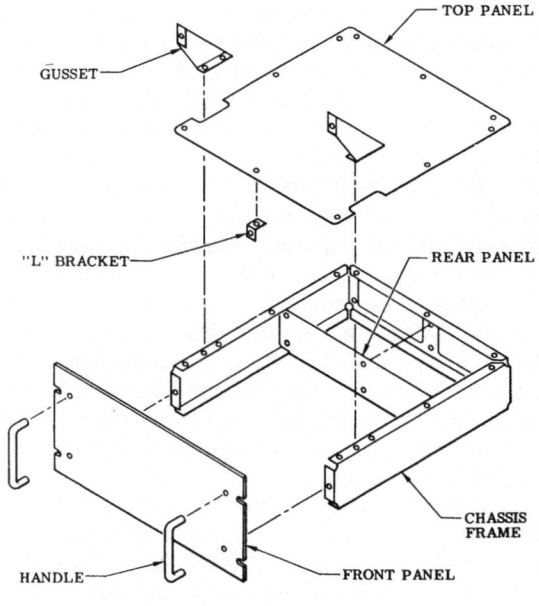

STANDARD CHASSIS BREAKDOWN

Fig. 1. Standard chassis.

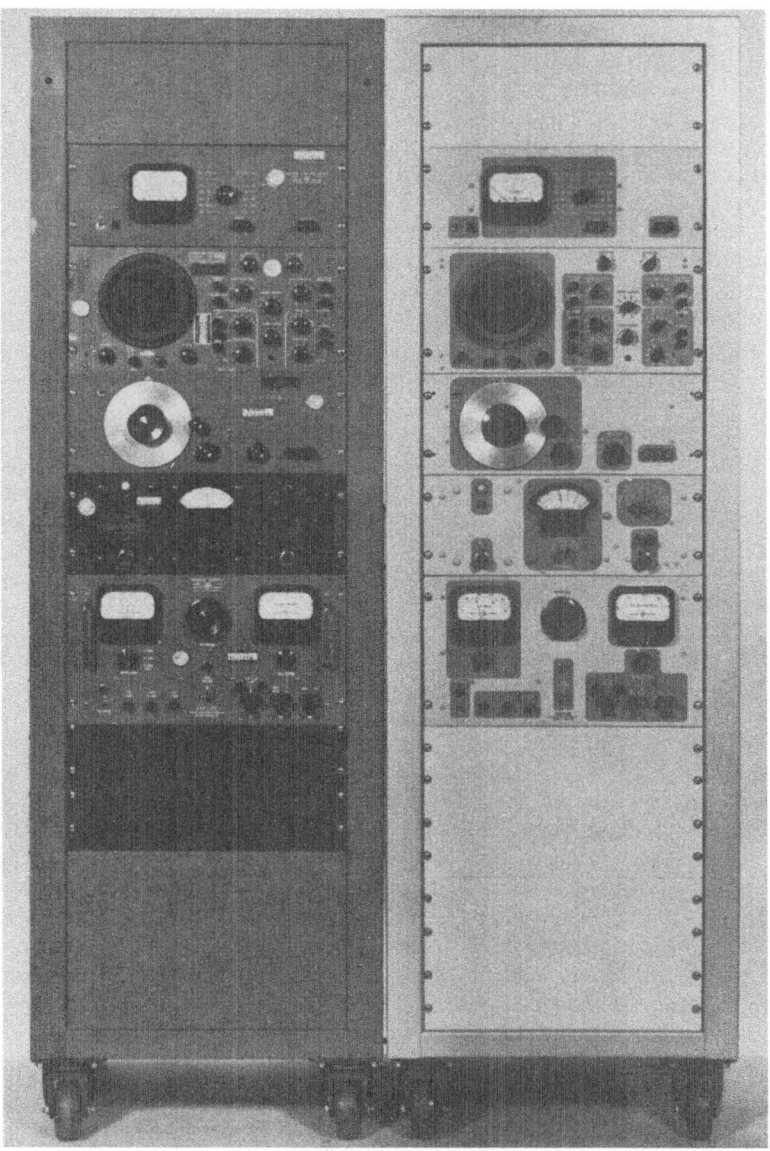

Fig. 2. Comparison of reworked commercial unit.

ing. Improvements were made in the chassis over a period of time, and it was found that two basic sizes would satisfy a majority of cases with the addition of minor bracketry to handle the extremes.

The development of this chassis provides several advantages. The adaptability of the chassis to most R&D packaging jobs permits large scale production of the chassis although only a few are used on each individual test set. The position of the chassis in relationship to the front panel allows the chassis to be supported by angle runners within the enclosure, and if slides are desirable, there is sufficient space available at the sides for their attachment. Of equal importance, the combined attachment of the chassis and the handles on the panel face presents a controlled starting point for the design of the panel face.

A panel face which is compatible with the enclosure and adaptable to many different electronic packages was the final step in the analysis. Because electronic equipment of this type is often quite complex, it is necessary to use human-factor techniques for panel arrangement. However, idealizing the human-factor aspects of controls and display arrangement is not always possible. Engineering considerations such as wire routing, heating, and RF shielding of wires and groups of components often take design precedence. The panel arrangement, therefore, is a compromise between the requirements of the circuitry and human factors. In addition, panel faces must be standardized so the finished design will have continuity of appearance. From a mechanical standpoint, there is only one variable between panel faces which cannot be standardized. This is the hole pattern which is necessary to mount controls, displays, and indicators. Other variables, such as panel size, hardware, including handles, mounting devices, and finishing material, can be chosen to provide the best appearance compatible with the manufacturing limitations. Commercial units such as oscilloscopes, vacuum tube voltmeters, and power supplies are often used with units manufactured "in house." This adds the requirement that the in-house units be compatible with purchased items. Several solutions to this problem have been explored. An early answer was to simply paint all the panel faces one color. While this gave some semblance of order, it failed to provide a truly satisfying appearance. The human-factor technique of placing functional pads around groupings of instruments and displays was investigated to see if a panel design incorporating this technique could be used.

A group of commercial units was randomly arranged in a standard 6-ft enclosure. A second set of identical equipment was repainted using two colors to separate the functional pads from the background. The result, as shown in Fig. 2, was a marked improvement in appearance, but use of this technique required extensive coordination between designers and packaging engineers. Attempts by packaging engineers to use this system without the aid of an appearance designer often resulted in a haphazard appearance. Further work with division of functional areas from panel background revealed that if the distance between functional areas rather than the size of pads themselves were fixed, the entire problem of arrangement could be closely controlled. Starting with the handle position, the face of the panel was divided into two main areas. The first was the background and contained the handles and slots for fastening the panel to the enclosure. An area at the top was set aside for the panel title. The remaining area could be subdivided into a series of small subareas, separating the control and instrument groupings by function.

An instruction sheet (Fig. 3) was drawn up and distributed to packaging engineers. When there was sufficient time, these same instructions were provided to the manufacturers of commercial equipment to be used in certain test sets. Where this was not feasible, the commercial equipment was painted the same color as functional pads and segregated within the enclosure to minimize the contrast. An example of the results of this effort is shown in Fig. 4.

This test set uses a commercial enclosure with standardized chassis and panels in the upper portion. The lower portion contains power supplies which were purchased and painted to contrast as little as possible with the control panels above them. It was necessary to produce only two of these sets, and all of the previously mentioned restrictions were encountered. Yet they were manu-

STANDARD PANELS

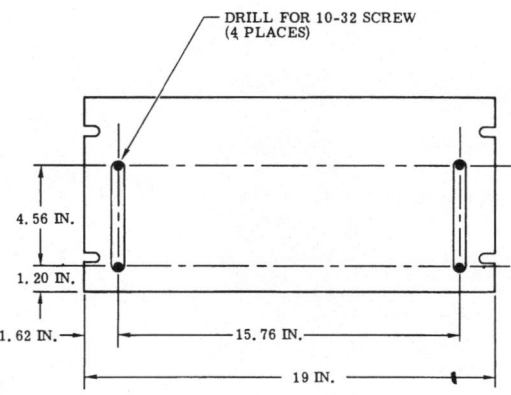

DIMENSIONS ARE TO BE USED ON
ALL PANELS 7. 0 HIGH & LARGER.

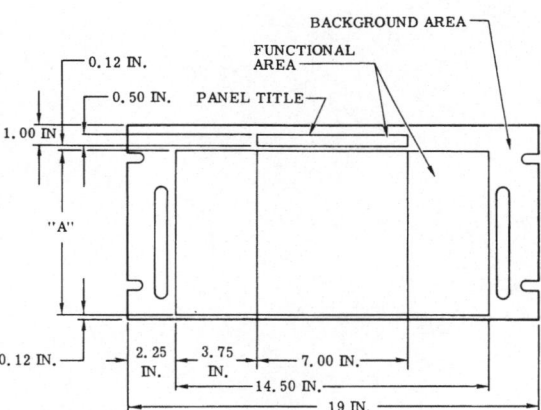

DIMENSION "A" VARIES WITH PANEL HEIGHT.

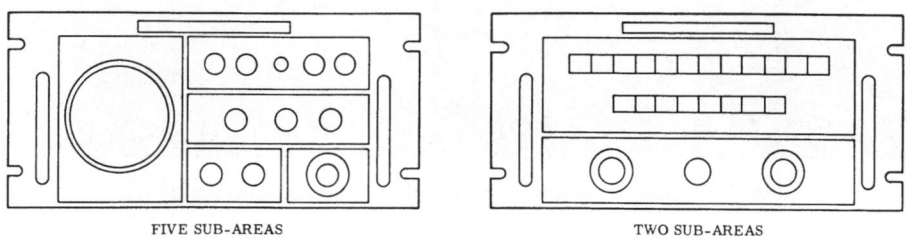

FIVE SUB-AREAS   TWO SUB-AREAS

Fig. 3. Standard panel configuration.

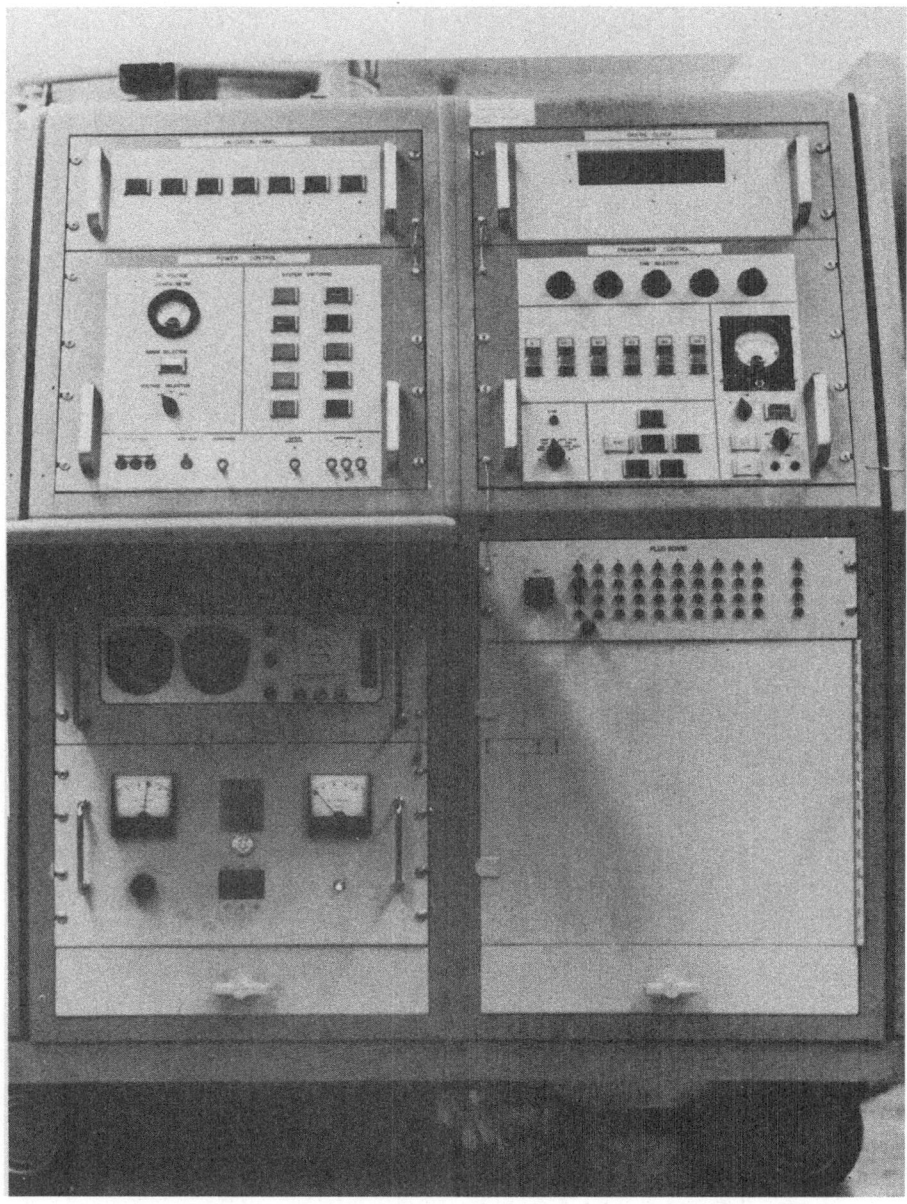

Fig. 4. Checkout set built with standard chassis and panel configuration.

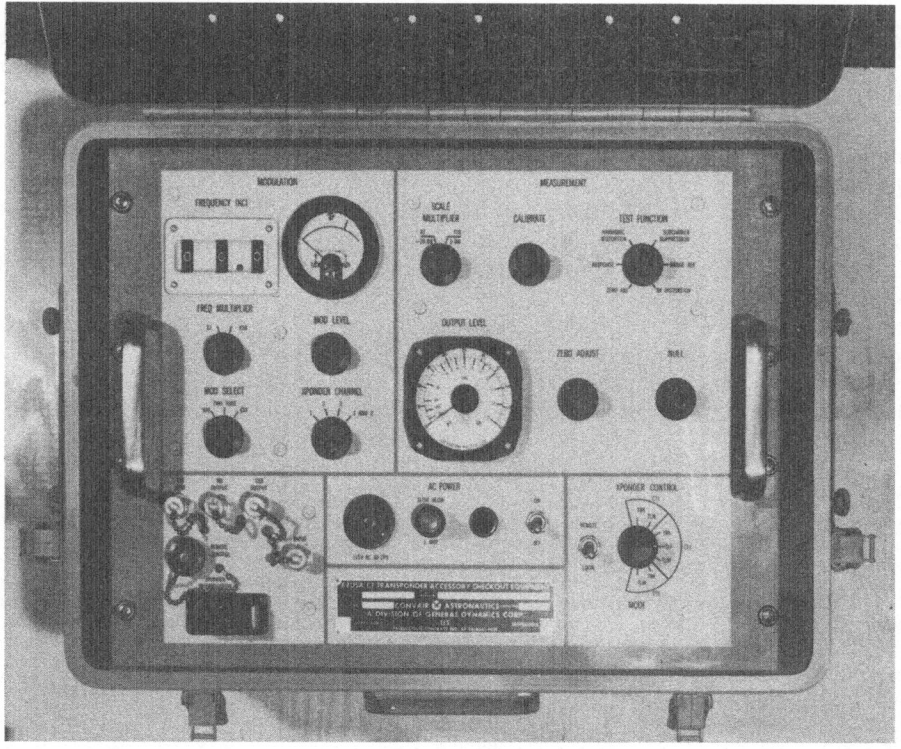

Fig. 5. Small test set using standard panel configuration.

factured rapidly at low cost, and the final appearance is a distinct improvement over equipment of this type previously produced. It is also interesting to note that the four upper panels on this test set were designed by four different packaging engineers. This panel arrangement is easily adapted to a variety of equipment as shown in Fig. 5. The same panel arrangement, with only a change in the handle position and elimination of a panel title block, is used.

Besides the previously mentioned advantages found in appearance design, there are several intangible but nevertheless obvious values in the application of appearance design. Since they are difficult to measure and are often overlooked, they might be considered hidden values. How many electronic equipment manufacturers have placed sufficient emphasis on appearance so their company can be identified without consulting the nameplate on their equipment? You can probably name only three or four firms whose original design projects a corporate image. These same companies are invariably the ones which set the trends in styling and whose products represent excellent quality in the electronic equipment field. The growing competition within the electronic industry has forced advertising budgets into astronomical figures. Yet many companies still have not recognized the fact that their best advertisements are their own products. A corporate image created through the use of appearance design not only increases the quality of a product, but enables the product to be its own best salesman.

The care a customer gives electronic equipment is often proportional to the maintenance required during the life of the equipment. Proper care and handling is usually outlined in a small instruction sheet which is lost soon after the equipment is uncrated. Why is it certain equipment is given proper care while other of a similar nature is abused? One reason is that equipment by its appearance either demands or rejects care. A piece of equipment whose appearance reflects a quality product is usually given the care a quality product deserves. At the same time equipment which has a poor appearance is accorded an equivalent amount of care.

Finally, pride and enthusiasm with which people within an organization perform their work has a definite effect on the equipment they produce. In this day of exotic machines the average worker and engineer contribute only a small portion of the finished complex mechanisms of the electronic industry. If the appearance of the final product reflects the care and skill of an entire group, the pride of all those who contributed to its completion will be bolstered and eventually appear in better work and superior products. Quality equipment should be the goal of every electronic manufacturer, and appearance is the mark of quality.

DISCUSSION

Question: P. J. Guillot, RCA, Van Nuys. Would you care to comment on the black knobs and the black hardware and all this equipment you have as part of the human factor of engineering?

Answer: Yes. The black knobs and handles and the various colors shown here are all Mil spec colors. All this equipment was designed within military specifications. The knobs are Mil spec knobs, the color schemes are Ballistic Missile Division color schemes which are called out in military specifications. We adhere to these specifications and still, as I hope I showed here, improved the appearance of our equipment.

Question: K. A. Allebach, Nortronics, Los Angeles. How do you feel about the use of 100% push button controls rather than knobs?

Answer: As far as the use of push button controls as opposed to knobs, working with our human engineering department at Astronautics, when we design a particular type of equipment, the function to be performed by the equipment dictates the type of knobs to be used. In the illustrations I have here, we use both the push button switches and rotating knobs. Once again this is the function of the equipment and is worked out in conjunction with the human factors department.

Question: Do you think it's worthwhile to attempt to use more push buttons, however?

Answer: From a standpoint of appearance it's excellent.

Question: Don Gardner, Melabs. What have you done in the panels by way of engraving? Is it photoetch?

Answer: The various panels shown here include quite a spectrum of panel finishes. Originally at Astronautics we did all photoetching on panels. However on test equipment, due to the short time and low budget, we have found that silk screening or merely masking and spraying is sufficient. Some of the panels you saw were merely leroyed and then sprayed over with a lacquer to protect the ink on the panel. Some of the panels you also saw here were done with a new process which is a proprietary process by one of our vendors called the Wrinlay process. They use an epoxy base paint (Perma-Resin) which you cannot scratch or mar. It is the finest finish we have found yet. We have had a great deal of success with it.

Question: How did this leroy get by on Mil specs?

Answer: On the test equipment to be used "in house" we are not required to meet the same Mil specs required for operational equipment.

Question: Beck, Bendix Research. Would you repeat the source for the epoxy paint?

Answer: It's called the Wrinlay process using Perma-Resin paint. It is a proprietary process of the Photo Chemical Co. located in Santa Monica, California.

Comment: Joe Bachus, Medtronic, Minneapolis. I might say an alternate way to get a panel finish is to use the silk screen and cover it with clear epoxies. We have done this quite a bit and it works out very well.

Answer: I have no idea what the cost of this would be. Oftentimes our finishes are dependent on the budgets of the particular job.

CRUCIFORM PACKAGING—A GENERAL SYNTHESIS FOR AIRBORNE ELECTRONICS

J. C. Rubin
General Electric Company, Philadelphia, Pennsylvania

SUMMARY

Accelerated development schedules, combined with a tremendous increase in number and complexity of design factors, have pointed out the need for a general package design adaptable to a wide variety of environmental conditions.

Cruciform packaging has been developed in order to optimize reliability, maintainability, size, and weight, with capabilities of meeting any specified environment of temperature, humidity, shock, vibration, and radio noise. Starting with an internal structural skeleton, in lieu of an outer casing, the synthesis proceeds structurally to diagonally braced plug-in module assemblies of selected cross section and continuously variable length. Reliability target of duplex connections is achieved by mechanical and electrical connections at each component and dual lead support in fixture boards and encapsulation. Maintainability at any level is assured by plug-in structure, plug-in modules, and chemically dissolved encapsulation. Size is kept minimal through high-density cordwood arrangements of components between parallel fixtured boards; as component parts of smaller size become available, they can be immediately accommodated without extensive design modification. Solid circuits and thin-circuit subassemblies are likewise acceptable component parts of cruciform modules.

Weight can be minimized consistent with environmental requirements by straightforward machining operations in the metalwork and substitution of foamed urethane for loaded epoxy encapsulants.

Heat transfer by free or forced convection, radiation, conduction, or any combination thereof, can be readily employed to suit customer requirements.

The over-all result of a generalized design with extreme versatility in size and weight with positive environmental margin has been the transferring of the design engineer's packaging effort from synthesis of a required package to analysis of the generalized package in light of specific requirements. This analysis, based on actual test data obtained from the generalized design, results in simple straightforward adaptations to meet specific needs. In two years' utilization at the General Electric Company's Armament and Control Section, Johnson City, N.Y., the cruciform has appeared in actual delivery, or in proposals for airborne, shipborne, and jeep-borne equipment, and for primary and ground-support equipment, for all Armed Services. In each case, the configuration and customer requirements dictated the design specifics, but the general concept has remained essentially intact throughout.

INTRODUCTION: STATEMENT OF THE PROBLEM

Presenting a paper such as this, dealing with a solution to problems of electronic packaging design, would have little meaning, unless we can first agree on what the objectives of such design should be. Last year's Packaging Symposium was enough to convince me, however, that unanimous agreement on objectives, even among those well versed in this branch of engineering, is quite impossible. (It may even be recalled that certain members of the audience spent as much time out of their seats as in them, discussing and debating basic approaches suggested by the speakers.) The best I can do, therefore, is list for you certain objectives of electronic packaging design which were satisfied to some degree by the synthesis described in this paper. You are encouraged to fill in your own special objectives, determine the applicability of the procedure to them, and comment accordingly.

Engineering Objectives

1. To design to meet customers' requirements;
2. To effect a smooth transition from circuit engineering breadboard to completed design;
3. To meet contract schedules;
4. To stabilize work loads for engineers, technicians, draftsmen, and shop personnel, thereby minimizing overtime charges and costly crash programs;
5. To obtain sufficient environmental margin to protect the circuitry from field failure, without excessive penalty in weight, space, and cooling power.

Manufacturing Objectives

1. To facilitate production of hardware meeting specifications;
2. To meet contract schedules;
3. To stabilize manufacturing engineering and labor work loads, thereby avoiding overtime and crash programs;
4. To obtain the cost advantages of standardization.

Marketing Objectives

1. To obtain a distinctive saleable hardware design of proven reliability and customer acceptance;
2. To obtain design information in the form of written and pictorial descriptions, models, and samples of typical package designs, for use in proposal and presentation efforts.

The major difficulty in reaching these objectives appears to be a composite of two factors. First, the design cycle normally places packaging design in a "series" position, following release of circuit information (see Fig. 1). Packaging is not a continuing but an intermittent process, with large gaps in its engineering work load of often months or even years, with little historical continuity thereby established. With normal departmental turnover, this often places the packaging design responsibility largely in the hands of personnel unfamiliar with previous designs, evaluations, and trade-off decisions. Second, the time allowance for military contracts has been decreasing, while the design complexity has been increasing. Environmental factors have been increasing in number and severity also. This has had the effect of requiring a maximum of design in a minimum of time.

It can easily be seen that these two factors combine to combat every single one of the objectives listed above. Results will include one or more of the following:

1. Poor packaging: failure to meet environmental requirements in test or field; poor transition between breadboard and final design.
2. Expensive packaging: crash programs in engineering, manufacturing, and test; forced decisions to make rather than buy due to tight schedules, extreme variability in engineering, drafting, and shop workloads.

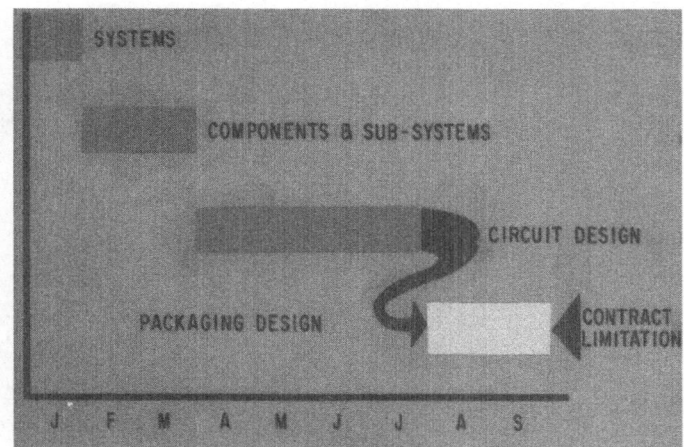

Fig. 1. The packaging squeeze.

3. Little historical value for future designs; little standardization.

To all of the above considerations, I would like to add another factor—a personal or professional one rather than a corporate or business one. This factor is the professional status of the packaging engineer, i.e., his recognition as a continuing active contributor to equipment program efforts, rather than a "fireman" to put out a once-a-program blaze. I have felt more and more strongly, over sixteen years of association with electronic packaging design, that this factor is an important one in attracting and retaining top technical talent in the packaging field, a factor that is unfortunately overlooked, or considerably minimized in many equipment programs I have seen.

PURPOSE OF PAPER

The body of this paper contains, in an abbreviated fashion, the details of a single, simple design which has been synthesized and applied at General Electric's Armament and Control Section at Johnson City, New York, over the past two years. It has appeared, in one form or another, on programs and proposals for airborne, shipborne, and jeep-borne equipments on a wide variety of programs. But the objective of this paper is not the presentation of a panacea, a cure-all for equipment packaging design, nor are the design details to be presented important in themselves. They are employed to illustrate the working arrangement and procedure which generated them. Features of this procedure are as follows. Also refer to Fig. 2.

1. Establishment of packaging engineering as a continuing design effort, adequately funded, and independent of contracted programs;
2. Execution and regular review of generalized package synthesis to meet anticipated needs for two, five, ten, or more years in the future;
3. Follow-up efforts in design, fabrication test, and evaluation on these generalized packages, again adequately funded;
4. Dissemination of results to design engineers (initially by consulting contact, later by product seminars, and finally by formal documentation in specifications and standards, complete with detailed application data, charts, and curves).

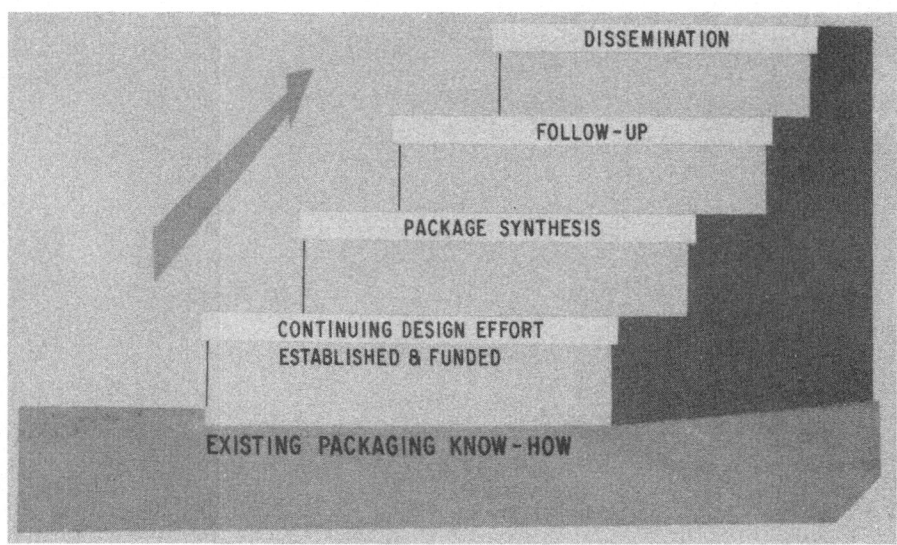

Fig. 2. Package design—steps involved.

ORGANIZATION OF THE PROGRAM

The original concept of the cruciform was evolved by George Carris, Value Engineer, with design work being done by Tom Tehan, Design Engineer, both of General Electric, Johnson City. Continuing efforts were carried on in Advance Design Engineering within the General Design Engineering Operation.

Advance design engineering specialists are primarily concerned with advancing product design, incorporating new design features, and examining product designs for areas of improvement, including reliability, cost, and standardization. As such, their efforts are closely tied to their associate groups in General Design Engineering as well as the product design engineers in all groups. These specialists are largely, if not entirely, funded on an "unapplied" basis; i.e., their time is not charged directly to the customer, but appears only indirectly in the engineering overhead. It is apparent that these contributors are relatively free from the pressures of contracts and schedules, and can be expected to concentrate on long-term design objectives without the usual qualms about contracts "running out" before completion.

The original engineering synthesis discussed in the paper was arrived at in a rather routine series of discussions between advance design engineering specialists, design engineers, and other members of the general design engineering organization. Engineering effort was provided without special procedures of any kind; drafting, fabrication, and testing were covered by development authorizations requiring signatures of top management. Very little delay was experienced in handling such items as the procedure is routine. Evaluation of results by advanced design engineering was followed by product seminars to introduce the concepts to design engineers (again a routine procedure), and finally by formal documentation. It was rather interesting to note the reactions of the various groups handling the test items; somehow a noncontract item on which one could proceed at a routine, leisurely pace was a welcome change.

Although specifically solicited in the work orders, the comments from manufacturing appeared to reflect individual personal interest in the design. All of this indicated that a continuing long-term packaging effort was able to achieve several side-effect, intangible benefits as well as its principal design objectives.

SYNTHESIS: EVOLUTION OF THE DESIGN

Where to Start: Structural Section. The first step in the particular synthesis was selection of a structural form. Leaning heavily on the developments of civil engineers, selection was based on the equal-leg rolled angle, long shown to be an efficient structural section in building construction. (See Fig. 3.) Advantages of the selection were:

1. Extremely open for acceptance of plug-in modules;
2. Convenient variability:
   a. In the plane of the section by selection of modular sizes (trimming if necessary);
   b. Perpendicular to the plane of the section (infinitely variable by cutoff);
3. Rigidity—high stiffness-weight ratio for high natural frequency in vibration;
4. Producibility in case of fabrication by casting, rolling, and/or machining.

First Modification: Cooling and Interconnection. On examination of the equal-leg structural angle, it became evident that separation was required to accommodate cooling and interconnecting cabling. This was achieved by the expedient of adding a strip at the end of each leg (see Fig. 3).

Second Modification: Machined Surfaces. In order to obtain good thermal contact between plug-in modules and package structure and to facilitate assembly, machined surfaces were specified and built up to avoid expensive over-all machining (see Fig. 3).

Combination of Sections: The Cruciform. The single angle developed so far could well have been considered as the complete synthesis. Variability to meet over-all configuration requirements could be obtained by resorting to variability in length (normal to section), variability in leg length, and unequal-leg angles (see Fig. 4).

Of these, only the first variation was generally acceptable to meet over-all configuration, the other two being reserved for special requirements. Later, however, a means of combining, or back-to-back angling, was used.

Figure 5 shows the variations used. The design at the top is employed for minimum cross section, maximum length configurations. The cover plate forms a cooling plenum and cabling channel. The assembly of two angles, or the tee, shown at the lower left, is used in place of the unequal-leg angle where 2 : 1 ratio section is applicable. The assembly of four angles, or the cruciform, shown at the right, is used in place of the elongated-leg angle. Figure 6 shows a production drawing for the assembly of two angles.

MODULES

Modules to fit the predicated design are generally described by the following:

1. Shape: Square parallelepiped
2. Size: Square variable in standardized increments; length variable indefinitely above width of one standard connector.

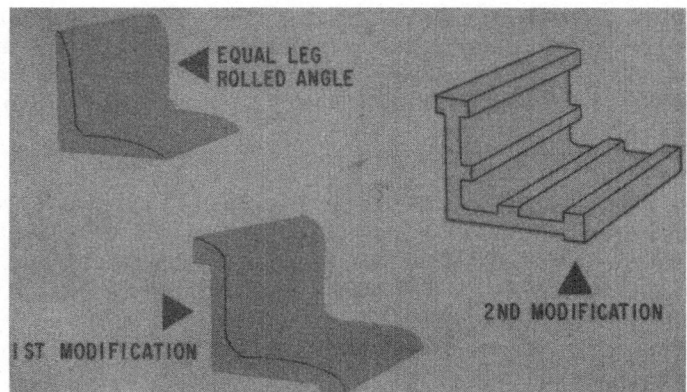

Fig. 3. Structural section.

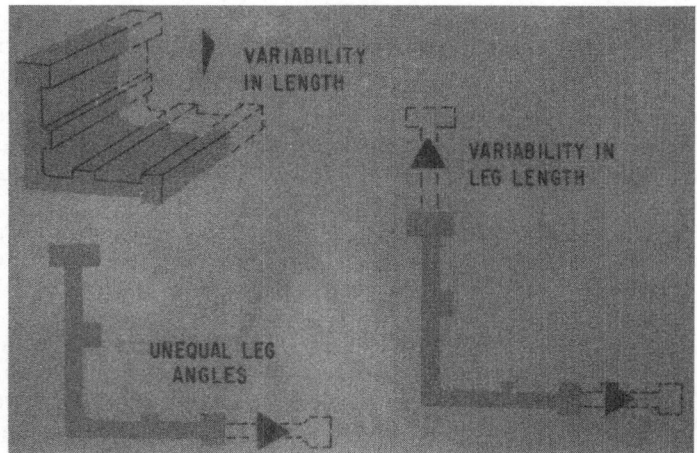

Fig. 4. Structural section—variations studied.

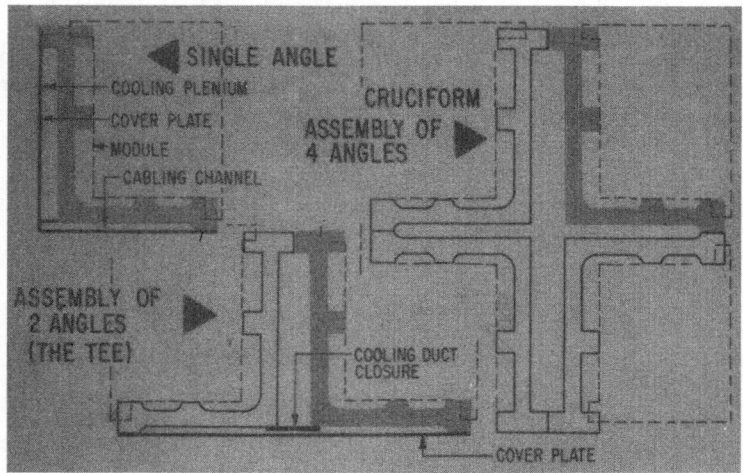

Fig. 5. Structural section—variations.

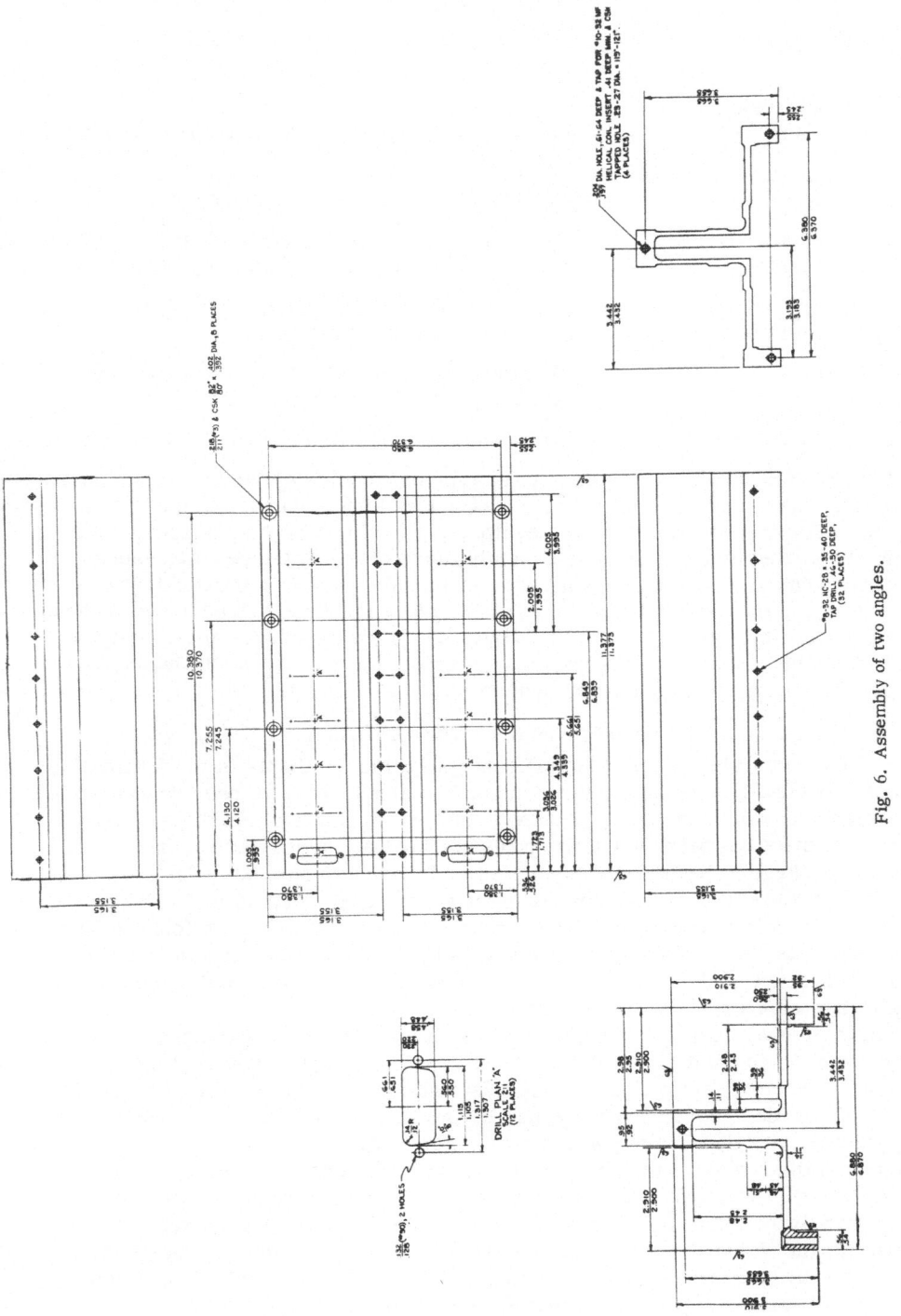

Fig. 6. Assembly of two angles.

3. Outer case:	Square tubing incorporating internal and external mounting areas. May be cast or extruded.
4. Wiring boards:	One or two per module (square, longitudinally mounted).
5. Electrical part orientation:	Parallel or perpendicular to boards or mixed.
6. Dimensionless assembly drawings:	Applicable, using grid system orientation.
7. Coatings and/or encapsulants:	Optional and variable; if used, should be chemically removable without damaging parts.
8. Variable parts and special test points:	To mount directly to outer case of module on faces opposite the angle.
9. Heat sinks:	To mount directly to outer case of module on faces adjacent to the angle.
10. RF shields, gaskets etc.:	To mount directly to ends of the outer case.

Figures 7, 8, 9, and 10 illustrate the type of modules employed by designers using the cruciform synthesis as a basis for design. These include a control amplifier for jeep-mounted survey equipment, a relay tree for ruggedized shipborne support equipment, an oven assembly for extremely temperature-sensitive electronic parts, and the original dummy module used in vibration testing. As can be seen, the desired degree of standardization has been achieved without infringing on the versatility or creativity of the designer. A typical equipment employing this concept is the platform control electronics and 3-channel reset integrator for an artillery survey system shown in Fig. 11.

EVALUATION AND DESIGN DATA

An important step in the over-all concept of this synthesis is the fabricating of test models for evaluating the design. To date, tests have been conducted in the areas of vibration, heat transfer, and encapsulation. Perhaps, the most striking of these has been the vibration results. We have found natural frequencies of 30-lb packaged electronic assemblies as low as 100 cps (or lower). Gains at resonance in the 100 to 500-cps region have run as high as 10 to 15 : 1 in some cases, and limiting inputs of 5 g maximum are not uncommon. At Johnson City, vibration tests were conducted on the cruciform under the direction of Daniel Stern, Vibration and Stress Analysis Engineer. The results indicated a much improved characteristic in each of the above areas.

Thermal evaluation of the cruciform synthesis is being conducted. Preliminary analysis indicates extreme versatility in high-efficiency cooling by conduction, convection, and radiation methods, and combinations thereof. By suitable design of the duct cover, baffling may be introduced to optimize pressure drop for forced-convection cooling with air or other media.

The detailed results of all tests are incorporated in company design standards, and application data sheets. Based on this information the design engineer is able to analyze the design model in terms of his customer requirements, and determine the variations required. Instead of the time-consuming design effort, his problems are reduced to a series of straightforward decisions, such as:

1. Which cross-section combination should be used: the standard 1, 2, or 4 angles, or some special grouping?

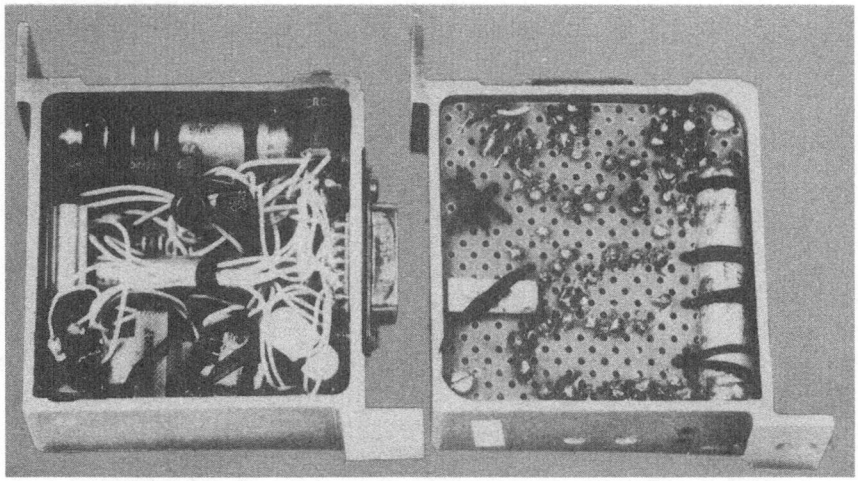

Fig. 7. Control amplifier for jeep-mounted survey equipment.

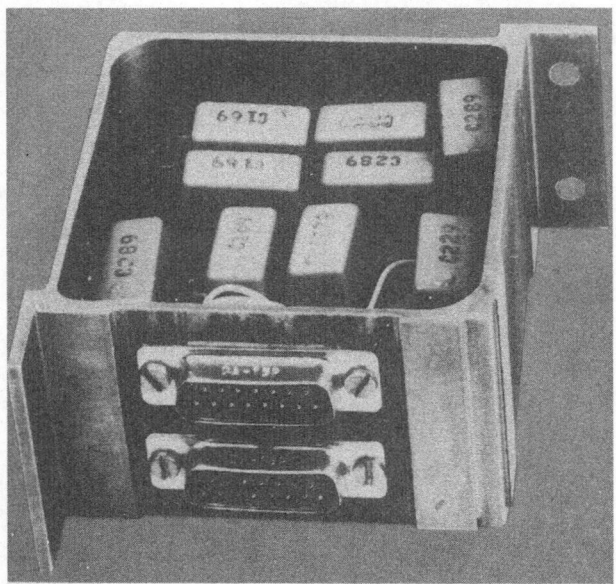

Fig. 8. Relay tree.

2. Which standard cross-section size should be selected from the available standards, or is a special size required?

3. Does the module case weight plus structure exceed weight limitation? If so, how much rigidity and strength margin can be sacrificed for weight reduction?

4. Must modules be encapsulated or coated, and which of the standards shall be employed, considering mechanical and electrical properties listed?

5. Which cooling technique of those described in the test data and analyses is best applicable to the design?

6. Is shielding or sealing of modules required?

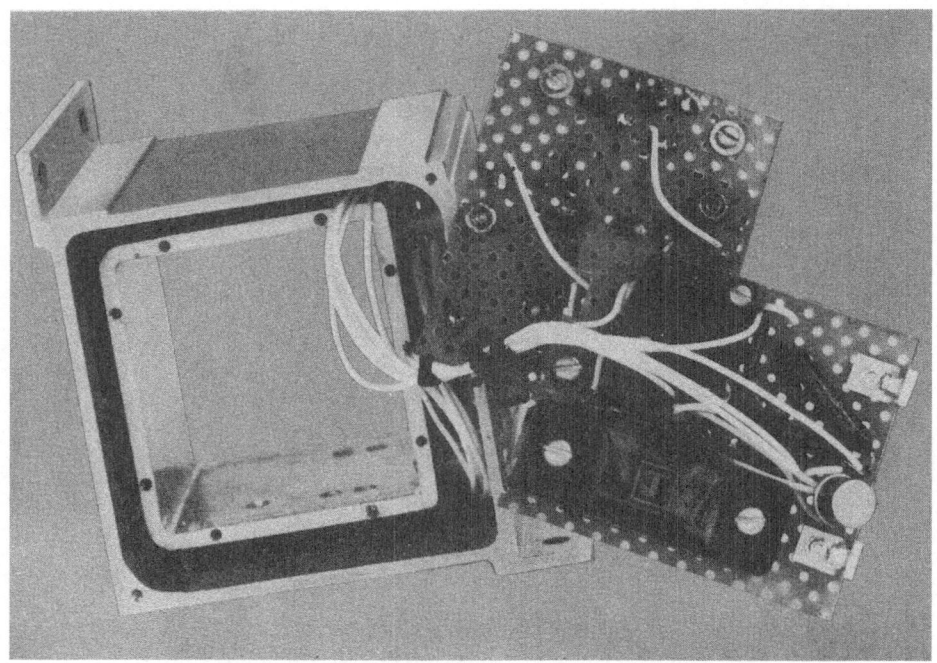

Fig. 9. Oven assembly for extremely temperature-sensitive electronic parts.

Fig. 10. Original dummy module used in vibration testing.

After reaching decisions on questions such as the above, and designing mounting adapter plates and primary cabling to suit the installation, the designer is ready to proceed with his layout.

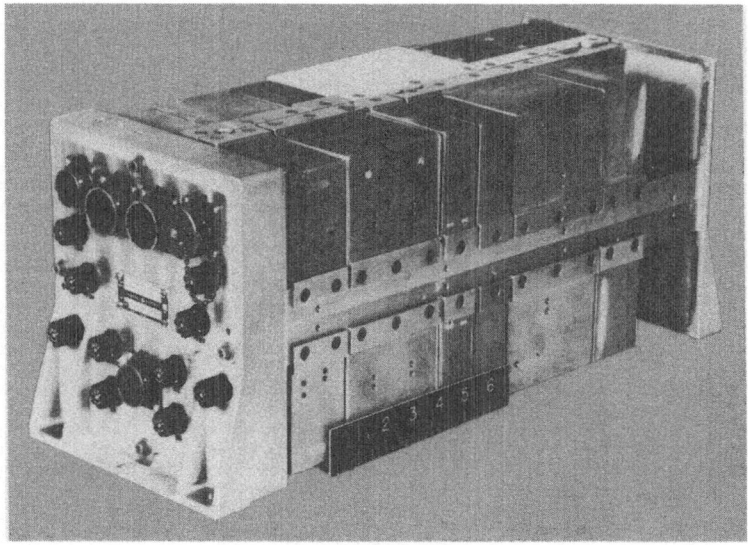

Fig. 11. Cruciform electronics packaging.

CONCLUSIONS

As previously stated, neither the particular design selected, nor the general method employed in the selection, is presented as a cure for all electronic packaging problems. What we have attempted to show is that a packaging investigation leading to one or more generalized design syntheses can serve many useful functions in an electronics organization. For example:

Engineering—The synthesis technique provides information to the design engineer facilitating the establishment of package designs of high quality in a reasonable time period.

Manufacturing—The synthesis procedure effectively introduces the design into the prototype shop long before it would normally appear. Detailed shop problems are discovered, make-or-buy and cost information obtained on a preliminary basis, both internally and with potential vendors. Standardization helps to minimize inventories, and obtain predictable shop workloads.

Quality Control and Test—Preliminary information on testing, connecting, and fixturing is available before the actual design cycle begins. Environmental tests are actually performed long before circuits have been designed, thereby providing valuable test experience.

Marketing—Drawings, models, and actual tested hardware are available to assist marketing personnel in their presentations and proposals to potential customers. Preparation time for proposals is minimized in the packaging area without sacrifice of quality.

DISCUSSION

Question: P. J. Guillot, RCA, Van Nuys. Is there any reason you kept the cabling and air cooling separate? Why couldn't they run together?

Answer: In many cases they can be run together; in the cases where there is a very specific need for specific cooling, they can be easily separated. In some cases, the individual designing the cooling actually wants to put an intricate little deflector in that area; and there again, this works out quite well.

Question: Singletary, Texas Instruments. I have two questions. The first is: Could you give me typical pounds per cubic foot of a complete package using this design? The second question is: What would be the percentage of the metal weight to total weight in such a design?

Answer: This is a very good question because weight was one of the main restricting conditions; the solution was to overdesign metalwise, with the added provision of being able to remove metal very easily. For example: we had a design made up of a single quadrant (the L) which was used on the West German F-104 fighter aircraft for a complete control amplifier for the sight. The total amplifier, including the L castings, 5 modular subassemblies, and a power supply assembly, weighed three pounds as originally drawn. The potting compound used was RTV silicon rubber. The material used was aluminum, and the fabrication method was casting. We decided to do a weight study to see how we could possibly cut down the weight of a three-pound amplifier, and we found, incidentally, this thing took 20g's of vibration over the full range out to 2000 cycles. It was quite successful. We were able to cut down the weight of that three-pound amplifier to a little over $1\frac{1}{2}$ pounds by cutting down metal thickness of the module part and by a substitution of foam polyurethane potting compound for the silicon rubber. I don't want to avoid specifically answering your question (typical cruciforms run 30 to 60 pounds per cubic foot) but what I do want to say is, the way it's set up, the metalwork is provided deliberately high in the general synthesis, as much as 70% of total weight, and then the engineer cuts it away by straight milling operations as he desires. He can get it down to module wall thicknesses, we estimate, of 12 mils or less in specific areas without significantly destroying the vibration capability which is evidenced by this particular design.

Question: Bob Condon, Boeing. I am quite interested in your approach to the aspect of developing such a program for synthesis. However, one thing is going to be necessary before I start. Can I get some figures from you on approximately the time scale and manpower involved in that development program?

Answer: Again, this is a question I very much like to hear because we were able to accomplish this by using, you might say, nonexistant effort in many areas. While the direction of the program was with people directly concerned in the General Design Engineering activity, much of the other contribution to this program was in areas where people had free time between programs, and could devote some time to it. So it's very difficult to assess the total dollars or total manhours that went into this activity, especially since various funded projects are picking it up and starting to use it, and you are getting feedback from them. The nominal full time manpower and effort put in was extremely small, two manyears over twelve months. The materials cost was extremely small. The net effect of the program in terms of those who ended up working on it, thinking about it, and contributing to it, was very much larger. I don't know if this is a satisfactory answer to you but I would like to emphasize one thing; you can set up a massive packing effort with 30 to 40 people full time put them in a corner, and have them design the best packaging in the world, and you will never sell it to the rest of the people in your plant. This "not-invented-here" thing is an extremely important factor. The thing we aimed at in Johnson City was getting a program that people in the design area would pick up, would use, enjoy using, and make suggestions about. I've seen other programs elsewhere. much better funded than ours was with many more people than we had; and it turned out that the engineers developing the program spent three times as much time selling it after they got done, than actually doing the work. So I propose an integrated program, starting out with a small knowledgable group in a Standards area, in a General Design area, (or whatever you may call it in your company) set aside to do this work but closely coordinating, consulting, and working with the package engineers on or between funded programs. Does this answer it?

Question: Partially. We work the same way you do, I think. However, I am interested in getting some kind of feel for the over-all calendar time period involved in the particular development.

Answer: I think cruciform started in mid '59, conceptually, with active work going on through early '60, at which point it had already been picked up by other people. Then the work continued, especially the evaluation phase, which just keeps right on going. One thing I must mention: many other things were coming out of the same group at the same time. The same individuals working on the starting point of this synthesis were the ones working with AMP on MECA, working with Utica's Tom Telfer on the welded-wire matrix and a number of other programs, all within the same time interval. But normally there were different people in the design areas coming into the picture. To get a good time figure on it is very difficult. It's part of a full-time effort for a few individuals for an indefinite period. As far as I know, the evaluation is still going on, getting some of the detailed data on different forms of the same synthesis. It's a continuing effort. It isn't something that goes on for six months and stops, or at least it shouldn't be.

Question: Harry G. Frankland, Ryan Electronics. Can you tell us how the cruciform is attached to the front and rear panels?

Answer: It simply bolts on, into helicoil, rosan, or stainless-steel inserts. If you will come up front you will see that the structural connections are simply straight bolt-through. It could be adapted to virtually any type of fancy interconnection. Bear in mind that what we have here is a building block, the starting point, from which the engineers adapt and make their own individual variations. The purpose of the standard synthesis was not to stifle initiative but to channel it into constructive areas.

THE SYSTEMS APPROACH TO ELECTRONIC PACKAGING

Henry Chrystie

Consulting Engineer, Pomona, California

The hardware that represents an electronic system is usually in the form of a group of packages, generally referred to as units, components, or "black boxes," together with racks or other mounting, cables or harnesses, and miscellaneous input and/or output devices. All too often, this assortment is an accumulation of odd sizes and shapes, not well coordinated from a design point of view, resulting in a hodge-podge of poorly arranged items that are costly to manufacture and difficult to service.

A systematic modular design approach to the packaging of the components and subsystems of an electronic system (whether this is a military weapon system or a commercial accounting or computing system) has many advantages over the use of a random arrangement of packages of unrelated sizes and shapes. These advantages can be clearly demonstrated in terms of improved reliability, easier maintenance, lower manufacturing costs, reduced size and weight, etc. This concept of modular design has been applied to a number of military and commercial systems and affects all the equipment in these systems.

When a system is evolved, it is usually necessary to choose designs that will provide a working system in the shortest possible time. Because of this, expedient arrangements are used or just grow on us, even though we know very well that there are much better ways to accomplish the desired result. These better ways require more time than is available early in the program. As we proceed with the further development and refinement of a system, we must take advantage of the opportunities to stand back and take a fresh and critical look at what we have, or what we are doing, and occasionally set it aside and start over with a new, clean, fresh, and unrestricted approach.

We should not let ourselves fall into the trap of continuing with an approach that our study and development programs have shown may be wrong or not very good, but must recognize that sometimes we cannot determine until after considerable effort that the approach we have taken was not the best choice. If we have doubts about a program, we would be well-advised to run one or more parallel approaches until we have enough data to make a wise choice. On the other hand, occasionally engineers have had the opportunity to start early enough in a development program to lay down suitable ground rules by which to evolve a packaging system in which all units have common features of size, shape, assembly methods, construction techniques, wiring methods, installation, etc., and which can be made on simplified common tooling, and can be tested on standardized test equipment. This paper describes some of these latter methods and shows the advantages in higher reliability, easier servicing, and lower manufacturing and maintenance costs.

Let us consider the following definitions:

Parts: Bits and pieces such as condensers, screws, transistors, cores, dials, handles, etc.

Subassembly: Minor groups of bits and pieces, such as a terminal board with resistors and condensers, a dial mechanism, a filter, etc.

Assembly: A major or functional group of bits and pieces which are a part of a unit, such as an assembled printed-circuit card, a chassis with electronic parts wired together, a motor with gear train, an optical readout, etc.

Module: A basic building block shape of package used in multiple to form units. One or several fastened together may be units, or modules may lose their identity when combined into units.

Unit: A box, (black box) or package, complete within itself, that performs a specific function, such as a guidance platform, guidance computer, power supply, battery, etc. This may be several modules fastened together.

Subsystem: A unit or a group of units that are normally used together to accomplish a particular task necessary to the over-all performance of an over-all system. A subsystem may be an over-all system or several subsystems may be required to complete a system. Examples are radio receiving equipment, missile guidance equipment, fire control computing equipment, flight telemetry instrumentation, detection radar, etc. The first two examples may be either subsystems or systems.

System: A group of equipment, or a group of subsystems, that can perform or complete an assigned task without any additional outside contribution (except the operators, of course). Examples are a complete missile of the self-guiding type, a banking computer which sorts checks and does the bookkeeping, etc.

Weapon System: Everything which is a part of or has anything to do with the successful performance of a weapon. One example: a group of complete missiles, plus a fire control system, plus ground handling equipment and test equipment, etc., plus training devices, shop and field manuals, instruction books, etc. In other words, the whole works having to do with one kind of weapon.

Component: This word means many things to many people, so I avoid using it.

There are many groups working in electronic packaging who consider the items in this list all the way from parts to the weapon system, but there are many more packaging engineers who consider only parts, subassemblies, assemblies, modules, and units—and there they stop. It is this last group in particular to whom my remarks are directed, in the hope that they will broaden their point of view, and thereby do a more effective job. It requires some digging to find out the facts which affect a packaging system above the level of units, but the improvement in the over-all result will be well worth the extra effort.

Some objectives which are sometimes overlooked in the systems approach are given below, and this is by no means a complete list.

Reduced Costs
    a. By common features such as size, shape, plug type, and location, etc.
    b. Independent manufacture, testing, and stocking of subsystems making up a system

   c. Minimum rehandling of completed and tested apparatus

   d. Identical equipment in all similar systems regardless of the ultimate use

   e. Simplified system checkout by mating pretested subsystems

Improved Access

   a. All equipment easily accessible in factory, depot, or field, through openings in equipment container, section, or cabinet

   b. All equipment easily removable and replaceable by one man without special handling equipment

   c. Any equipment removable without disturbing other equipment

   d. Easy access to all adjustments

   e. Adequate access space for personnel

Improved Performance

   a. By separation of power, signal, and instrumentation wiring, to reduce crosstalk and interference

   b. Minimum length of signal and heavy power leads

   c. Provision for low impedance signal ground buss

   d. Adequate electrical bonding (independent of signal ground buss)

   e. Higher reliability by having no blind installation details

     (c, and d, must be done early in the game. Are you who are evolving new systems doing this?)

Improved Maintenance

   a. Harnesses easily changed and carried as spare items

   b. No special tools for servicing

   c. Minimum of special test equipment

   d. Minimum of training or skill to test or replace units

The systems approach to electronic packaging is really the application of value engineering to the entire project right from the start, where value engineering is the obtaining of maximum performance in all areas of the project for minimum cost. Application of value engineering after the design is completed is becoming too widespread. We must stop this locking of the barn after the horse is gone. Maximum performance is measured in reliability, operation, technical adequacy, competence, percentage of time available to perform, etc., and cost is measured in man-hours and dollars of original cost and man-hours and dollars of continuing cost. The continuing costs come from such things as preventive maintenance, prevention of obsolescence, cost of training and maintaining operational and repair crews, cost of mobility if this is a factor, etc.

The time to apply the principles of value engineering is right at the beginning, not when the weapons system is so far downstream that there are operational systems in the field. It is the project director's responsibility to see that this is done, and the long-term savings and improved performance are some of the obscure but very profitable by-products of good project management. Much more can be accomplished by carefully directing the effort in the first place, but all too often this part of the task is neglected because of the urgency in competition for a contract and the resultant staging of a good show for the benefit of those who must be impressed.

This analysis of values, to be most profitable, must be so thorough and far-reaching that it covers every single bit of the entire system. To do all this in advance of any hardware or designs requires a great deal of thought, study, and imagination, and requires a group of highly skilled and broadly experienced people. This is not the sort of task that can be turned over to just any engineer. This in turn means that important people must be taken off other important tasks, so the powers-that-be must recognize the long-term gains in doing this. It is not easy to convince unimaginative people who may be in decision-making positions what the studious application of imagination can do for them in saving of dollars. After all, this is the application of techniques for improving hardware, but done before the hardware is designed, so the improvements must be made to nonexistent hardware which is only in the visionary stage. This, incidentally, is what every equipment designer does during the process of thinking out how he is going to design any device, except that the equipment designer seldom has the opportunity to do this on an entire system and only the more experienced designers have developed this capability.

There are many advantages in making a thorough advanced study of the packaging scheme and method to be used in an over-all system. From this study will come a set of "ground rules" which will guide the designs from then on. The design tasks then become easier because major design details such as size and shape of modules and their mountings have been predetermined. An incidental result will be a consistent appearance in the system which will make it look as if everything were designed under a common director, instead of just accumulated from here and there. What is far more important will be the gains in reliability from studying the various methods and choosing those which will produce the best results. This will tend to eliminate the pet ideas which so often come from people with limited experience and a tendency to go overboard on some hard-to-manufacture novelty.

Packaging schemes for missiles and airframes may be based on rectangular (orthogonal) units or modules or may use odd shapes such as pie sections, or may use a combination of both, but schemes for commercial equipment are almost always rectangular. The rectangular schemes are often cheaper to produce because advantage may be taken of standard adjustments already provided as part of common shop tools, such as backstops, fences, etc. This eliminates the need for the jigs, fixtures, or dies that are more likely to be required in the manufacture of odd-shaped items.

A carefully thought-out design approach and program early in the conception of a system will allow an equally early concept for testing, and this well thought-out and planned test program will, in turn, allow sufficient time to design and procure test equipment, whether it be manual or automatic, so that the test equipment will be "debugged" and ready when needed. All too often the test equipment becomes really ready to use well downstream, and we have to improvise to test the first systems that are completed. This can be awfully aggravating and frustrating.

Some descriptions of typical systems are given. Because this is an unclassified session, the systems described are only representative examples and not actual cases, although the ideas presented here can be and have been applied to complex military systems.

FIRE CONTROL SYSTEMS

Fire control systems, because of the nature of some of the devices making up the system, are more apt to be an accumulation of black boxes, especially since the major units may come from different sources of supply scattered around the country. These usually present a very difficult design coordination problem. It must be remembered in a case like this that we will not obtain a

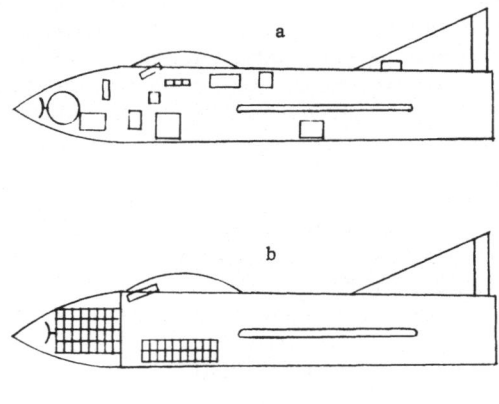

Fig. 1

coordinated design unless we specify, and we are not in a position to specify unless we have made a preliminary study; therefore, this study must be made very early in the game, long before any circuitry is firm. This can and has been done and has resulted in standardized modular construction which is readily available to a variety of installations with minimum rearrangement.

This technique is illustrated in Fig. 1. Sketch a shows the hodge-podge that is obtained when we round up a lot of units and try to put them into an airframe in which the design is already firmly committed. This is a chaotic arrangement in which treatment of such factors as vibration absorbing, heat removal, ventilation, radio noise and grounding, cables and harnesses, etc., was not planned in advance, but generated to fit the situation at hand. Since with this approach there are inevitable compromises, there cannot help but be a reduction in performance and reliability over what might have been accomplished.

Since each new fire control system is an extension or extrapolation of an older existing one, we have enough detailed information on hand on the circuits and functions that we can make a reasonably precise estimate of the details of any proposed new system; certainly there are enough data so that we can conduct a careful study of various proposed ways of packaging and installing such a system. We know, for example, just about what we must put into a radar for tracking and intercept purposes, what we must give a pilot for communications, what other electronic instruments he must have for navigation, and what kind of computer must be aboard to usefully combine all of these data; we can thus propose standardized geometric modules which will contain any or all of these devices in identical cans or boxes which all fit into a common type of mounting. Referring to Fig. 1b, these modules might be arranged in a nose section, a belly section, in wheel wells or wing sections, or elsewhere in an airframe, depending on the size, shape, and purpose of the aircraft.

The nose or belly sections shown in Fig. 1b may be made entirely removable and may contain most if not all of a fire control system. It is a fine idea to be able to rapidly remove and replace an entire elaborate system like this but it is quite costly to do so considering the complications added into the airframe structure, plus the need for special, large ground handling equipment both at the factory and in the field. This would be a great convenience during the development phase of a system or with a system that is quite unreliable and must be changed often like early World War II radar, but doesn't make much sense with our present reliability and performance. The things that go wrong are usually within a package, or are in external instruments, and we rarely need to change harnesses, mountings, cooling ducts, etc., so there is no point in paying for an expensive feature we do not really need. We can have the access we need with large doors, and should be able to change any package easily, without disturbing any other package.

The reason I mention this possible removable portion of an airframe is to illustrate how far I think a responsible packaging man should go on a major project. I think he should have the authority provided that he can prove that the project or system will benefit from it; but he must be able to back up his ideas with facts, and these facts can only be obtained from a careful study.

LARGE MISSILES

The same general comments on fire control systems apply to the equipment for large missiles, except that this equipment is more likely to be specially designed for the specific application. What usually happens, however, is that a survey is made of available spaces in a proposed or experimental airframe, these spaces are listed, and are then assigned on a volume basis adequate for each expected piece of equipment. The designer of each equipment is then told what the dimensions are of the space assigned to his unit (and there are often no two spaces alike). The designer then proceeds to create a box which exactly fills up all of the space assigned to him, because he does not dare to make a smaller box for fear that he will not be able to fit everything in. He doesn't worry for a minute about the other fellow's box size or external details such as kind of connector, etc., since that is not his assigned problem, and he is not about to go looking for more problems.

The right approach is to determine the total space needed and see that it is made available in the airframe, then proceed to divide this space in a uniform standardized manner, even if the airframe must be changed to accomplish this. Airframe changes are easy to make early in the game, in spite of the traditional screaming and yelling that issue from the airframe designers.

This approach must be made by a competent staff who has the authority to set up and direct the execution of this plan, and it must be done soon enough so that the energy expended in design is all in the desired direction. This desired direction must be clearly understood by those people who must carry out the tasks, otherwise things will drift away from the over-all objectives. The best way to control the objectives and convey the desired requirements to the designers is to produce a full scale mock-up which clearly demonstrates what is wanted and why, and is available from then on for reference. This mock-up should then be presented to those who must take on the design tasks so that they can see what their objective is. Remember, Confucius say, "One mock-up worth ten thousand pictures."

An interesting approach to a standardized packaging system for large missiles is shown in Fig. 2. In part a, the randomly arranged odd items are attached to the missile structure in a sort of patched together scheme. A careful study shows how standard packages may be made. These packages are not necessarily the same width, but are the same height and depth. They are made from a stamped shallow pan end plate, in which the same end stamping is used for both ends.

The top, rear, and bottom are a U-shaped wrap-around cover, and the front panel is extended above and below to provide flanges for mounting. These packages slide into a frame consisting of upper and lower sheet metal shelves held together by separators which are identical to the stamped ends of the boxes except larger. Thus, complete control of tolerances is obtained in a packaging system with variable box sizes, using only two special stamping tools, one for the box ends and one for the separators. All the other parts of the boxes and shelves or frames are simple sheet metal pieces made on ordinary shop tools, yet this system is flexible enough so that any normal electronic circuitry may be contained. The exception usually is the inertial guidance platform, which defies ordinary packaging and usually requires a specific location, but the related electronics easily fall into the standardized scheme.

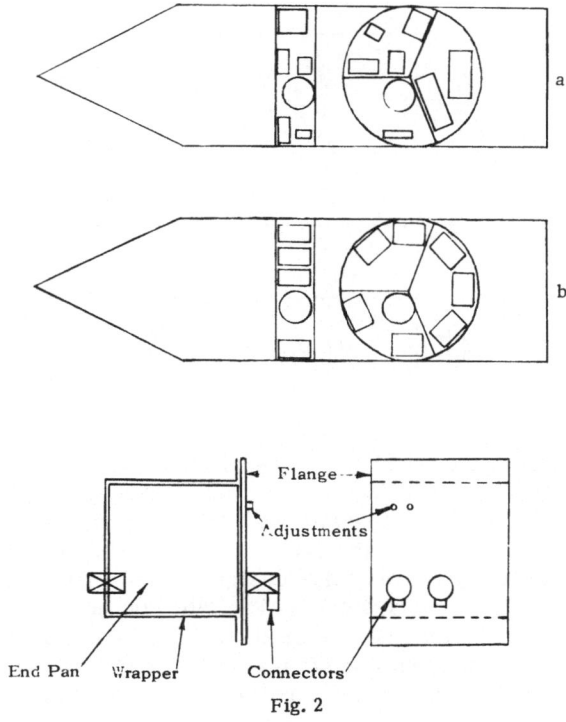

Fig. 2

SMALL MISSILES

There are several important factors that affect the packaging in small missiles. One is that because of the application of these smaller missiles, they are usually made in relatively large quantities compared to large missiles or fire control systems. Therefore, it is important to design for high volume or semi-automatic production. In large missile system packaging, we try to design small quantity items so that we can produce large quantity details, such as a box end for example, where there may be ten boxes and if we can use the same type of end for all of them, that is, twenty box ends per missile, we achieve a multiplication of production quantities by twenty. This same idea can get us up into production details in the thousands when we are making small missiles in the hundred lots.

Now we can afford to go into more elaborate tooling, and can stamp out little cans for modules, or mold little bases for welded wire assemblies, or make IBM card- or tape-controlled printed-circuit machinery, etc.

Figure 3a shows large odd-shaped units which are expensive to make, awkward to handle, and difficult to service, but which suffice until we have time to do better. Figure 3b shows the use of smaller modules designed for volume production as throwaway items, such as welded wire modules, encapsulated or hermetically sealed cans, etc. These are put together into uniform assemblies which may be quadrants, sectors, or slices (wheels).

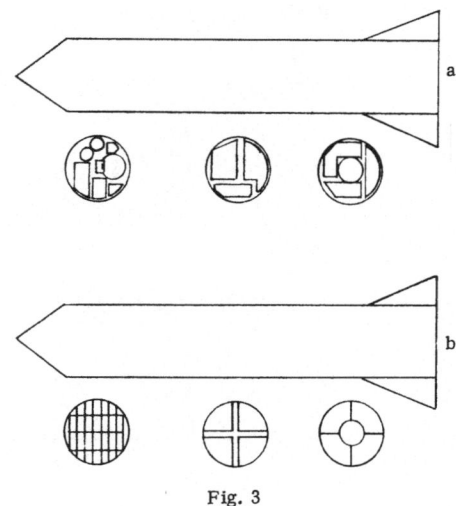

Fig. 3

To sum up, we have known about some of these ideas for a long time. One wonders if we are applying these ideas to our space efforts.

The schemes illustrated demonstrate ways of designing, building, installing, and servicing the electronic equipment in various complex systems. These schemes have the advantages of modular design but still maintain the flexibility that is needed in a dynamic system which will continue to change, develop, and improve. This illustrates the direction to be taken in our future designs, for which a study program should be set up early in the study phase of the project to allow exploration, expansion, and elaboration of the best electronic packaging methods.

In some of our satellite and space vehicle work, we are starting off in the same old ways, namely, to use or adapt what is available, and along with it, to build a few specialty items, which do not have much by way of common shape or form. This is for the same old reasons: urgency in getting started and the need for small quantities. In other of our satellite and space vehicle work, we have been able to apply the systems approach to packaging with a resulting uniform coordinated appearance and a general orderliness that indicates good thinking and proper planning. This, of course, is the way to go, and a systems approach will become more important as we get into larger, more complex systems and into larger quantities.